127113
#49.95

THE GEOPOLITICS OF SOUTH ASIA

South Asia is one of the world's great cultural heartlands, and home to nearly one quarter of the earth's population. It is the cradle of two of the world's oldest and most important religions, Hinduism and Buddhism. It has also embraced Islam, and there are now more Muslims here than in the Middle East and North Africa combined. This enthralling volume tells the story of how the earth's shifting continents moulded this cockpit, of how the great Hindu, Mughal and British Empires in turn came to dominate it, each leaving their marks on the geography and cultures of the region, and on their ideologies and political structures.

The analysis is woven around the three major forces of integration; 'identitive' forces – the bonds of language, ethnicity, religion and ideology; 'utilitarian' forces of common material interest; and 'coercion'– the institutional use or threat of physical violence. By studying the interplay of these forces over time, set against the geography of South Asia, Professor Chapman shows how the organisation of territory – as states and empires, as monarchic realms and representative democracies – has been central to the region's cultural and economic development. Anyone who is planning on carrying out research in South Asia or indeed anyone who simply wishes to understand more about this cultural heartland should read this book.

To the memory of Peter Gould,
inspirational geographer and supportive friend

The Geopolitics of South Asia
From Early Empires to the Nuclear Age
Third Edition

GRAHAM P. CHAPMAN
Centre for Advanced Study, Oslo, Norway
Professor Emeritus of Geography, Lancaster University, UK

ASHGATE

DS
341
.G48
2009

© Graham P. Chapman 2009

Published by
Ashgate Publishing Limited
Wey Court East
Union Road
Farnham
Surrey GU9 7PT
England

Ashgate Publishing Company
Suite 420
101 Cherry Street
Burlington, VT 05401-4405
USA

www.ashgate.com

British Library Cataloguing in Publication Data
Chapman, Graham
    The geopolitics of South Asia : from early empires to the
    nuclear age. - 3rd ed.
    1. Geopolitics - South Asia 2. South Asia - Politics and
    government
    I. Title
    954

Library of Congress Cataloging-in-Publication Data
Chapman, Graham.
    The geopolitics of South Asia : from early empires to the nuclear age / by Graham
Chapman. -- 3rd ed.
    p. cm.
    Previously published: 2nd ed. 2003.
    Includes bibliographical references and index.
    ISBN 978-0-7546-7298-2 (hardcover) -- ISBN 978-0-7546-7301-9 (pbk)
1. South Asia--Politics and government. 2. Geopolitics--South Asia. I. Title.

    DS341.G48 2008
    954--dc22

                                                                    2008030037

ISBN 978 0 7546 7298 2 (Hardback)
ISBN 978 0 7546 7301 9 (Paperback)

Mixed Sources
Product group from well-managed
forests and other controlled sources
www.fsc.org Cert no. SGS-COC-2482
FSC © 1996 Forest Stewardship Council

Printed and bound in Great Britain by
TJ International Ltd, Padstow, Cornwall

Contents

List of Figures

List of Tables

Foreword

This book is an introduction to the Indian Subcontinent and its three largest countries – India, Pakistan and Bangladesh. I hope and believe that it will be of use to students in history, geography, development studies, sociology and politics – and also to the lay reader who simply wishes to understand a little about this cultural heartland. I have long felt that students in many subjects and many departments are asked too soon to get down to the details of this period or that, this part of India or Pakistan, or that, before they had some understanding of the broader context. I have tried to provide that broader context – to give the kind of launching pad from which those who are interested in learning and researching in more detail can then move on. I have tried to do so as much from an emotive point of view as a detached and 'objective' point of view. What I have therefore tried to do is to tell a story chronologically – beginning in the geological past and continuing to the present day. At different points in this chronological story I take time to spin out some of the threads that might otherwise have been dealt with systematically – for example on caste, or on the management of water resources. The organising lynchpin of this story is a geo-political one. The theme to which the book constantly returns is the relationship between society and the organisation of territory – as states and empires, as monarchic realms and as representative democracies – and the development culturally, linguistically, economically of the ever-changing pattern of states. For this reason, this book could not have been published by an Indian, Pakistani or Bangladeshi author in his/her own country. Both peer pressure and legal constraint forbid the telling of disputes from an opponent's point of view, and prohibit the publication of maps with counterclaims shown clearly on them.

Human society exists because of what the word society implies. It implies a degree of co-operation and order or, in other words, the antithesis of anarchy. But anarchy can and does break out, at any scale from violence within the family to the outbreak of war between states. Throughout history social organisations of increasing scale have tried to control anarchy within, thereby implicitly leaving a residual anarchy without. Family units are grouped in tribes or clan units, traders and artisans are grouped in leagues and guilds, amalgams of both are subjected to empire. Empires then both wax and wane. In Europe a new system of states emerged after the anarchy of the Thirty Years War, when the Treaty of Westphalia was signed in 1648 and the final vestiges of the Holy Roman Empire were dismantled. The 'modern' states gained internal strength and cohesion, but were competitors with each other, and quite frequently at war. For this reason the international system of states which Europe has more or less imposed on the world has been called *The Anarchical Society* by Hedley Bull (1977), an analyst of international relations. The story of South Asia is a story partly about its indigenous political and cultural forms, and partly the story of the invasion and implantation of foreign forms, including most recently the European idea of sovereign states. These successor states have engaged in competitive warfare – in other words have resorted to international anarchy. The book, therefore, is also concerned with South Asia's engagement with Europe – Britain in particular – and the external world.

I cannot conceal the fact that I am British and that that fact on its own could explain why I became interested in South Asia in the first place. The British involvement has left permanent marks in British culture as much as the British left marks in South Asia. Quite a few words in common currency in English have their origins in India. Bungalows are houses of Bengal: their inhabitants may sleep on divans which take their name from the Mughal tax officials – the Diwans – via the cushions they sat upon – and perhaps asleep on their divan people wear their pyjamas – the 'leg garment' of Punjab. The juggernauts that oppress small rural roads in England are named not after Prussian militarists, but the Lord Jagganath, whose image is taken through Puri in Orissa (one of the states of India) in a wheeled chariot, beneath whose wheels his devotees would throw themselves to be crushed. Thugs originally committed Thuggee, ritual murder, in the name of the Gods of destruction. Weekend sailors sail in Catamarans, though not the 'bound wood' of the original Tamil. The army wears khaki ('khak' means dust) uniforms. The Labour party has Holy Cows, expert commentators are pundits, and we hear of the Moguls (Mughals) of British Industry. The involvement has of course not ended. Since the 1950s Britain has received several waves of immigrants of South Asian descent, directly from India, Pakistan and Bangladesh, and also indirectly from the Asian communities expelled from east Africa. The South Asian diaspora in Britain is a large one, contributing much to the new multi-cultural Britain. I do not know, but I suspect, that more mosques, Sikh gurdwaras and Hindu temples have been built in the last two decades than Christian churches (some inhabiting the disused churches and chapels – a kind of architectural apostasy?). There are now more people of South Asian descent living in Britain than ever there were Britons in India. And both the marriage trade and the tourist trade sends a steady stream of Indians, Pakistanis, Bangladeshis and Britons in both directions. The British in general seem to have fallen in love with Indian curries – and interestingly the title 'Indian' restaurant retains the imperial idea of greater India – since only a few style themselves as specifically Pakistani or Bangladeshi cuisine. But for me the most enduring connection is the number of close friends I have made in India, Pakistan and Bangladesh and with immigrant friends from South Asia living in Britain and elsewhere. This book is written in warmest appreciation of these friendships. I hope I have written something that will help those who have no contact with, and little understanding of, South Asia to learn something of the outlines of the history and culture and politics of this cultural heartland – home to one quarter of the world's population.

It might seem as if the current political map of sovereign states has some permanence: but much of that permanence is in the perception of it, not the reality. The United Nations charter subscribes to the idea that there are such things as sovereign states, and that interference by other states in the internal affairs of others is 'illegal' – although actually the level of interference by multi-national forces in the affairs of 'weak' states has increased markedly since the end of the Cold War. The European student knows that for a 'long time' there have been such countries as 'France' or 'Britain', and yet a moment's reflection reminds us of the changing German-French border in the region of the Rhineland, or the Schleswig-Holstein question and the demarcation of the southern Danish border, or the acquisition by Hitler by treaty before the Second World War of the Sudetenland from Czechoslovakia. More recently the USSR has dissolved, and Yugoslavia has fractured violently – the latter being one illustration where there has been intervention by both the UN and NATO.

Figure F.1 shows a map, which depicts the frequency with which certain borders have been used in South Asia. We may argue about the exact status of a border – whether it was or was not between truly sovereign states, we may argue about the meaning of 'state' at different times, but the generality of the map still conveys a point. To the European there may always have been only one 'India', but it is a sub-continental name. There is only one North America, but within that there have been many states and empires too, and the persistence of a unified Canada has sometimes been in doubt. Today there are three major states in South Asia – India, Pakistan and Bangladesh (the others are Sri Lanka, Nepal, Bhutan and the Maldives). These three have been born out of the India which the British ruled more or less as a totality – but even this is a mighty simplification.

First let us look at the land which the British finally controlled within their Indian Empire (Figure 4.2) and compare it with Figure F.1. It is immediately obvious that the valley of Assam has rarely been part of the South Asian system, and that the 'most frequent' border of the north-east is drawn across the mouth of the valley where it joins the Bengali plains. The people of Arunachal Pradesh and Mizoram and other north-eastern states share less in common with the north Indian heartland than many other people of the modern Republic, and there are continuing political tensions between them and the Government in Delhi, resulting still in armed insurgency. And when the British came to leave, they did not divide their Indian empire into three, but into two – the state of India, and the state of Pakistan, which was in two geographically separated parts. The Eastern Wing of Pakistan was formed by partitioning Bengal more or less along one of the lines which is shown on Figure F.1 as a minor and 'infrequent' boundary. In turn this Eastern part of Pakistan achieved its independence in 1971–72 and became Bangladesh. Was that the end of the story? This book is not a futurology, and so the only truthful answer to that question is; I do not know. But, we do know that an extremist group in Punjab took to armed insurgency to try to force the independence of 'Khalistan' for the Sikhs. The future of Kashmir is unsettled, as a bitter struggle between Kashmiri groups, India and Pakistan shows little sign of resolution. We do know that there is tension between the different regions of Pakistan. And although it is not a topic discussed in any detail in this book, we do know that the problem of the Tamil North of Sri Lanka has not been resolved, and that India has been involved because of its own Tamil population, and that the Dravidian South of India also has its independence movements.

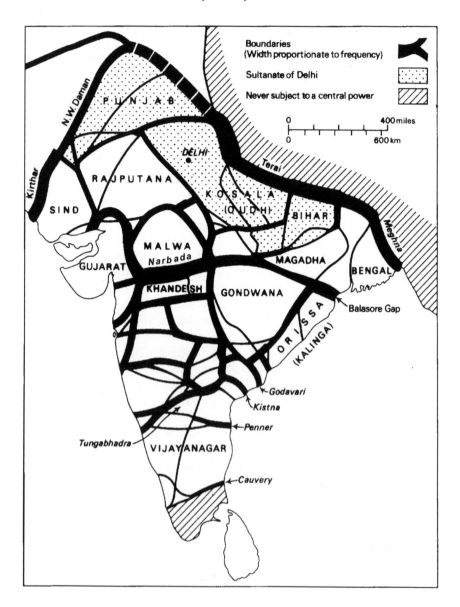

Figure F.1   The Relative Permanence of Former Boundaries in India

Source: Day (1949)

Note: This map is subjective and illustrative rather than 'proven'. At some times, for example, Assam was not as isolated as implied here. The general point, that some boundaries do recur, is valid, and the thicker lines are reasonable indicators of regional divisions.

Where do we start with such a story? Clearly we can start by saying something about the area under discussion. It is geographically distinctive, a subcontinent of plains and hills and rivers, bounded by massive mountains and by the sea. It is what the political geographers call a geo-political region, that is to say, a region derived from geographical features which give it a unity within which cultural, political and economic processes of integration can occur. There is no determinism in this, but a simple statement about the backcloth against which history is played out. There is no determinism which ordains that there will be a single cultural area, but within such regions cultures do often integrate and assimilate. Indeed it is the pervasiveness of Hinduism and its syncretic abilities which has provided a cultural unity to much of South Asia. Islam came of course (in historical terms as a recent cultural invader) and in resisting integration within the manifold of Hinduism sowed the seeds that led to partition and the creation of Pakistan. Even then the six or so centuries of Muslim supremacy resulted in a distinctive Indian Islam – Islam-i-Hind. So the division of South Asia into the current states is very much a matter of division within the region, and still the political and sometimes military struggles between these states is very much a regional affair, even if it concerns external allies. In response to this fact the states of the region have formed their own South Asian Association for Regional Co-operation.

If there is no determinism to state formation, what is it that enables areas to be integrated in one sovereign state? The forces of integration may be conveniently classified under three headings, with perhaps a fourth subheading. Again and again we will return to these three groups of forces, and in the summary at the end I shall retrace the path of history again expounding them more fully.

There are 'identitive' forces, those forces of common identity which link people together and which bring them to a tacit agreement that they constitute the political group or arena within which final arbitration occurs. These are the bonds of language, or ethnicity, or religion, or ideology. Sometimes the identity is almost a tautology – as with the case of nationalism. Identitive bonds need not be congruent – the Swiss have a very clear national identity but neither a common language nor a common religion. In modern India there is religious continuity between north and south which crosses an extremely wide linguistic gulf.

There are 'utilitarian' forces – those bonds of common material interests which mean that it is in the interests of the different component parts of an integrated unit to stay integrated. There are now strong utilitarian bonds between the British and the rest of the European Union. Most of the British are aware of the damage that would be done to the UK economy if they were to withdraw. Through history utilitarian bonds have sometimes been immediately obvious to the populace, as with the case of dependency on river waters from adjoining areas, but quite often they are not so obvious. And of course they have changed with time: the integration of the global economy has a strength now which was unimaginable a few centuries ago when the process first started as European traders first reached India by sea. Technological change has changed the nature of utilitarian bonds faster than any comparable force has been able to change the nature of identitive bonds.

There are 'coercive' bonds. Simply put, force can be used to carve out a state and to hold it together. But force alone is expensive: the costs of maintaining armies of occupation and the sluggish economy of a hostile populace combine to ensure that as soon as possible other bonds must be developed to maintain the hegemony, and that can lead to political dialogue and perhaps changing identitive bonds. Commonly, in history, force has been an initiator of integration – but rarely has it been the prime integrator for long periods.

The fourth type of bonds I group as a sub-class. They are those which I attribute to administrative technology and administrative skill. There are examples from South Asia, as from elsewhere, of Empires which appeared remarkably large for their historical epoch – often short-lived – but which cannot be explained by the identitive bonds of the mass of the populace, least of all – given primitive transport technology – by utilitarian bonds, nor solely by coercive bonds. Behind them were some of the great imperial figures of history, who devised political systems of accountability and land and resource management that extended their empires far, and enabled the ruling classes to reproduce themselves over a few generations, sometimes fostering their cohesion with new identities. The ways in which these bonds can by reinforced are often complex and subtle: local leaders can be recruited by an alien élite who reward them with rank and security.

As I have said, my approach in elaborating on these themes in this book is mostly, but not completely, chronological. The early chapters give some understanding of the environment of South Asia, and of the early history of settlement and culture which I use as a base to explain something of the contemporary form of Hinduism. Similarly, the chapter on the Muslim invasions is used as an excuse to say much about contemporary Islam in South Asia. The involvement of the British is considered mostly chronologically, but there are certain issues in the changing nature of the Indian economy and of resource development which are dealt with in a more systematic way. Independence and Partition are looked at from several geo-political perspectives, which then widens out to consider international relations both within South Asia and with the outside powers. There the book ends – but not of course the story which can be followed daily in newspapers and magazines as it continues to unfold.

A Note on Dates

Dates Before the Christian Era are given as BC. For dates since 0 AD, the AD is omitted.

A Note on Spellings

For many terms taken directly from Indian languages, there has been no standard transliteration into English. Over time the fashionable ways of making transliterations has changed – so over time the transliterations may vary greatly, the changes often loosely tied to tacit notions of political correctness. Versions used in the British imperial period have by and large been substituted now with more 'up-to-date' versions. This means the reader may have to use his/her imagination when perusing the literature to realise when two spellings refer to the same word. As examples consider: Muslim, Moslem, Moslim – any of which in older texts may be replaced by Mohammedan, Muhammadan, Mahometan, Musselman etc. The Mughals may even by Moguls or Moghuls. The Kalifat movement may be referred to as the Kilafat, Caliphate, Kalifate movement. The River Kavari is also the Kaveri or Cauvery; the Jumna the Yamuna. The ritual self-immolation of a widow is suttee or sati. Poona is now Pune; Bombay has become Mumbai; Calcutta, Kolkota; and Simla, Shimla etc. More confusingly, Madras has become Chennai (officially – but many Madrasis will not use the word.) In this book I have tried to be consistent with spellings, but I have probably failed. In general I use the spelling of a place which was common at the historical time about which I write.

A Note on Maps

This is a monotone book and not a colour-printed atlas with large plates; therefore, some of the maps shown are not necessarily the 'best' or most authoritative. The maps of early Hindu Empires and the Mughal Empire, taken from Davies (1959), are simple and effective in suggesting the broad ideas, but may be misleading in some respects. The reader is referred to Schwartzberg (1992) for detailed material.

Foreword to the 2nd Edition

The subtitle of the first edition of *The Geopolitics of South Asia*, published in 2000, was: *From Early Empires, to India, Pakistan and Bangladesh.* That there is a second edition with a changed subtitle so soon afterwards reflects two things. The first is the critical welcome given to the first edition, for which I am very grateful. The second is the American "War Against Terrorism" waged against Osama bin Laden and the Taliban in Afghanistan, and the threat of war between nuclear-armed Pakistan and India over insurgency in Kashmir.

I am glad to say that the main framework and arguments of the first edition have remained intact, and that the recent events have not caused me to change any of my previous conclusions. I have taken the chance to up-date different parts, and to write a little more on Kashmir. More significantly I have written a completely new chapter on the North-West Frontier and Afghanistan, drawing attention to the repetitive, disastrous, and damaging pattern of external incursions into that benighted country from the earliest of times. So, although I can say that current events have not changed my arguments and conclusions, they have compelled me to spell out much more clearly the geopolitical significance of the north-west, and to underline some of the religious and cultural problems of the subcontinent.

It is of course natural to dwell on 'Recent Events', as they knock history from this course to that. Sir Richard Temple's *Survey of India* published in 1880, which is cited in the main text, starts his 29th and concluding chapter (see Table 4.3 below) with "Recent War in Afghanistan", and ends the chapter with "Virtues and merits of the character of the Natives – Their general contentment under British rule – Happy prospect rising before them".

By contrast, I have chosen not to concentrate on the Recent Events in Afghanistan and the American-dominated coalition war against the Taliban and al-Qaida. Instead, my treatment of Afghanistan and the North-West Frontier expands on the security issues of the 19th and 20th centuries. The relevance of these to the contemporary situation is very clear. As for the recent events of 2001–02, I note them and make some acerbic comments about the failures of Western policy, but it is too early to discern whether current action (or inaction) will break the mould of history and lead to the establishment of an integrated, democratic and modern Afghanistan. My hunch is that it will not, that the Americans are as self-deluding as Sir Richard Temple, that they will leave without knowing how to replace tribalism and warlordism, and without either the will or means to invest in economic development. I see little chance of a 'happy prospect rising before' the people of South Asia until and unless both the internal nations and the external powers realise that security is not just a matter of reactive and cripplingly expensive military 'defence' against force, but a matter of pro-active education and development. Just a few modern swords can make so very many ploughshares.

Foreword to the 3rd Edition

There is a 3rd edition partly because too many people were unable to buy a copy of the 2nd edition, which sold out, and which turns up rarely on Abebooks, second hand, at 'inflated' prices. But it is also a 3rd edition, rather than a reprint of the 2nd, because there were still gaps in the story which I felt needed filling. The biggest single difference between the 1st and 2nd editions was the addition of the extra chapter on the Northwest Frontier and Afghanistan. The biggest single difference between the 2nd and 3rd editions is again an extra chapter, this time on India's Northeast Frontiers. In a way, the first edition accepted that this unique-subcontinent was the geopolitical arena to be examined, without saying enough about the borders through which invasions have come. The subcontinent is separated from Asia by the dent that, following the movements of the earth's tectonic plates, the Deccan Block has made, in the process throwing up mountains in the west, north, and east. While I had dealt with the Northwest in the 2nd edition, I had said almost nothing about security and the geopolitics of the Northeast. Besides this chapter, there is new material on water resources, on the New Security Agenda, and a substantial number of updates.

Acknowledgements and a Lament

It is usual to start with the acknowledgements, and to end with a disclaimer on behalf of the named persons, that all remaining fault and error is the author's alone. However, in a book of this historical and inter-disciplinary sweep, the chances of residual error are high, and the occurrence of contesting interpretation certain. So I wish to start by accepting responsibility for all fault and error. Then, without in the slightest implicating any of them in these faults and errors, I wish very sincerely to thank the following. Firstly, the many students I have taught over the years, who have challenged my ideas, and provided me with new clues and references. Their feedback has been of inestimable value. Next I wish to thank John Harriss for reading and commenting on an earlier version of the script, and Joe Schwartzberg for very generously going through the current text with a fine tooth comb. Next I wish to thank Mike Young and Arthur Shelley, of Cambridge University Department of Geography, and Catherine Lawrence and Claire Ivison of SOAS Department of Geography for their preparation of many of the maps; and above all Chris Beacock of Lancaster University Department of Geography, for completing the maps and also for painstaking efforts in preparing camera ready copy. Finally, I would like to thank the Indian Institute of Advanced Study at Shimla for bestowing a Fellowship on me, during which, amongst other things, I undertook the work on International Harmony and Stability repeated in Chapter 14.

Now the lament, which will be understood by most of my British colleagues and some overseas. I first went to India early in 1970; thirty years ago and after I had completed a PhD on the unrelated topic of Systems Theory. I was funded by a 'Hayter' grant financed by the UK Government through the UGC to preserve a stock of area expertise within Britain. When I came back I was so enthused by South Asia that I proposed this book to Academic Press, and signed a contract to write it. My early drafts were justly savaged by unknown referees, and Academic Press in the interim published a book stemming from my thesis on Systems Theory. Some ten years later the commissioning editor of Academic Press moved, and while emptying his files found the old contract which he returned to me for disposal. But I had not stopped writing the book. In 1988 I moved from Cambridge to SOAS (at interview I promised this book was almost ready for the press), where my colleague Tony Allan observed that the pay-off from the Hayter investment was very long term (there were others who went to SOAS who had had similar support in other disciplines). At SOAS the regional expertise around me put some justified further dents in drafts of the MSS which I shared with colleagues. Then in 1994 I moved to Lancaster, where, after further elaboration, finally I have decided that I am probably not going to improve substantially on what I have written.

Not only has this book been thirty years in the making, I now think it had to be thirty years in the writing. To the extent that I have been successful in integrating different disciplines over such an historical time span, then I know I could not have done it as a short-term exercise.

My lament is that in these days of the Research Assessment Exercise held at circa five-year intervals, few young academics could expect the toleration of their Head of Department for spending time on such a long-term project. In some fields it might not matter: in the areas of cross-cultural understanding, which are the most significant problem areas confronting human-kind as a species at present (all others including global environmental problems boil down at implementation stage to cross-cultural understanding) long term investment in expertise is essential. So the lament is actually the greatest acknowledgement of all. I thank the trust, true creativity, farsightedness and collegiality of the British university system as it was, before oxymoronic 'Conservative reformers' engineered its downfall.

Acknowledgements for the 2nd Edition

My first thanks go to Val Rose of Ashgate for proposing a second edition. My next thanks go to my father K.F.Chapman, who has lost none of the proof-reading skills he had in his forty years as a technical editor, and who at the age of 92 has shown a much more acute grasp of spelling, grammar and punctuation than I have ever had. Thanks also for extensive comments and feedback from Bob Bradnock and Christina Coleman. I also wish to re-iterate the thanks expressed in the first edition to Chris Beacock, who has again had to labour with preparing maps and the camera-ready copy. The quality of Figure 5.1, which he has prepared from original material, makes me wish he had re-drawn all the illustrations. But that would have taken too much time.

Acknowledgements for the 3rd Edition

In addition to those acknowledged above in the 1st and 2nd editions, I wish to thank Barbara Harriss-White and Kunal Sen for thorough feedback on the new chapter on the Northeast, and to Simon Chew of Lancaster University, for extra illustrations and graphic advice.

PART I
Introduction

Chapter 1
Brahma and Manu: Of Mountains and Rivers, Gods and Men

The Land

Our understanding of the formation of the continents has been revolutionised over the last few decades. The Earth's outer crust is now thought to be made up of fairly rigid plates – of differing sizes and irregular shapes – which have been moving relative to each other for at least 1,000 million years. About 400 million years ago these were grouped together in one piece – a super-continent called Pangea. This split into two parts – a Northern part known as Laurasia, and a Southern part called Gondwanaland. Between the two lay what is known as the Tethys Sea – which remains now in the remnant string of 'middle earth' seas – the Caribbean, the Mediterranean, the Black Sea and Caspian and the trough of the Ganges Valley. Both the Northern Laurasian part and the Southern Gondwanaland part then split. One split common to both runs down the middle of the two Atlantic oceans – so the fact that the eastern coastline of the Americas 'fits' the western coastline of Europe and Africa is not an accident. The Atlantic split is not the only split that occurred in the Southern part – Gondwanaland. The Antarctic broke off and drifted towards the south pole, Australia drifted Eastwards (relatively) and about 200 million years ago what was to become the Deccan block broke off, and began to drift across the Indian ocean (Figure 1.1). For more than 100 million years it was isolated – hence its flora and fauna could evolve in a distinctive manner similar to that of Madagascar and Australia. Then about 80 MA (million years ago), it struck into the Southern flank of Laurasia (Figure 1.2), and began to push the edge up – lifting the Tibetan plateau and beginning the process – which has not ended – of pushing up the Himalayas, partly from the sea-bed rocks of the Tethys Sea. So what had been a fairly straight line from the Straits of Hormuz at the mouth of the Persian Gulf to Malaysia became severely dented. The Deccan Block is still pushing north at about 6 cms a year – so the Himalayas are also continuing to rise – at an average of between 1 and 9 cms a year. The front ranges, the Mahabharat that separates Kathmandu from the Indian plains and the Siwaliks, began their uplift 200,000 years ago – well within the time period of human settlement – and so recent in terms of Earth history that if the earth has lasted for one year, then this event started in the last 23 minutes.

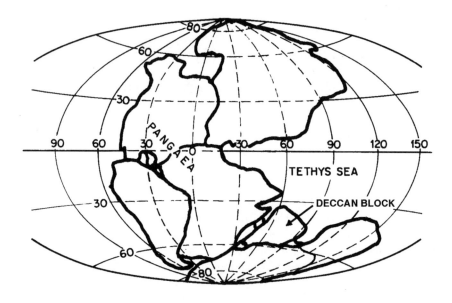

Figure 1.1   Pangaea

Source: Lapidus (1987)

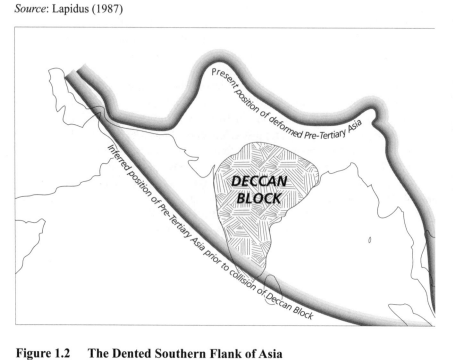

Figure 1.2   The Dented Southern Flank of Asia

Source: Tapponnier (1986)

The raising of the Himalayas has meant that the northward seasonal march of the meteorological equator known as 'the zone of tropical convergence' gets delayed through the long, dry and very hot summer, until it suddenly bursts at the beginning of what we call the monsoon. Because the mountains have been raised so high, the monsoon no longer reaches across the mountains into Tibet – which has become an arid as well as a high plateau. But the amount of water that the monsoon releases on the mountains and plains is exceptional even by equatorial standards. The depth of moisture bearing air is about 6,000 metres – three times as deep as in the other Asian monsoons. South Asia boasts the world's wettest place – Mawsynram, just north of Bangladesh on the southern flank of Meghalaya block. This amount of water falling on the world's highest and youngest mountains inevitably means that natural rates of erosion are extremely high, and the volume of silt brought down to the plains by the Indus, the Ganges and the Brahmaputra are amongst the highest river-borne loads on earth. The Indo-Gangetic plains are the world's largest riverine lowland deposits – in places perhaps up to 5 kms thick – and they result in two massive deltas – the Indus and the Ganges-Brahmaputra. The latter is – one almost feels like saying 'of course' – the world's largest, with sediments up to 22 kms deep – and deposition which continues 2,000 nautical miles into the Bay of Bengal. Other massive rivers, tributaries of these river systems, show equally dynamic behaviour. The Kosi in north Bihar has migrated 100 kms west across its own fan of deposits in the last 250 years: the Beas, one of the five rivers of the Punjab, was captured by another, the Chenab, in 1790. In the Bengal delta the flow of water has shifted over the last few centuries from the western distributaries to the eastern, reducing the water flow past Calcutta – a fact which will be seen to be important in later chapters of this book.

When the tide recedes from a sandy beach, the many small rivulets that form and drain the small springs and rock pools shift their courses many times in an afternoon, capturing each other, forming braided patterns and deltas, and leaving abandoned courses etched on the flat surface. Indra, the God of the rains, from his loftier vision of time, must have watched such a ceaseless patterning of the plains of India. In the mythology of India the Holy River Sarasvati ran to the sea, and there is little doubt that once there was a river that left the Punjab and reached the sea at the Rann of Kutch. The divide between the Indus and the Ganges is so slight that it may well have been the Yamuna (Jumna), which was later captured by a tributary of the Ganges. At any rate, there are marks left on the landscape, and the dry bed of the misfit Ghaggar peters out after it has left the Punjab, now in places reinvigorated by the canals of modern man, like streams on the beach reinvigorated by scheming children with spades and buckets. In Pakistan the Hakra marks a continuation of the old course. The rivers soak into the plains, and provide the ground water for the wells of time immemorial. Upon the rivers and their waters and their silt, so much of life depends and has so long depended, but always subject to their wanton floods and shifting moods, that it is no wonder that they have become revered and holy in themselves. And the land whence they originate is truly the abode of the Gods. That so many of them issue from sources

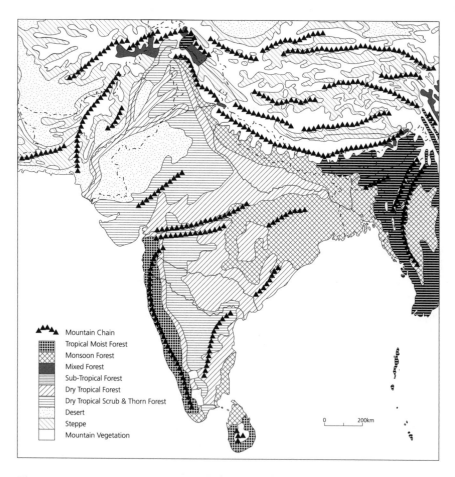

Mountain Chain
Tropical Moist Forest
Monsoon Forest
Mixed Forest
Sub-Tropical Forest
Dry Tropical Forest
Dry Tropical Scrub & Thorn Forest
Desert
Steppe
Mountain Vegetation

0   200km

Figure 1.3   The Natural Vegetation of South Asia and the Principal Mountain Ranges

so close to each other must surely show where heaven reaches down and touches the earth. Could this be the centre of the world?

Thus it is quite clear that the sub-continent is aptly named. It is part of Asia, and yet it is not – it is a geological and climatic region in its own right, cut-off from mainland Asia on its landward sides by high and difficult mountains. These form one of the three major elements of the sub-continent's geology. They together with the second zone – the great riverine lowlands – constitute one of the most dynamic physical regions on earth. The last major element is the old Deccan block, peninsular India, thought to be rigid and stable – though there have been some major and unexpected earthquakes within the block itself. This is the fragment of Gondwanaland – which takes its name from the Gond tribe of Madhya Pradesh, where some of the geology was first studied. Even despite the different history of human settlement and the different climate, the Deccan block remains sufficiently like East and South Africa (see Morgan (1993) for an examination of the similarities) for it to be used in hoodwinking the public: when Richard Attenborough produced his film *Gandhi* he shot all its 'South Africa' location scenes in India.

The physical regionalisation of South Asia (Figure 1.3) is not quite completed by the demarcation of these three geological zones. Although the monsoon may give the East of the country and parts of the south extremely heavy precipitation, this is not true of the north-western parts of the Deccan, where rainfall is unreliable, and least of all of the north-west of the subcontinent – in the middle and lower Indus valley (Sind) and Rajasthan in India – regions straddled by the shifting sands of the Thar desert.

The obvious physical distinctiveness of South Asia is partly the cause of its equally distinctive culture. The physical geography has created barriers that surround the sub-continent, defending it against all but the most determined invaders. West of the Himalayas are the adjacent great mountain systems of the Hindu Kush and the Pamirs, stretching over the nexus of boundaries separating Afghanistan, Pakistan, the Central Asian republics, India and China. And south from the Hindu Kush run the mountain ranges that skirt the west flank of the Indus Basin in the Northwest Frontier Province and Baluchistan. It is a wall around India's fertile garden, but not quite as impenetrable as the Himalayas. Beyond the wall there lie two great cradles of the human race and their civilisations – the Iranian Plateau and the foreland that is now in the republics of Central Asia. The Europeans can recoil even now from their long historical memory of the latter, the hearth of the invading Huns and Gengis Khan, and the many other waves of invasion that have emanated from its core, almost absorbing western Europe into the Eurasian Heartland. But in learning our history thus we tend to overlook the fact that these same hordes could also have twisted south and then east, to penetrate the fabled fertile garden beyond the jagged dry mountains of India's northwest. Like a badly-thatched roof, they can inhibit average rainfall, but not downpours and deluges.

In the north-east in Assam, beyond Bengal, the border with Burma is defended by young mountains that are often covered in a nearly impenetrable jungle. Here the thatched roof is even better proof than that in the northwest – although in this century it was nearly breached by the Japanese whose invasion of India in the Second World War was finally stalled at Kohima and Imphal.

The deltas and river lowlands present a different kind of boundary to the outside world. There are fewer easily navigated waterways in the Indus delta than the Bengal delta – and of course the delta leads nowhere except into a desert. On the other hand in Bengal the sea, the rivers and the land inter-digitate with each other over an area of 30,000 square miles. And this land is highly productive too – and the rivers naturally full of fish. Here is an area where sea-borne trade has been known for centuries, and which can admit a naval power of sufficient strength to gain a toe-hold.

For the Europeans, Africa south of the Sahara was the last great unknown continent partly because it is so hard to penetrate into its interior. Most of the continent is a plateau, with steep escarpments near the sea on all sides. It is not possible to navigate up the rivers such as the Congo or the Orange, and even in the railway age the escarpments have presented great difficulties. The Deccan of India is part of the same terrain. It is skirted on the west by the savage escarpment of the Ghats which overlook the Arabian sea. The coastal strip is narrow and does not lead inland easily anywhere. For the most part the drainage of the Deccan is east to the Bay of Bengal, threading through the lower and disjointed Eastern Ghats, sometimes in impressive gorges. Here on the eastern coast there are several bigger coastal plains and productive deltas – of the Mahanadi, Krishna-Godavari and Kaveri – but the peninsular rivers lack water from snow-melt in the hot season which the Himalayan Ganges and Brahmaputra continue to receive. Hence these deltas do not provide the same kind of rivers and the same possibilities for navigation. Further, there are no attractive anchorages on the east coast.

The peninsula is divided from the northern plains not by one but by several boundary lines. In the north, two contrary rivers flow from east to west – the Tapti and the Narmada – which make a line on the map which has been redrawn by many kingdoms and empires, dividing the Northern empires of the Plains from the Deccan Plateaux. Further south, there is another such line along the Krishna, which demarcates the most recalcitrant parts of the southern peninsula.

The People

It seems likely that the earliest still identifiable groups of people in the subcontinent were what are now called Veddoids (a name taken from the Vedda tribe of Sri Lanka and not to be confused with the Vedas (texts) of the Aryans), sometimes termed proto-australoid because of their similarities with the Australoid type. The first major group to invade, settle and to remain important today were the Dravidians, whose descendants in 'purest' form are in the south of modern India. The Dravidians are dark-skinned and short – neither factor being without contemporary significance. It is not possible to work out in any detail how different racial groups have migrated into India since then, since there is a continuum of mixing of characteristics throughout the sub-continent, and already by the first millennium BC the population was possibly 100,000,000 strong (somewhere near 10 per cent of the current figure), implying a massive gene pool from which backward extrapolation would be extremely difficult. Attempts by physical anthropologists such as Risley (1915) to use cephalic and nasal indices to distinguish between higher and lower castes are now largely discredited, but there are physical anthropologists who identify visible regional differences. The people of Ladakh and many other mountain areas are Mongoloid. In the north-east in Bengal and Orissa it is thought the original Dravidian type became mixed later with mongoloid blood to form the basis of the Bengali nation. In the north-west (e.g. Punjab) people are on average fairer skinned, and a small proportion have blue eyes.

Our understanding of early Palaeolithic, Mesolithic and Neolithic cultures in the sub-continent is weak. The archaeological record for this period, for whatever reason, is not as strong as in Africa. Our understanding of prehistoric South Asia improves dramatically from the time of the Indus Civilisation. Recent interest in desertification and the changing margins of many of the world's deserts has coupled with the explosion of interest in Indian archaeology that followed the discovery after the First World War of the cities of the Harappan civilisation, whose greatest city, Mohenjo Daro by the Indus in modern Pakistan, revealed a magnificence of early urban life in the subcontinent previously undreamt of.

A tentative chronology has been proposed by Goudie, Allchin, Hegde (1973) for this corner of India which suggests that this civilization flourished in a wetter phase than now prevails. The culture at its height extended from the Aravallis and the Punjab (many sites have been found by the course of the Sarasvati/Ghaggar) east to the Ganga – Yamuna Doab, to the seaward end of the Narmada (sometimes Narbada or even Nerbuda), and across the mouths of the Indus far into Baluchistan. It is now evident that throughout the period floods were a recurrent problem at many sites, but despite such vicissitudes the civilisation achieved a high form. In Mohenjo Daro buildings on a regular plan were built of regular baked bricks – and the city was served by proper sanitary drains and public granaries. Huge tanks, skilfully proofed, stored water. In the third millennium BC the population of some cities may have been as much as 35,000. And it is evident that it traded with the Mesopotamian cities by land and sea.

The sophistication of the mature Indus civilisation is well attested in the work of Allchin and Allchin (1982). Their concern is more to examine the nature of the civilisation than to document its history. Most people agree that a combination of changing environmental factors – and possibly even tectonic activity that might have caused the rivers to shift – would have weakened the society to the point where it may have been open to external assault. The more colourful view is that the end of Mohenjo Daro was as rapid and sudden as the apocalypse of Judgement Day. Sir Mortimer Wheeler summarised the shock by saying that it ended 'at four o'clock on a Wednesday afternoon.' But in the view of the Allchins, though the end of the city itself may have been cataclysmic, the civilisation and its technology survived, mostly in rural production and rural areas, to become the seed of the Ganges Civilisation of 500 BC, even if that newer civilisation was dominated by a new people who learnt from the older ways.

In the second millennium BC, or perhaps earlier, the people we call the Aryans began to migrate out of the heartland of Central Asia. They moved north and west towards Europe, and also south and east to India. Some were fair-haired and blue-eyed, they spoke a language which was derived from a lost stem which is the base of all Indo-European languages. In northern Europe some of their descendants became the Scandinavian tribes, whose mythology shares much in common with the mythology of the Aryans of India. Linguistically, in Europe, Lithuanian is closest to Sanskrit. They represent the archetypal Aryan that Hitler wanted to re-establish as a master race – and from India he adopted the 'swastika' – a symbol both of 'luck' and sometimes of the sun – as the emblem of his hegemonic project.

In Asia Minor (modern Turkey) the Aryans became dominant in the later stages of the Hittite empire (c. 2000–1000 BC), and ultimately their descendants and their languages became the basis for Greek and Roman civilisations. In India their descendants gave India its classical language, Sanskrit, once thought when first encountered by 'modern' Europeans to be the original Indo-European stem language, but since proved to be another off-shoot of the lost stem. It took indeed several generations of contact between the 'modern' and 'enlightened' Europeans and India to realise the extent and depth of these relationships, partly because the 'orient' had become associated with barbaric power, which must by definition be alien, and because during Europe's renaissance the great powers of Islam had driven a wedge between far Asia and Europe. The 'otherness' of the East is of course the basis for Said's (1978) critique of 'Orientalism' – the implicit disparagement of the East in much western scholarship.

The Aryans were a pastoral people, and they reached the Indus plains at a time when the Harappan civilisation was waning. Bolstered by the traditions of their oral history as recorded in the Vedas (the 'Truth' or the 'Wisdom'), the sacred hymns of the Aryans (very, very roughly analogous to the Jewish psalms) for a while the view was held that they absorbed little of what was there before them. But it now seems likely that even such important Hindu Gods as Siva and perhaps such common Hindu practice as phallus worship, and certainly the technology of the ox-cart, were adopted from their predecessors. Thapar (1992) has summarised how our knowledge and interpretation of early Indian society is constantly reviewed.

We have some idea of these early times because the priestly caste of the Hindus, the Brahmins, have jealously guarded the Vedas. There are four sets – the oldest known as the Rig-veda, was composed about the time of their first arrival in the Punjab. From it we can get an idea of the geography of the land of the Sapta-Sindhu (they recognised seven (sapta) rivers in the Indus where now we know the Indus and the five of Punjab (Punjab meaning Panch (five) ab (river)). It was a land of cold winters and hot summers, and unreliable rain which could fall in either winter or summer.

South lay the land of Vitra – the land of drought – the present lower Indus Valley and the Thar Desert. To the east lay the land of both Vitra and Indra – drought and rain, or in other words quite clearly the monsoon lands of the Ganges basin.

The Aryans pushed east to this land of Vitra and Indra and intermarried with the indigenous Dasyus (Dravidian people and tribal groups), but also treated them as beneath and below them. So we now come to the highly contentious and charged topic of the origins of caste – whether it is in a sense theological, or functional, or racial. In India sometimes the word *varna* is used instead of caste, and it means 'colour'. On average it is probably possible in any region to identify a colour gradient from fairer-skinned higher castes to darker-skinned lower castes. The marriage adverts of present Indian newspapers frequently refer to the 'wheaten complexion' (fair-skinned face) of a high caste bride. In other words, the caste may have originated in the limitation of intermarriage between conquerors and conquered who had different racial characteristics. This is a topic to which we will have to return.

The Aryans also began to adopt the ways of the agriculturists of the plains, in the days of the Rig Veda cultivating barley in the Punjab, and later in the Ganges lowlands to the east and south both wheat and rice. From the Vedas, Saxena (1968) has been able to sketch a convincing picture of their agriculture, and for me one of the exciting details is to learn that it was already organised according to the *nakshatra* (literally 'asterisms') an astronomical calendar of 24 divisions (as opposed to our twelve months). Farmers were enjoined to sow in Rohini, exactly as even now in Bihar farmers use the same calendar and still sow in Rohini (Chapman, 1983).

It was not an easy task to tame the landscape. The tropical deciduous forests (losing their leaves in the hot dry season) thick with teak, sal, simul and sisoo, and in Bengal with bamboo thickets as well, was stocked with a fearsome fauna – tigers and elephants (often more dangerous), snakes and boars, and disease. The forest demons were not far from the early wooden stockades.

Society Crystallises

In the thousand years from the first of several Aryan waves of invasion to the Mauryan Empire that followed, Hinduism (as we now know it) began to take its early shape. It seems there was little or no political unity or cohesion between the many small states, and indeed even the idea of 'state' or 'polity' may be inappropriate. Society was organised more on clan lines. But these tribal origins became associated with segregated lines of marriage and descent, with guilds and crafts, and ultimately crystallised into a caste system. I use the word 'crystallise', because it carries with it some idea of permanence – a settling out into a final form. Clearly nothing in human culture does ever achieve a final unchanging form – but nevertheless although the expression of caste may have weakened and strengthened, and changed in other ways over time, one of the distinguishing marks of caste in India has been its remarkable persistence and longevity as a system. It was a problem that bothered Karl Marx – who predicted that society would move from feudalism through a class conscious phase of capitalism before achieving socialism – and here was a society that seemed to stand outside of this progression.

The explanation of caste can take many forms. To some it can be explained as *sui generis* – a thing unto itself – unlike anything at other times or in other cultures – and therefore it can be explained by Hinduism in its own terms. For others, it has to be explained comparatively, that it is the Indian realisation of a form that can occur in other guises elsewhere – and sometimes the black–white divide in the USA is quoted as a parallel. Historically we need look no further than to Japan in the early 19th century before the Meiji Restoration to recognise another rigid caste society – which is still not completely erased from the cultural record. In Japan the 'problem' of untouchability still adheres to the descendants of the untouchable (cleaners and sweepers) caste – although it is something which receives little publicity. Go back further in time and in Europe in the Middle Ages there were hereditary crafts and guilds, and much further back in time unequal citizens and slaves in Greece, and another caste system in Egypt.

Within Hinduism there is a theoretical or theological explanation of what caste is – although this theory is a simple and high level abstraction which often bears little relation to caste in its practical form. Therefore, in saying the following, be warned that I am generalising and simplifying perhaps excessively. In the next chapter the whole subject of caste and Hinduism will be greatly expanded.

In theory there is a four-fold division of varnas – at the top there are the priestly castes of the Brahmans, followed by the warrior caste (Kshatriya), then the merchants and landowners (the Vaisya), and finally the Sudras who are labourers and menials. There is a fifth group of outcastes and untouchables – but the distinction between them and the fourth group is complicated and varied, and we will deal with this in the next chapter too. The reality of the system is that there is a definition of what constitutes a caste group which is separate from this theology. Groups called castes exist: these are explained by mapping them on to this four-fold system – which is, I repeat, essentially a theoretical abstraction. The definition of caste is based upon the pillars of marriage (it is only possible to marry within one's own caste – it is therefore heavily tied up with lineage and genetics), of occupation (a caste in theory will have only a traditional occupation, and in practice will certainly exclude many demeaning occupations from any consideration), of pollution (members of a caste may be polluted by eating food prepared or touched by other castes, and the higher the caste the more stringent the separation), and of ritual hierarchical status (some castes are superior to other castes, a fact which is acknowledged by virtually any two castes in relation to each other).

This empirical definition when applied to India will indicate that there are thousands of castes, all perhaps theoretically identifiable with the four-fold division, but separated by region and language as well as marriage and other barriers. Currently it is commoner to term the four-fold theoretical division to be 'caste' and the actual 90,000 or so endogamous marriage groups, grouped in perhaps 3,000 sub-castes, to be *jatis* (*jati* = birth).

There is one fact that must be stressed here, because it may well escape the reader's attention otherwise. The south of the subcontinent retained Dravidian languages (Figure 1.4) – which in the current day are known as Telugu, Malayalam, Kannada, and Tamil. The greatest linguistic divide between Europe and the southern tip of India is therefore in India itself. English is more akin to Bengali or Hindi, than either of those to Tamil. Despite this and the political separation of north and south throughout most of history, the south is thoroughly Hinduised, and indeed there is in many ways a stronger Brahmin presence and tradition in the south than in the north. It is a remarkable fact, and of course this cultural continuity across such a linguistic divide is one of the reasons why India is 'India'. It is a fact which has not been well explained, but which could in part be a phenomenon strengthened in the last 800 years by Muslim pressure in the north.

The Epic Ages

The Mauryan age (321 BC to 185 BC) coincides with what is known as the age of the Epics. It was then that the events described in the epic texts (akin to the Iliad and Odyssey) of the Mahabharata and the Ramayana were written down. Both concern warfare and the struggles for dominion over lands and kingdoms, and the instatement of lawful dynasties. There is little doubt that they have their origin in historical fact but the narration has assumed the proportions of spiritual dialogue at the highest level. They contain discourses about the nature of man and his relation to God, and of man's role in this world. The Mahabharata includes within it the Bhagavad Gita – the Song of the Lord. It is not a revealed holy word like the Koran or such as many Christians believe the New Testament to be, but its status for many Hindus is very nearly that – and in hotel rooms where in the west one might find a Bible, in India you are likely to find the Gita.

In it, Arjuna argues with the God Krishna, who at first is in disguise, as to whether or not he should wage war against his relatives and his former friends over a disputed inheritance. He is told that it is the duty of the warrior to fight in a righteous war. The man who seeks God must also seek detachment, must accept both good and evil, heat and cold, praise and blame, be tied to no-one and despise no-one, but above all fulfil his duty and obligation.

Arjuna went forth into the dreadful battle at Kurukshetra, and won. Kurukshetra is near Panipat, a town a little to the north and west of Delhi. It is on the divide between the Indus and the Ganges, between the erratic climate of the Sapta Sindhu (the Punjab) and the land of Vitra and Indra, the monsoon lands wherein India's agricultural heart has always lain. At Panipat many centuries later the Muslim invaders fought to break into India.

The philosophy of the Epic age stems from the Upanishads, (which originate in the period between the Vedic and Epic), a collection of discourses taught from teacher to student (who sat upright facing his mentor) about the nature of the universe and man. The world and the universe is unitary, in it all is Brahma, beyond definition and impersonal, but found through contemplation and meditation. It is an abstract view of cosmic completeness. This philosophy is not equally accessible to all men since their intellectual abilities vary – and hence the deification of so many of the heroes of the epics, who became allegorical representatives of the abstract ideals. The philosophical importance of the Upanishads continues to this day: they became the basis of the Vedantas (that which concludes the Vedas) which are the basis of contemporary Hindu philosophy.

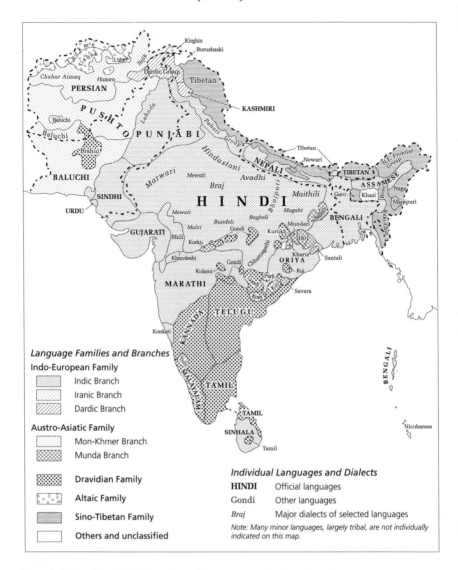

Figure 1.4   The Distribution of Languages in South Asia

Source: Schwartzberg (1992)

Before we jump too far, let us stay for a moment in the 6th century BC. Although we recognise now the legacy of the philosophy of the period which by now included the doctrine of reincarnation, there were also for the ordinary people the rites, sacrifices and prayers of everyday religion, and the strife and violence of society around them. Much of it was what we would now call pagan or near pagan, and much of it very similar to the practices of the Scandinavian tribes, distant

Aryan cousins. It was in this real world that there occurred two reformist revolts, Buddhism and Jainism.

The New Religions

Buddhism was founded by a Kshatriya Prince of North India, Gautama (sometimes Gotama) who lived from c. 562–483 BC. He strove for eight years in meditation under a peepal tree at Bodh Gaya (in modern Bihar), and finally achieved enlightenment. He became the Buddha, and preached his gospel of detachment, of peace and love, of the Path to Nirvana to the people of North India. The second was systematised by Mahavira, known as Jina, the conqueror, whence the name Jainism. Like Buddhism it, too, preached the sanctity of life and the path of non-violence and detachment. Both were also anti-caste, believing in the essential equality of man. Both found adherents more from the non-Brahmanical castes than the Brahmanical ones. Buddhism spread throughout India, and might perhaps even have become the dominant religion of India, except that the Brahmans fought vigorously to keep alive the doctrine of inequality and their own ritual superior status. It then spread beyond India, into Afghanistan (but whence later it was vanquished by Islam), Central Asia, Tibet, China and Japan, Indo-China and Sri Lanka. Jainism never spread so far. Its strict enforcement of non-violence went so far as to exclude even agriculture, since the plough damaged insects. For most Jains it became a sect associated with trade and traders. The Jains are now mostly found in West India, mostly in business and trade, but also still with a strong intellectual and ascetic tradition.

We are now approaching the beginnings of recorded history, and the first of the great Indian Empires. But before we consider this period, I wish to reflect a little more on what has been said so far. And at the risk of confusion I wish to introduce yet another of the great literary works of early India. These are the Puranas, the many folk histories and the genealogies (many invented or 'improved') of the early kings of the Vedic people. The Puranas were transmitted, and therefore presumably adorned, orally for many centuries, and were written down at various dates between 500 BC and 500 AD. But they encode and encapsulate so much.

The first King of India was Manu Svayambhu (the self-born Manu) born directly from Brahma, the god of all, and (s)he was hermaphroditic. From him there sprang a line of descendants who gave the earth its name, and cleared the forests, cultivated the land, and introduced commerce and cattle breeding. But the most famous Manu was the tenth, who was warned by the God Vishnu of a great flood that was to come. Manu built a boat to carry his family and the seven sages to safety, and the boat was towed by Vishnu, who had taken the form of a fish, to a high mountain, where they waited till the floods subsided and Manu's family could re-people the earth. From his hermaphroditic son there issued the two lines of descent, the Lunar and the Solar, which include the names of all the King's until the Epic Age, so that the list finally fits with Rama the hero of the Ramayana who lived at Ayodhya (the site in recent years of the struggle between Muslims and Hindus over the replacement of a mosque by a Hindu temple).

The story of the flood is so common to so many religions – the parallels with Noah and his ark are obvious – that one cannot help wonder whether or not it has a common beginning in Mesopotamia, or whether it is simply a recurrent theme among the early civilisations, many of which were in riverine valleys in highly seasonal climates.

Even more than the Vedas do, the Puranas contain a great deal of geographical information, much of it fascinating. The world is like a lotus (as one version depicts it) with at the centre the great Mount Meru, and around it the eight petals of the eight lands of the world, one of which, to the south, is Bharat (India). On the summit of Meru is the vast city of Brahma, enclosed by the Ganges. The river issues from the foot of Vishnu above, washes the Lunar orb and falls here from the sky, encircles the city and then divides into four mighty rivers flowing in four directions, north, east, south and west. These are the Sita, the Alakananda, the Chaksu and the Bhadra. The first flows east through the country of Bhadrasva to the ocean, the Alakananda flows south to Bharat, and then divides into seven and flows to the sea. The Chaksu flows west, and the Bhadra north across Uttarakaru to the northern sea. The Sita has been identified with the Yarkand, the Chaksu with the Oxus, the Alakananda with the Indus certainly, and perhaps the Indus and Ganges (one of the two major head rivers of the Ganges is currently known as the Alakananda). The last is unidentified. The rivers flow through regions whose people are described and named. Obviously identification becomes even more tenuous here, but clearly some haunting early memory survives. The name Bharat which is used in India to identify itself (it is what is written in Hindi on postage stamps) is said to derive from king Bharat, a descendent of the mythical king and law giver Manu Svayambhuva (see Chapter 2 on Hinduism). As a name it therefore evokes explicit cultural and religious origins.

The great Mt Meru may refer to the Pamirs, the pivot of Asia, where Tibet, India, Central Asia and Afghanistan grasp each other. From this mountain stronghold the rivers flow in many directions to many seas, some inland, like the Aral sea, some to the oceans. The Aryans came from somewhere beyond the Oxus and crossed it on their way to India. It is a curious river dribbling to the Aral Sea, but it crops up in many stories of the Aryans themselves, the Persians, and the Bactrian Greeks. Another version would have Meru and the city of Brahma at Lake Manasowar, 700 miles east and south at the origin of the Sutlej and near the source of the Brahmaputra-Tsang-po.

What is clearly of most importance is not the accuracy of the memories found in the Puranas, but the deeply ingrained memory of an origin near these life-giving rivers, which spout from the home of the Gods where the great snows fall from the sky. These memories are part of the history and mythology of Hinduism, part of the blend of legend and reality, and reach into our own age. No wonder the Ganges is still Holy, that the Himalayas are revered by the people of the plains who have never seen them, and that Kashmir is not a trifling piece of land to be given to whosoever demands it.

It is evident that the authors of the Puranas knew much about the people of the Deccan, but equally that the Vindhya range had prevented them from penetrating far south. This folk history is, then, North Indian.

The First Empire

Oil that is spread on water often splits and regroups into separated fragments. After the Aryans came to India their contacts across the North West frontier (the leaky thatched roof) broke, and they became an 'Indian' people, much as the Mughals would do millennia later. So the passes of the North West became a frontier to be defended against other incursions. In 516 BC the Persians under Darius conquered the North Punjab and proclaimed it the 20th satrapy (province) of their empire. In 334 BC, Alexander the Great crossed the Hellespont into Asia Minor and within 11 years had subdued the Persians, taken Egypt and Syria, marched into Afghanistan and passed by Kandahar, Ghazni and Kabul north to the Oxus. He went beyond it to found on the Jaxartes (Syr Darya) the city of Alexandreschate, 3,500 miles east of his native Greece. Having secured these lands of Bactria (the name survives as Balk in modern Afghanistan) and Sogdiana he decided to complete the Empire by taking over the Persian 20th satrapy. In 326 BC he reached the Indus, which he crossed 16 miles north of Attock. He then entered the city of Taxila near modern Islamabad/Rawalpindi. What the Greeks found amazed them. Here was a great civilisation, but replete with some barbaric sacrifices, and customs such as suttee (the immolation of surviving widows) which horrified the British 2,000 years later, and replete with marriage markets and the non-violent naked ascetic Jains.

Pressing east again, Alexander found his crossing of the Jhelum contested by the army of Paurava, superior in numbers and equipped with what must have seemed like tanks to the Greeks – fighting elephants. But the Macedonian cavalry outflanked the elephants and stampeded them with trumpet blasts, so that in the end they proved a greater danger to their own side. Alexander won, and presumably stood on the gateway of India. Indeed he wanted to continue down the Ganges valley, but his army refused. Instead he withdrew south down the Indus valley, leaving behind Greek Governors in his new domains. His army then returned through the horrific desert wilderness of the Makran coast, suffering privations of hunger and thirst. He himself died of a fever at Babylon. He was aged 33.

Dying so soon after his triumphs, it seems that his empire did not for long hold together, but yet there remained in Afghanistan, central Asia and North West India a Greco-Buddhist culture that left an evolving imprint over centuries. The archaeological site of the city of Taxila dates from this period. To the Indian kings the event seemed peripheral and little is said about it. But it proved an event which transmitted considerable knowledge of, and fascination about, the fabled lands of India to the ancient Europeans.

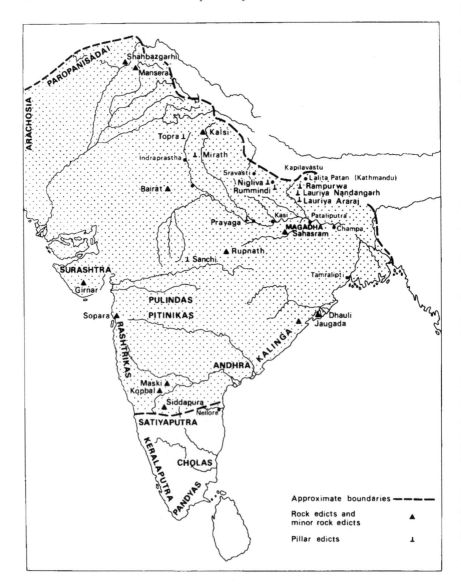

Figure 1.5   The Empire of Kala Asoka

Source: Davies (1959)

In north India in 323 BC political power was seized by a young man called Chandragupta Maurya. He united many of the petty kingdoms, chiefdoms and republics of the Ganges valley in one empire, based on his capital city at Patuliputra near present Patna. His empire was well administered, though inevitably given the communications technology of the time, significant power and responsibility was devolved on the five viceroys of his five provinces. It was a rich empire, trading abroad with China and Egypt. It was well armed, with a standing army of 600,000. His fabled capital was, though, built of wood: for on the great young river lowlands stone is hard to find. (Even today road builders bake bricks and crush them to produce a hard-core foundation.)

Chandragupta's son extended the empire as far South as Mysore. He in turn was succeeded by his son, the great Asoka. The empire (Figure 1.5) he inherited linked north with south, yet there was a hole in its eastern flank, which hindered coastal communication in the Bay of Bengal. In Orissa the Kalingas stubbornly remained independent and hostile. Asoka subdued them in a bloody, relentless campaign that is reported to have caused directly 100,000 deaths, the deportation of 150,000 and the deaths of hundreds of thousands of others through starvation. From this he gained the epithet of Kala (Black) Asoka. The experience was traumatic. In his triumph, Asoka renounced violence and embraced Buddhism, and his own concept of a new moral order.

The empire he ruled was the greatest in extent of all until the peak of the Mughal Empire, and the peak of the British. It was effectively administered, with a hierarchy of officials that reached down to accountants in each village. The taxes raised were spent on roads and irrigation, tree planting and well-digging, and the maintenance of his army. He is reputed to have had an effective state police-force too, effectively an internal political espionage system. The exact extent of his empire remains conjectural. He left behind many stone pillars on which were carved edicts about his concept of Dharma, his concept of obligation and duty. It is these which are used principally to assess the extent of his empire. (The surviving capital of a stone pillar from Sarnath, a Buddhist monastery erected where Buddha preached his first sermon, is capped by the heads of four lions looking outward. This is the emblem of the government of the modern Republic of India.) But of course many neighbouring princes who were effectively independent might have humoured the imperial power by agreeing to their erection. We cannot be sure then of the exact extent of his power. We do know that after the war with the Kalingas he negotiated treaties with the remaining independent states in the south and did not try to subdue them militarily. And we also know that he sent missionaries abroad to spread the Buddha's gospel – to Sri Lanka, and west to Egypt and Greece. This has led to a suspicion that many of the stories about Christ may have been borrowed from or influenced by stories about the Buddha. There are many similarities – including wilderness and temptation, revelation and their respective sermons on the mount.

At the end of his reign it seems likely that the empire was already faltering. The reasons why are not known, and are presumably many and complex. In a way though, to write of the decline of empire is perhaps wrong. The empire of this period, given the distribution of the population, the wild jungles, the difficulty of transport particularly north to south, must be the exception to be explained, not the norm whose absence requires explanation. The technical mastery of iron production may be one key. Clearly administration is another key, and clearly the development of an extensive and effective administration was due to remarkable men. But it was bound to face a problem that was common to the Mughal empire later, and was shared by the Spanish Empire of the New World too, and which also in Rhodesia-Zimbabwe plagued the last days of British imperial power in Africa.

In the struggle for conquest the élite beneath the Emperor needed to stand shoulder to shoulder and support each other. But once each was alone as deputed ruler of his own vice-royalty, his interests would become literally rooted in his own area. Pressures could develop to suggest that local interests are different from central ones, and that the remittance of revenue to a central authority should be scaled down in favour of local accumulations of wealth and the associated patronage and power. There were probably also strains at the centre simply emanating from the scale of public expenditure.

The perpetuation of empire required loyalty to a central concept of state. The state was embodied in the emperor, who had a reciprocal obligation to uphold the social order. This is not mere surmise since political theories of the responsibilities of the monarch were debated and recorded. One treatise, the Artha-Sastra, believed to have been written by Chandragupta's prime minister Kautilya, displays in great detail the workings and principles of the empire. Principles of foreign policy – my enemy's enemy is my friend for example – were clearly enunciated. The right to depose a monarch who does not uphold his contract is well attested. But the social order which the monarch should defend was still basically the order of caste, and this was local and fragmentary. There were certainly continuing struggles between the various castes, because of the asymmetries of economic, ritual, and political power. The Vaisyas developed economic power, and struggled against their political inferiority to the Brahmins. Although there is an old adage 'divide and rule', there is always the possibility, as perhaps later the British realised, of 'divide and it will fall apart.'

Whether it was because Asoka's successor lacked his outstanding statecraft, whether no man could have maintained the cohesion of the disparate social groups, or whether it was economic decline, the fact is the Mauryan Empire shrank soon after Asoka's death back into the northern river plains, and even there it shattered into many smaller states. This was presumably part cause and part effect of yet more invasions from the northwest. The Bactrian Greeks were driven south and east from the Oxus by the Scythians, and after them came the Kushans, a branch of the Yueh-chi tribe(s) (originating in China), whose greatest ruler Kanishka established a vast empire centred on Peshawar. He, too, became a Buddhist, although by now (c. 150 AD) that faith had become one of icons and temples and the many incarnations of Buddha.

The Hindu Empires

The next truly Indian empire to arise in the Ganges plain was that of the Guptas, which reached its greatest territorial extent under Chandragupta II (AD 385–413).

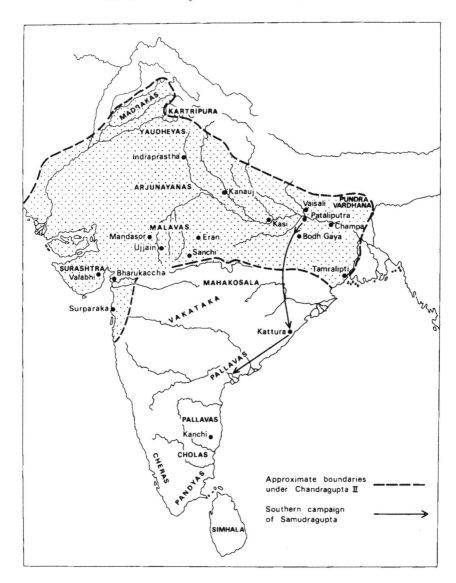

Figure 1.6   The Gupta Empire
Source: Davies (1959)

This again dissolved back into a pattern of petty states, until a new dominant ruler, Harsha emerged from c. AD 606. (Figure 1.6). As had Asoka's empire, this imperial power also interested itself in science and learning. Arybhata in 499 AD put forward the proposition that the earth was a sphere rotating on its own axis, and that the shadow of the earth falling on the moon was the cause of lunar eclipses.

Iron working reached a famous peak – a stainless steel pillar, now erected near the Qutb Minar outside Delhi, is unrusting to this day. How it was made is still not known. And a thousand years before Voltaire parodied the ridiculous simplicity of western science's notions of cause and effect, Hindu philosophers grappled with the problems of multiple causation. Though it was firmly a Hindu empire in which the Brahmins codified Hindu law to a greater extent than previously, Buddhism was still strong and sponsored the great University at Nalanda in Bihar. It attracted scholars from many lands, and much of what is known about the empire has reached us particularly through the writings of Chinese pilgrims and travellers.

But it was not a time of peace. The Huns were constantly warring on the western margins, and after Harsha the empire collapsed. From then until the Muslim invasions North India was a shifting kaleidoscope of warring states. It stands as a Dark Age about which little is known. One group of invaders does, however, need special mention. These were the Rajputs, who established themselves as Hindu rulers in what is now Rajasthan. Their rule was legitimised by the guardians of the social order, the Brahmins, who went to great lengths to establish for them a Kshatriya status, by inventing a genealogy that gave them a descent in the lines of the sun and the moon from King Manu of the Puranas. They were, and are, a martial people, who had elaborate rituals of chivalry and arms, not unlike the knights of our middle ages. Like the Scottish clans they waged war against each other, and seldom did they present a unified alliance to the outside world. They proclaimed themselves the swords of Hinduism, but their endless vendettas prevented them from expanding south and east.

In the south in the Deccan, indeed new and independent Hindu kingdoms arose, often dominated by the Dravidian Tamils. The armies of Rajendra Chola campaigned as far north as Orissa, and the majesty of Vijayanagar and its huge capital city Hampi was probably unrivalled at its peak anywhere else in South Asia. The ruins still visible certainly constitute one of the world's great archaeological sites.

Concluding Remarks

The material in this chapter has already touched on many of the main themes that thread through this book. Firstly, it is clear that we are dealing with a well-defined region, within which there is the possibility for extensive and interconnected settlement – particularly in the Ganges valley. Just how distinctive this area is globally can be discovered by considering Cohen's (1963) geopolitical account of the world – to which we will return again in the concluding chapters. Briefly, the world is divided into two major geo-strategic zones, corresponding somewhat with Mackinder's famous division first published in 1904. On the one hand there is the heartland of the old world – the land empires of Czarist Russia and China, succeeded by the Soviet and Communist Chinese empires, and more recently, of course, by the Commonwealth of Independent States. The other region is the region of maritime trade and movement – Europe and Africa and the Americas, where settlement is largely coastal in orientation. These two geo-strategic regions provide different possibilities for movement and trade – the silk roads on the one hand, the trade winds on the other. They are subdivided into geo-political regions, zones of geographical contiguity which provide a framework for the common evolution of culture and for common economic and political action. This is not deterministic – in the sense that common action has to result – it is more permissive and suggestive of what might happen. In this schema South Asia – the only so-styled 'sub-continent' on earth – has a unique designation. It is a geo-political region in its own right, belonging to neither geo-strategic region. It is a zone where culture and power can develop in their own right, with only infrequent interference by the forces of either of the two geo-strategic regions – and then usually incorporating the intervention rather than being subjected to it.

Given the possibility that this geo-political region will be integrated, we can then ask how the different forces of integration might work. In the early phases of pre-history and history to which we have referred so far, it seems that transport technology was so weakly developed that there would be little chance of developing much utilitarian integration. We may say this even with respect to the Mauryan empire, in which roads were developed, but not everywhere, and as much for military reasons as for trade – a parallel with the Roman roads. Certainly the rivers of the Ganges valley provided then the best possibility for major movement – and it is the Ganges valley either in its lower parts near current Patna or its higher parts near Agra and Delhi (on the tributary Yamuna) which provide the core region for urban development and empire building in the post-Harappan phases. Next we come to the idea of identitive integration. This is a slippery concept to apply to these early stages – we do not know enough about the sentiment of the common people, nor the unarticulated assumptions of acceptable social order and the modes of expression of fealty and power. We can say that a quite remarkable continuity of Hindu culture seemed to have developed through the region, even if with many local variations – but that culture was itself divisive, a society of differentiated groups which in some sense might make an organic whole, but in other senses provides all sorts of possibilities for divergent interests. Language also provides little scope for identitive integration. Perhaps we may interpret Asoka's attempt to impose Buddhism as some kind of quest for identitive integration – but it did not take root nor work in that way. This leaves us, therefore, with coercive integration – and that was common to the founding of the early empires. But force is expensive, and when the possibility of new conquests diminishes, so the costs of maintaining the armies falls back on the existing tax base rather than new plunder. At this point the ability of an Emperor to manage efficiently the affairs of state becomes critical – it is the fourth (sub-category) of the forces of integration. There is no doubt that in the Mauryan and the Gupta empires these skills were for a while evident. But in the long run it seems as if these empires could offer no 'value-added' by remaining intact as opposed to breaking apart – and indeed perhaps the reverse, that maintaining their integration was an inflationary cost that exceeded the benefits that flowed.

Let me emphasise that what I am saying is highly conjectural – but the questions raised by these themes are valid, and as the book progresses it will become easier and less conjectural to answer the same questions the nearer we get to our own time.

Chapter 2
Hinduism: The Manifold of Man and God

An Unrevealed Truth

Precisely because it appears so different, based on ideas and ideals which are quite often very different from those of Europe, Indian civilisation for the Europeans has always had a certain mystique about it. These values are, of course, partly bound up in what the European likes to term ' religion' – contrasting the Christian basis of European culture with the Hindu- and Muslim-dominated culture of India. In the next chapter, which deals with the Muslim invasions of India and the establishment of Islamic empires, we will introduce some of the ideals underlying Islam. In this chapter I try and give the reader some empathetic understanding of Hinduism – but let me stress immediately that the idea that it is a 'religion' in the same sense as Christianity is not helpful at all. Hinduism is often been described as a way of life – it is not separate from secular life – indeed the concept of secular life is very difficult for many Hindus to embrace. Hindu society does have its immediate and observable characteristics which attracted the attention of European travellers – the ornate temples, some with erotic carvings, the division of society into castes, a few of which were associated when the Europeans first contacted India with horrific practices such as *suttee* (the self immolation of a new widow upon her husband's funeral pyre). But these 'characteristics' say little about Hinduism itself, no more than the clothes a person wears reveal their state of mind. What matters more is a basic understanding of the principles from which such apparently 'strange' customs could emerge. I assume the reader has an understanding of western Christian culture (even if not a practising adherent). I will take great liberties in order to try to expose by contrast with Hinduism some of the fundamental tenets on which this 'Christian' life may be based.

The Western mind is peculiarly partitional. It likes to put things in boxes, to call this sacred, that profane; to label some things religious, and some secular or temporal. Hinduism is therefore labelled by Westerners a religion, as though that were a separate box in the scheme of life. But to a Hindu, the word religion would have no particular meaning. 'Hinduism' simply means the ways of Hindu society, a society or societies of people found living by and beyond the Indus river, whence the name originates. As we have already seen these people have many languages, and are descended from many invading groups. They have a long folk memory and history, stretching back to the Aryans and before. It goes without saying that they have their ways of life, and that there will be many ways of life in South Asia, and that many of these ways will have influenced each other, as neighbouring cultures often do. What they do not have is a single dogmatic revealed source of truth. They do not have 'a religion', a 'privileged revelation by God to man on earth'. The Christians do have a privileged source, the testament of Christ. They have a creed, which adherents profess in order to demonstrate their acceptance of the dogma. They can do this because God spoke directly to them, by placing his son on earth, God Incarnate. For any other man to claim he is God would be blasphemy. The Muslims do not have access to God via God Incarnate, since Mohammed was the Holy Prophet of God, but not God himself. But in other ways his function was similar to Christ's. He enabled God's dictation of the Koran to be written down, so that it is God's word, a direct revelation, and not simply Mohammed's own inspiration. The two revealed religions have similar accounts of the origin of the world, in the Christian case in the Old Testament book of Genesis. A 'religion' which has no revealed dogmatic truth still must confront the questions of Genesis, but its answers are not constrained by a single revelation. The next section will take a closer look at Hindu concepts of genesis.

Cosmologies East and West

The Christian Old Testament has a simple account of the origin of the Universe. God is separate from and above and beyond the world, and he made it. The world has a material existence, it is an objective reality. The last of God's acts of creation was to place man (and woman taken from man) in the garden of Eden. The creation is there for man to enjoy, it is his birthright. But the Garden does have its dangers, and Evil tempts man away from righteous living. The existence of evil in a world purpose-built by God is of course a theological problem which is never resolved. The purpose of man's struggle on earth is to enable his soul to achieve salvation in heaven in the afterlife.

Note that this Universe is basically objective: that there is a material world which can be explained by the laws of physics and chemistry – themselves part of God's universal design. God is separate, beyond and outside of it. Our souls, when they become detached from this physical world, will go out and beyond it to this other place where God resides. Each of us has one soul – it is an individual soul, which even in death does not lose that individuality.

Modern scientific cosmology does not take the earth to be the centre of the universe, and all but the most extreme would accept that our planet is part of a solar system, itself part of a galaxy, itself part of a galactic cluster, itself part of the universe. Most people who have a view on the matter currently accept the big-bang hypothesis of the origin of the universe. Once there was a singularity, all matter was compressed in a single point. This exploded, throwing matter in all directions. Locally, under the force of gravity, some matter drew together, and under pressure began nuclear fusion, to gives us suns and radiant energy. Other smaller concentrations of matter have formed planets, which can support life in the right conditions near the suns.

What we do not know at the moment is whether the universe will expand for ever, or whether these forces of attraction will eventually start drawing the galaxies together, and then all matter back in on itself to form another singularity, and another big bang. The cosmologists are working on the data collection and theoretical models that may be able to answer this question.

The general process is easy enough to grasp. The Bang produces chaos and random scattering. Out of that chaos order emerges. In that order life is created, but by a peculiar twist. The concentration of matter produces a sun which will eventually die out, but while it continues to radiate energy (low entropy), life can be supported by degrading that energy (to high entropy – or disorder). In doing this life is not breaking the second law of thermodynamics – that everything is going to a more and more degraded state, that eventually there will be no free energy left for work, that the universe is running down. What life does is roughly analogous to what a kettle achieves while gas is burnt beneath it: it gets hotter, and for a while it may be hotter than its surroundings, although if the gas is turned off, in the end it will lose heat to the surrounding area and the kettle and its surroundings will have an equal temperature. So for a while there was a local structure – a hot kettle against a cold background – which represents a *local* 'reversal' of the universal trend to heat death. But that could only be achieved by degrading the gas from a useful state (unburned) to an unusable state.

Plant life creates structure by degrading solar energy. Animals eat that high-order vegetation, and by degrading it build their own structures – which may be food to yet other animals in the food chain. Life is structure and life exists by degrading structure. Life is the summit of the balance of the forces of construction and destruction.

If there is in Hinduism no revealed total truth, no central dogma, but a variety of beliefs and customs, why call it one religion? From the philosophy of the Vedas and the Epics, which has never stopped evolving, there are some ideas which have a consensus appeal, and some sort of universal validity for all Hindus. Although the word God is wrong, I will use it as the best approximation, to say that many, perhaps most, Hindu's believe there is one God, or Life Force, and one alone, and that this God is Brahma, the force of the Universe itself, or sometimes simply pure 'energy'. Brahma is not separate from the Universe, not outside it. 'He' is it. The whole creation is one, and all is Brahma. Brahma wakes and sleeps, and the Universe goes through cycles of re-creation and collapse. Order is born out of chaos, to collapse into chaos again. In all there is a struggle between creation and destruction. And both are necessary. They are two aspects that are split apart as order is created, and re-unite as disorder is re-established. These two aspects are often associated within Hinduism by the God in the form of Vishnu the Creator – or the form of Shiva the Destroyer.

This cosmology has numbers attached to it. Each cycle of the Universe, from one big bang to the next, is 100 years of Brahma's life. One day of this cycle is equivalent to a Kalpa, itself equal to 4,320 million earth years. Further subdivisions of subdivisions of the Kalpa take us to the current Kali Yuga (Black Age, an Age of

Destruction) in which we live. It will last 432,000 years, out of which 5,071 have already passed.[1] (Wheeler, 1973: 466).

Everything in this Universe is Brahma, even you and me. Brahma is in everything. There is no distinction between the sacred and the profane. Everything is holy. If you say to me you are God, this is self-evident, not blasphemy. There is no Garden of Eden in which man and woman were placed last: they and the animals are all parts of the creation. And within it destruction is as important as construction.

We do have souls that can outlive our current incarnation – but the individuality of them is expressed in another way. We are fragments of the Brahma, and want re-absorption with the Brahma, that is the loss and extinction of individuality. At a simple level the path to absorption is expressed by many Hindus as a path through many re-incarnations, each being better than the previous one if merit has been earned. The highest achievement is escape from the wheel of rebirth.

The Three Paths to God

Most Hindu thought stresses the fact that there are many ways to God – and although some sects may stress one rather than another, each can tolerate and sympathise with other views. The path to God through true knowledge, *jnana*, is often associated with intellectuals and ascetics, with philosophical contemplation, and with high Brahminism. There is a path through *karma*, right action, with those involved in the physical practicalities of everyday life. It means that man has a duty and action to perform, and that the completion of such dutiful action is in harmony with the divine working of the Universe. Rejection of right action is against universal harmony. Since, as we shall see, there are duties associated with each caste, *karma* has often been associated with the duties of caste and the achievement of a better incarnation in the next life. Then there is *bhakti*, devotion or ritual, which is the praise and thanksgiving due to God. Many groups, perhaps particularly of the poorer and lower castes, celebrate God joyously with singing and music, for hour upon noisy hour, partly inducing changes of consciousness through the repetition of sounds. Ritual of such sort can, of course, bring enormous solace to the human soul.

In the Christian Church we can draw some *very approximate* parallels between these three ways and parts of a single service. There is knowledge as enshrined in the reading of the lessons and the sermon of the priest, there is action of a dutiful and practical kind in almsgiving, and there is ritual in the singing of hymns and the chanting of prayers. What is very different in the Christian church is the whole concept of a service, that the whole group of devotees come together at a fixed time, and in an enclosed building usually. The Hindu goes to the temple certainly on particular feast days, but in general as and when he or she feels. It is an individual unorchestrated act. The temple usually has little space for more than a few individuals at a time to approach its sacred shrine. And 'religion' is not something that belongs

---

1 There are other Yugas – for example Ages of Construction – and other lengths.

to the temple: all life is religious. Bathing in the morning has a ritual value. Religion is not as partitioned off as with most western practitioners.

An aspect of Hinduism that attracted derogatory judgement from many Europeans is its apparent polytheistic nature – that there is an enormous pantheon of Gods – from the Goddess of Wealth Lakshmi, revered by traders; the Goddess of Knowledge, Saraswati, revered by students; and the Gods of strength such as Ganesh, the elephant-headed God; and Gods of Cunning such as the monkey God Hanuman – to so many others. By contrast Christianity is supposed to be monotheistic like Judaism and Islam – which is possibly the most incisively monotheistic of all. But Christianity starts by confusing God with a Trinity of God the Father, Son and Holy Ghost, and continues by adding quite a number of orders of Angels and Archangels whose exact theological status has provided legendary grounds for dispute, although no-one doubts that angels are half-bird, and all appear rather androgynous and are rarely properly sexed.

Hinduism is monotheistic – in that all is Brahma – but there are many ways in which different aspects of the whole may be revealed, including by the different avatars or (loosely) 'incarnations' or manifestations of God – including those just mentioned and many more besides. Different village communities will have different local deities who in turn may be different manifestations of some of the better known Gods. Sometimes the symbol for the god may be an almost abstract form – a small phallus representing energy and life – or, like Ganesh, may be much more representational.

The link between Hinduism and the geography of India is explicit – it is perhaps as a religion more closely tied to its land than any other. To go overseas is forbidden by caste rules for many Indians – Gandhi broke with his caste's tradition when he went first to England then to South Africa. Mountains are holy, glaciers are holy, trees are holy, the rivers are holy, none more so than Mother Ganges – in whose waters all would wish their ashes to be scattered. Confluences are holy, none more so than the Sangam at Allahabad between the Ganges and the Yamuna. The great festival of the Khumb Mela occurs every three years, rotating between Haridwar where the Ganges debouches from the hills onto the plain, Allahabad, Ujjain and Nasik (the last is outside of the Ganges Basin). During the auspicious month of the 2001 Kumbh Mela, according to the Government of Uttar Pradesh, perhaps as many as 100 million pilgrims came to Allahabad, a city whose normal population is 1 million. On 24 January, 30 million people managed to bathe in the river at the holy confluence. It is the greatest number of people ever gathered together at one place on earth for one purpose – and, incidentally, a testimony to the phenomenal organisation of the Indian authorities.

Lineage and Caste

After the Roman invasions of Britain a new aristocratic stratum formed above the tribal British. When the Normans came they too formed an aristocracy above the complex of Celts, Saxons and Vikings. Suppose over the centuries the different peoples that had invaded Britain stayed distinct from each other, and only intermarried in their own groups, and different groups followed different traditional occupations, then the British would have had a caste system too. Many authors have said that the British used to have a caste system, and indeed still do have, meaning that we have a class system in which people marry within their own social groups, and are even recruited for education and work in their own social groups. Who you are and who you know may be more important for entry into a stock-broking firm than simple talent.

A purist definition of caste would exclude the British system, although in comparative sociological work the question has been asked whether the black-white divide in the USA is an early form of a caste system. The definition in India is of a group who by prescription can only intermarry (they practice endogamy), and who can eat with each other without polluting each other. Pollution can be defined as a contact which breaks the rules of that caste in a defiling manner, such that the polluted person has to undertake ritual cleansing. Do not think that I am trying to link pollution with sin, but it may help to understand ritual cleansing if for a moment one considers the Christian – particularly Catholic – ritual of Confession, which can bring absolution, a kind of spiritual ablution.

Because castes are marriage groups, a caste is essentially preserving a particular gene pool over time and through many generations. (This does not mean that these gene pools are greatly different from each other: it seems that over time there are periods of greater and lesser rigidity). In this there is a clue to the fact that in India individualism has a lesser value than in the egocentric West. The group matters more. Everything in India is group oriented, be it in an extended family, or in the wider caste or community. The sublimation of self to group is not necessarily to be seen as a 'cost' only, it can relieve people of the responsibility and tension of striving to mould their own destiny, and it can provide a supportive framework. The effect of this attitude is far reaching, and was little understood by the British for a long time. For example, in property matters the basic concept of ownership can be very different. A man does not own his land: he holds it on trust for his lineage. He, therefore, ought not to sell it – and this simple fact explains why there is such a slothful land market even today, particularly in some rural areas, and why prices for such land as is bought and sold appear economically irrational. It also prevents modern banks from taking rural land as loan collateral, since they have found that often forfeited land cannot be sold on an open market, when adjacent farmers know to whose lineage it belongs. Imagine that the Duke of Bedford in Britain sold Woburn Abbey, his ancestral home, to invest it in casinos in Las Vegas. Probably many Britons would think he did not have the right, that he did not truly own it as an individual, but that he held the estate in trust for his heirs. Loyalty to kin also explains what outsiders often see as corruption and nepotism in political life, when for example Mrs Gandhi promoted her favoured son Sanjay. But it would be disloyal for the family not to promote its own – something which the common person fully understands.

The Thousands of Separate Castes in India

Using the definition of the group that intermarries there are thousands of different castes (*jatis*) in India. Many are fairly limited in their distribution, being a group in some linguistic or geographical area. In modern life, though, some castes have scattered widely, indeed world-wide. Each caste is usually recognisable by the family name used – a little bit as though the Scots recognise the Frasers, MacDonalds and the Crawfords as separate clans by their names. Most caste members will still recognise each other and help each other, even if they are scattered world-wide, and sometimes there are strong caste associations, which maintain ties with their ancestral hearths.

In any one area in India there will be some number of castes, let us say twenty or thirty in a cluster of villages. These are ranked hierarchically, with some being at the top of the hierarchy and some at the bottom. Caste members will know what their position in this ranking is. The rank will be determined by historical custom, itself tied to occupation, ritual cleanliness, and local mythology. There will be some approximate ranking between the castes and the four-fold division of the Laws of Manu – that is to say castes will claim and be recognised as having the status of Brahmin, Kshatriya, Vaisya or Sudra. There is a complication with the last group to which I will return.

The first class is of course the Brahmins. The Brahmins are sometimes known as twice-born – not meaning that they are necessarily in their second incarnation (these could go on countlessly), but that in this life on this earth they are born once for real and are then born again when at the age of about eight they are ritually initiated in a ceremony lasting three days into the rights and privileges of their caste. Thereafter they may and do wear the 'sacred thread' made from fine spun strands of cotton, from left shoulder diagonally to right hip. Members of the Kshatriya and Vaisya classes of castes also often wear the thread, but are not initiated in the same way, and do not have the right to learn the Vedas as a Brahmin does. These first three groups are often referred to as 'Caste Hindus' – a confusing term which would be better rendered by the term 'High Caste Hindus'. The Sudras never wear a thread, and are low caste.

The complication to which I referred is that at the lower level of the hierarchy there are groups whose occupations or origins make them particularly inferior or polluting. For example, the sweepers, who clean sewers, are called 'untouchables', and some Brahmins may even feel defiled simply by seeing them. In many areas there is, therefore, a fifth class – the untouchables who are lower than Sudras. Another group who are sometimes put in this fifth class are the tribal peoples particularly from and near hill regions, who have themselves not accepted the Hindu framework of life nor the pantheon of Hindu Gods. They are often incorporated into modern economic life at the lowest levels. The exact distinction between Sudras and untouchables and others, or the incorporation of lower groups into the Sudra class, will vary from locality to locality, and is in any case an approximate mapping of real castes to theoretical classes. It would be the same

Table 2.1　　The Castes Present in 'Karimpur' (Wiser, 1936)

Rank	Name	Traditional Occupation
Brahmin	Brahmin	Priest/teacher
	Bhat	Bard and geneaologist
Kshatriya	Kyasth	Accountant
	Sunar	Goldsmith
Vaisya	none	
Sudra	Mali	Florist/gardener
	Kachhi	Vegetable grower
	Lodha	Rice grower
	Barhai	Carpenter
	Nai	Barber
	Kahar	Water bearer
	Gadariya	Shepherd
	Bharbhunja	Grain parcher
	Darzi	Seamster
	Kumhar	Potter
	Mahajan	Tradesman
Outcaste	Teli	Oil presser
	Dhobi	Washerman
	Dhanuk	Mat maker
	Chamar	Leather worker
	Bhangi	Sweeper and cesspool cleaner
	Faqir	Hereditary Muslim beggar
	Manihar	Muslim glass bangle seller
	Dhina	Muslim cotton carder
	Tawaif	Muslim dancing girl

(almost) if in Britain one tried to match families to Classes – e.g. working class, middle class. The latter are theoretical abstractions and rarely fit well, hence the endless qualifications of 'upper' or 'lower' middle class.

It is often said that the Hindi word for caste is Varna, which literally means colour. Those nearest the pure Aryans are the lightest coloured, and those nearest the Sudra or Adivasi (Tribal) base the darkest colour. And indeed there are colour gradients. In general, the highest castes in any area are lighter skinned. But there is also a regional colour gradient, from the lightest in the north and west to darkest in the south. There is no doubt that colour has played a part in the Indian perception of status. The Indians in East Africa were inclined to treat the Negro as inferior, from within the Hindu scheme of recognition. And it will not have escaped the reader's attention that the Europeans were seen when they arrived to have the physical characteristics of extreme high castes (but paradoxically in their personal habits they were basely unclean and they were meat eaters too). Modern India has legislated to remove the stigma of low caste status, and government has defined a list of the castes and tribes of any area which are to be given positive discrimination in education and public sector jobs. They are known as the Scheduled Castes and Tribes since the names are listed in two Schedules of the Constitution.

Each caste has been associated with particular occupations – in this sense the guilds of mediaeval Europe also came close to being caste-like structures. The list in Table 2.1 shows the castes noted by Wiser (1936) in a village he studied in North India in 1935. Together in any one cluster of villages (not in one village) there would normally be the range of occupations necessary to run the economy. Exchange within the economy was mostly, and sometimes still is, non-monetary. Instead a Barber, a Washerman, a Carpenter etc., will serve a local land-owning family, and their fathers will have served the fathers of the landlords in their time. Payment would be in kind (usually food) at harvest, and in customary amounts – so there is no economic 'market' in which a clearing price is set. The system is known as Jajmani – reciprocal obligation and service. It is often described as exploitation by high caste of low castes, but that is too simplistic, although undoubtedly there can be such exploitation. The patron has an obligation to his service castes, and if the carpenter dies, then he will bring the children up to their status in life as carpenters. Another criticism is that there is usually one dominant land-owning class in any one area, who control all activity through their control of land. Although often true, the origins of this are easily understood. Settlement of any area was not haphazard, but organised by a dominant caste, say Kshatriya Rajputs in some area of north India. They would organise the clearing of the land, and introduce the service castes to the area. Society itself then constitutes a totality, and all are sharing in the economy of the area. The Rajputs are then not so much owners as the trustees and managers of land. It is true that over time they may develop a more private concept of ownership, particularly if prompted by poorly-comprehending external institutions such as an imported English legal system, and the introduction of a money economy. The reason why the system may appear pernicious to the outsider is because the division of labour in society is associated with hierarchical values – that there are 'higher' and more powerful groups – who can exploit in the way that Marx saw the capitalist exploit the worker. The point is valid, because there is a ranking of castes, tied up with the concept of pollution, and indeed there are some jobs in society which are more onerous and more dangerous than others. It also struck Europeans as pernicious in another way. Justice and law were not universal. Each caste was responsible for maintaining the behaviour of its own members – the Panchayat was a council of elders (traditionally five in number) from a caste who deliberated on the behaviour of miscreants and meted out justice. But no high caste person could ever appear before representatives of lower castes. Contemporary village Panchayats in modern India use the same name but are in effect the lowest rank of local government, elected in at least a semi-democratic way and supposed to be much more inclusive.

Pollution and the Hierarchy of Caste

A westerner in India is often appalled at first glance by the squalor and filth of public places, particularly in towns. It is usually explained by the lack of garbage collection services, of sufficient sewers and sanitation systems, and by the pressure of population. Go back a little over a century in the UK and you would have found filth in the streets and open sewers too. But the modern Briton has come to expect that there are public lavatories, and that they should be used when in public. He/she has also come to rely on Public Health Acts which require kitchens in restaurants and pubs to be clean, and the utensils for public use to be cleaned. How would one behave if none of this were true?

The first and most obvious answer is to keep oneself to oneself. If there is little contact, there will be little disease transmitted. This is a lesson which can put cleanliness into the realm of morality. (This lesson is being re-learned now that there is a new sexually transmitted disease around. The safest way to avoid AIDS is not to have sex, or observe complete lifelong fidelity within marriage.) Among the precautions against the transmission of common complaints customarily observed in India are the following. Firstly, you should only eat food prepared within your own social group. Secondly, you should be more and more cautious of food which is potentially riskier and riskier, and, thirdly, you should be more and more cautious of groups who have more and more health-hazarding occupations. Take all these to extremes and you may find people who will only eat food they have prepared themselves, who are strictly vegetarian, and who will not even countenance being looked at by a more polluting individual. This is the kind of extreme in orthodox Brahminism sometimes caricatured in literature. Stories abound of Brahmins travelling (somewhat reluctantly) on a pilgrimage with all their own pots and pans, and taking flasks of their own well-water with them (from a well forbidden to untouchables).

The necessity of vegetarianism can of course spring from religious conviction – but there is also a practical side to that too. Meat in tropical countries which have no freezing system, and where the humidity can for some months of the year preclude drying, does, of course, putrefy quickly. The easiest way to keep an animal fresh is not to kill it: so there is a virtue in killing small animals, most of which can be eaten at one sitting. Hence, chickens will be safer than goats if part of the goat has to be kept a little longer: and goats will be safer than buffalo. The safety of meat will also be associated with the diet of the animal. The pig is a turd-eating animal that can clean open sewers of anything. It therefore can and does eat human excrement, and can thereby pass on such things as liver flukes and other parasites. Again, in the modern West we rely on health laboratories. Every pig slaughtered is inspected for infection before it is sold. To the meat eaters of the West, there has until recently been an accepted standard that meat eating, which is expensive, is the higher end of the dietary scale. But in India, for all the reasons given, a vegetarian diet is at the highest end. The sweeper who owns no land, cleans the drains, keeps pigs, because one of his only resources is the refuse of society. There is a theory that vegetarianism limits the extent to which agriculture can improve the value of its product, and that vegetarianism has played its role in the persistence of rural poverty.

The venerated cow also attracts attention – is it not irrational to think a cow holy? It is indeed irrational to have so many cows in India – it has the highest bovine stock of any nation and many are poor specimens and many are strays eking out a living on urban rubbish dumps. But the reason for the cow's sanctity is not hard to find. In an agrarian society such as India's, the cow and the bullock pull the ploughs, pull the carts on dirt roads, eat the weeds pulled from fields and other crop residues, and provide dung. The latter is of course often used for fuel and manure – but it also has another essential use. To thresh a crop (treading it with cows) requires a hard floor which is uncracked. A mud floor dried in the sun will crack, but a floor whose top layer is a mixture of dung and mud does not – it dries hard and smooth like cement. Without dung, threshing is extremely difficult. In other words, the cow is the pivot of agrarian society. It makes sense to revere it, and to accept the added bonus of the small yield of milk and ghee (clarified butter) it may give.

The origins of an elaborate hierarchy of pollution and of dietary custom can therefore be readily understood, and then in turn the relationship of this with other aspects of hierarchy. This gives rise to the possibility that status can be lost by breaking these customary rules, or even improved, by adopting higher ranking rules. I have known vegetarians, who have been advised by their doctor to eat eggs to improve their protein intake, be most particular to ensure their eggs are vegetarian ones – that is those laid by hens which have not been impregnated by a cock. The other way round there are people who reject meat which once they ate, in order to enhance their status. In general such altered behaviour does not happen at an individual level, but is followed by social groups. Over time and very slowly, there is mobility in the society. But it tends to be (outside of modern urban society) a mobility of groups – castes slowly inching their way up, over perhaps centuries. It is a process known as Sanskritisation.

Maya

It is often said that the Hindus have a sense of other-worldliness, and that they reject this world as an illusion – Maya. There are many different forms and explanations of Maya, so there is no simple single interpretation I can offer here. But, as we have noted above, the Christian view of the world somehow accepts the objective and external nature of the material world, through which the soul passes once in its incarnation. Many Hindus do not necessarily accept this duality, and maintain that the one God who created everything is everything, and the material forms are not separable from the rest of creation, that we cannot sense them as objective externalities, but imagine them. This can lead to an egocentric view – since if all is in the imagination, it is only in one's own imagination, but yet which is a fragment of the totality of the interconnected Universe/God. But there are also other interpretations that appeal to all of us. In the West there is a strong contemporary resurgence of interest in mind-body problems – particularly in the fields of the sciences of consciousness – which is being approached with great publicity from the material/physical side of Cartesian dualism. Ever greater attempts are made to understand how a physical brain can give rise to such a thing as self-awareness. In traditional Hindu thought more attention has been paid to understanding the mind-body relationship from the phenomenon of consciousness itself, one of the chosen paths being through meditation. If human beings try to strip themselves of all pretence, all mannerisms of behaviour, to find the basic essence of themselves, in the end there is nothing left to be a basic essence other than pure spirit devoid of intent, which is the denial of individuality and agency, but a locus of understanding. So our humanity is in a sense nothing other than mannerism and pretence, self-generated, and to that extent not real but illusory. Buddhism – which emerged out of Hinduism – has a very clear enunciation of the role of consciousness, its physical roots, and the doctrine of *Anatta* (the denial of an individual and separate soul):

> There is no unmoving mover behind the movement. It is only movement. It is not correct to say that life is moving, but life is movement itself. Life and movement are not two different things. In other words, there is no thinker behind the thought. Thought itself is the thinker. If you remove the thought, there is no thinker to be found. Here we cannot fail to notice how this Buddhist view is diametrically opposed to the Cartesian *cogito ergo sum*: "I Think, therefore I am." (Rahula, 1967: 26)

Caste and Hinduism in the Contemporary Era

This book is written in a semi-chronological manner – but in the introduction I pointed out that some issues would be developed further at convenient junctures. It is inevitable that at this stage I should say something about caste in more contemporary India. Many people have supposed that urbanisation and industrialisation would subvert caste and 'modernise' India. In the sense that a caste represents a marriage group, this has not happened. The vast majority of marriages are still arranged – very few couples manage to marry outside their caste. New technologies have created new jobs, but it is clear that in any small town particular castes will probably have monopolised any type of new job – being a secretary for example. In some sense there are even new 'castes' – such as (though this is highly contentious and reflects the debate about the colour bar in the USA) the descendants of mixed English-Indian marriages now known as Anglo-Indians, who virtually monopolised the middle managerial grades of the railways. The persistence of caste and its inegalitarian impact is recognised by legislation that reserves particular quotas of places in education for the Scheduled Castes and Tribes, and particular numbers of jobs in the public sector – this is often discussed in the current press by reference to enactment of the 'reservation' recommendations of the Mandal Commission. In attempting to make the less equal equal by making them more equal, the divisions of caste are written into the fabric of political life. There are backlashes from the middle-ranking castes, who feel excluded by the quota takers from jobs they are better qualified to take. Some of these backlashes have taken widespread and violent form.

In each age the expression of caste may have become weaker or stronger, and may be re-invented in different ways, as this Reservation crisis of the 1990s has shown. Each age too can re-interpret the significance of caste in the past. In this current age of the quest for 'sustainable development' one of the recent re-inventions is of an original Indian caste society which was ecologically benign. In this view (Gadgil and Guha, 1992) different castes had access to differing resources in the local ecosystem, so that competition for any particular resource was avoided; over-exploitation was therefore also avoided, and the stewardship of each resource that goes with a continuing lineage was well established. In their view, modernisation since the advent of the British has corrupted the system and rendered it unsustainable. There is, however, little evidence to support the imagination of a golden environmental past dependent specifically on caste structure. The castes of India have been associated with an agrarian and craft-industry type of economy which has for centuries both cleared new land, and abandoned older land when over-use led to diminishing yields.

The current political and ecological debates have one thing in common. India in the late 1990s has experienced a Hindu resurgence, and the secularism initially adopted by independent India is under threat. A Hindu revival is bound to open the question which Gandhi never wanted to answer – is it possible for Hinduism to survive without the institution of caste?

Concluding Remarks

The India that Islam encountered, first by Arab traders isolated wandering mystics – the Sufis – and only later by military invasion, was a world of all-embracing sanctity, of many Gods and none, of idol worship, of meditation, of world renunciation, of venerated cows, and of the inequality of man. Hinduism is highly varied and flexible in many ways, and is not dogmatic in matters of theology even if it is dogmatic in matters of caste. Indian society has indeed often been described as syncretic – able to absorb influences and even contradictory ideas from many diverse cultures and sources. Though this self-same Hinduism has given South

Asia much of its sense of cultural identity and indeed marks the region out as one of Earth's great cultural hearths, it ought also to be apparent that it does not really offer the kind of identitive bonding that would, in the absence of other bonds, lead to the political integration of the area. Its provides an over-arching system – which permits and celebrates difference and diversity, dividing as much as uniting.

It was to this world that came the new truth of the one true God. There is no God but God, Allah the Compassionate, the Merciful. The faith is known as Islam – meaning submission, to the will of Allah. The adherents are Muslims, 'those who submit'. Submission, not absorption, was what the Muslims demanded in their earliest raids, and which they revived from time to time throughout the period of their dominion.

Chapter 3
Islam: Submission to the One True God

The Prophet

In the period corresponding to the depths of the Dark Ages in Europe, the Arabian peninsula was also in disarray. The peoples of the area were divided by religion and by tribe, and dominated or exploited by the outside powers of Persia and the Greek empire of Constantinople. In many ways the factionalism might have resembled the divisions of modern Lebanon – not only were there different major religions such as Judaism and Christianity, there were also many different sects within these religions, and also many pagan and polytheistic faiths. It seems likely that one of the gods in Mecca who had popular support was called Allah, and indeed that among the polytheists Allah was becoming established as the leading God. But the greatest difference between modern Lebanon and Arabia in 570 AD was of course that there was as yet no Islam, and the Koran (Quran) had not yet been written.

Mohammed the Prophet, was born circa 570 AD at Mecca not far from the Red Sea. He was obviously interested in and influenced by local theological disputes, and by reformists in Mecca who were neither Christian nor Jew, but who were monotheistic. He was also a contemplative man, who spent time in meditation on Mt Hira nearby. He was forty years old or thereabouts when he received his first divine revelation from God via the angel Gabriel. Some legends would have it that there was a single revelation of the whole Koran in one night, but it seems much more likely that both the revelations themselves and the subsequent recording of them were more sporadic than that, probably over the length of Mohammed's prophetic period (610–630 AD). But whichever way, Mohammed was chosen as God's messenger on earth, to be the Seal of the Prophets – that is to say the last of the great line of prophets, after whom no others would follow. The Koran is divided into 114 chapters of unequal length, the shortest containing only 3 short verses, the longest containing 306 long verses. Most Islamic and non-Islamic scholars agree that the Koran has remained essentially unchanged, and is therefore still in the authentic original form, although recent discoveries of ancient manuscripts may undermine this view

The earlier prophets were the Prophets of the Jewish Old Testament, and the great prophet of the New Testament, Jesus Christ. Islam completes the line of evolution of these religions, and completes the revelation of God to man. There is no God but God, the Almighty, who was given the name which was most understandable in Mecca – Allah, the Compassionate, the Merciful. Mohammed revealed that the Christians had gone astray – that Christ was not God incarnate, though indeed a man of God, and that the Christians had erroneously and needlessly built an intercessory barrier between man and God.

Mohammed preached his gospel and gathered disciples around him. As with any other new sect that threatened orthodoxy, it was also seen as politically threatening, and it was not long before persecution started. In AD 622 Mohammed fled from Mecca to Medina. The Hijra, as the flight is known, is the starting date of the Muslim calendar. From Medina, Mohammed and his followers organised first armed resistance and then armed triumph over their enemies. The combination of the all-encompassing dogma of the Koran and the fanatical devotion of the new adherents proved an extraordinarily powerful mix. A new religion had been born, and soon unified the Arabs with each other and against the Greeks and Persians in turn, expelling them from Arabia. By the time of his death in AD 632 Mohammed had triumphed on earth to create a new kingdom for God, and had shown wisdom in its administration, courage and skill in its defence, and love and leadership for his adherents. His domestic circumstances were simple and unostentatious. His devout nature was beyond question. Where once tribe and kinship had created faction and division, there was a new and greater principle of faith in the one true God.

The Word of Allah

Allah created the earth and all things in it, and created man and put him in the garden. There is then an external earthly world, and we live in it as individuals but once. By serving God's will we shall attain heaven, a delectable place of beauty and leisure. By following Satan, the fallen angel, we will surely end in Hell. But God has revealed to us in the Koran precisely how we may follow his will and reach heaven. It is the comprehensiveness of the Koran that gives it such authority and appeal. It also should be learnt by heart by all Muslims at an early age – indeed learning the Koran by heart for many constitutes their sole formal education, even in places as far from Mecca as Bengal and Indonesia. And no matter where Islam has spread, the recitation is still in the original Arabic. The Book itself, even by those who do not accept its divine origin, is nevertheless credited with maintaining the literary beauty and integrity and homogeneity of Arabic over such a wide area of the earth. One of the mysteries surrounding it is how an illiterate and poorly educated man (or so legend would have Mohammed) could have written such a work – but the answer is that he was responsible for seeing that it was transcribed. In translation into other languages it loses the strong rhythms, poetry and cadences of the original, and therefore much of its beauty and power.

The Moslems refer quite often to the Peoples of the Book, by which they principally mean the Jews and Christians who follow the earlier prophets and their teachings. The Jewish prophets of the Old and New Testament re-appear in the Koran. Isa (Christ) son of Marium (Mary), is born of immaculate conception, and is revered as a prophet in the Koran, but he is not the son of God. The Koran claims that he was not crucified, but ascended unto God – so there was no earthly death and resurrection. There is therefore a kind of acceptance within Islam of the moral worth of these earlier but incomplete religions, and explicit provision for means of accommodating their adherents within an Islamic state. However, this is not a topic one hears much about in the current debates about inter-faith relations.

Islam is a proselytising religion, and in theory it gives those who are confronted with it three choices – to accept and submit (a Muslim is 'one who submits': Islam is 'submission' or 'faith'); to pay a financial homage but to remain unconverted; or obviously to resist and risk war, enslavement and execution. The carrot is as attractive as the stick is fearsome. Muslims believe in the equality of man, and in a strict ethical code. It has given people born both high and low a common brotherhood – and in the Islamic world over the centuries that followed Mohammed there seemed to be even greater social mobility than in many other contemporary societies, and many histories of rapid rises from the ranks to exalted positions. In terms of individual observance, Islam is said to have five pillars: 1) the Profession of Faith which must be made publicly at least once in an individual's lifetime: "There is no God but Allah and Muhammad is his Prophet." It defines the membership of an individual in the Islamic community. 2) Prayer or rather the duty of five daily periods of prayers. The first prayer is offered before sunrise, the second in the very early afternoon, the third in the late afternoon, the fourth immediately after sunset, and the fifth before retiring and before midnight, all facing towards Mecca. Special early afternoon prayers are offered on Fridays at the mosque, where the Muslim men congregate. 3) Almsgiving – the payment of zakat. This was originally the tax levied by Muhammad primarily to help the poor. It is now voluntary, but commonplace, and used to promote Islam as well as for poor relief. 4) Fasting during the month of Ramadan. (Because the Islamic calendar is lunar, the time during the solar year when Ramadan will fall changes). During the fasting month, one must refrain from eating, drinking, smoking, and sexual intercourse from dawn until sunset. At the end of Ramadan is the festival of Eid. 5) Pilgrimage to Mecca. Every adult Muslim who is physically and economically able to do so is enjoined to make this pilgrimage, the Haj, at least once in his or her lifetime. During recent years, air travel has allowed Muslims from all parts of the world to perform the pilgrimage, and the annual number is now in millions. Those who have participated often speak of the profoundly moving experience of participating with so many others of every tongue and colour and level of education, but all meeting as equals. On return to their homes, pilgrims thereafter may be deferentially addressed with the title Haji. The pilgrimage obviously promotes political solidarity in the Muslim world.

Besides these five basic institutions, other important laws of Islam include the prohibition of alcohol consumption and of eating the flesh of swine. In prayer much of the Koran will be recited – and the word itself means 'recitation'. The Muslim must make no idol, nor any image of any of God's creatures. This injunction is intelligible in its origin, but has had a possibly unforeseen effect – that much art associated with the Muslim world is abstract in pattern, and in the form of calligraphy associated with the Koran it has achieved spectacular magnificence.

The Muslim accepts the distinction between the physical world and the spirit of man (though believing in the ultimate resurrection of the body), but does not accept the difference between the spiritual and the secular. Whereas in Christ's gospel Caesar's government is accepted as lawful, the only state acknowledged in Islam is Islam itself- as so well displayed during Mohammed's life. The prophet was in many ways a man of peace and toleration – and indeed Islam permits the non-believer to pay his small tax and continue in his own faith. But what it will not do is allow the non-believer any political power or responsibility. This means in practice it can be difficult for an Islamic country to tolerate and administer ambitious non-Muslims, a point of importance to the story that follows. And it also means that 'law' – Islamic law – regulates more of an individual's life than civil law does in the West – where some aspects of moral law are voluntary.

Muslim Law: The Sharia

The Koran is detailed and specific on the laws relating to the economy and to society. Amongst the many codes are those that forbid usury, which means not the charging of excessive interest, but the charging of any interest at all. In the modern world there are Banks in the Middle East which work on Islamic principles, essentially becoming equity sharers in risk rather than secured lenders. The codes of marriage and divorce, and of property inheritance, and many other things, are all spelt out. The treatment of women is often picked out in Europe as being degrading. Women appear to be second class citizens, and certainly in a Muslim country women's liberation movements as known in the west would not be acceptable to the orthodox. But in Mohammed's time the laws he laid down were a great advance for the status of women. They were guaranteed rights in property and a share in their family wealth – and were no longer merely slaves and chattels. Muslim marriage contracts include within them the terms of the settlement on the wife in the event of a divorce. There must be many western women who wish our legal codes were so advanced.

Muslim law has two major sources – the Koran which is a comprehensive guide to righteous living, and the Hadiths. These are the customs and habits which the Prophet is known to have followed in his lifetime – graced by divine revelation. The collection of 2762 Hadiths by the early cleric Bukhari is venerated by Muslims as second only to the Koran itself. The interpretation of these together has resulted in the system of Sharia law which is recognised by most Muslim societies in some variant or other. In one basic respect it is profoundly different from, say, English law. Since they are part of God's revealed truth they cannot be changed by custom, and in essence no man has the authority to change them. Nor in theory will they ever be changed, since Mohammed was the last of the prophets. But of course in practice variation and flexibility in interpretation is possible, but usually the subject of fierce wrangling by lawyers and priests. In 2008 clerics in Turkey were trying to reform Islamic law to bring it up to date and more consistent with values within the European Union, which Turkey aspires to join. It is an experiment that is being keenly watched.

The appeal of this dogmatic religion has the same foundation as the appeal of dogmatic Roman Catholicism, or as once dogmatic Marxism had. Paradoxically, dogma may give people freedom. If we do not know what we can or cannot do, if we have few rules or bend those that exist, then we continually trip ourselves up by doing things which later we regret (if one's reach exceeds one's grasp, one is due for a fall), and we are continually hurt or frustrated in a society in which others may not adhere to expectations, and in our turn cause hurt and frustration. In addition, constant uncertainty requires constant attention and is ultimately fatiguing, and promotes withdrawal. By contrast if one is certain of the universe of behaviour, the bounds drawn around one's actions and responsibilities, and those of one's fellow men, then that certainty gives liberation, to act freely without fear or guilt in the known universe. The faithful who find this freedom with each other may find it difficult to be tolerant of those who do not live in this same universe.

The Spreading Fire

The armies of Islam, 'Fired by their new faith ...' (Robinson, 1982: 23), triumphed rapidly in North Africa and in the East. Within one hundred years of Mohammed's death the war cry of the Muslim warrior 'Allahu Akbar' (God is Great) – was heard at city gates in Spain and in Central Asia. The spread was rapid, but the society built was neither short-lived nor superficial. The Arabic Empire centred on the Caliphs (Khalifa – the Successor) in Baghdad became a bastion of patronage of the applied and theoretical sciences, and of music and literature. Much that Europe later believed it had rediscovered during the Renaissance of the learning of the Classical World was in fact knowledge which the Muslims had both kept alive and developed during the period of their dominance. Much of it was also knowledge that the Muslims gained from India and transmitted to Europe in mathematics, astronomy and medicine.

But not all was war without and tranquillity within. The Caliph, or Successor, guardian of the faithful, was supposedly elected. But after a few such elections, the post became hereditary, and the Caliph became in effect an 'Emperor'. But one group rejected the Caliph's leadership and claimed that it should be in the hands of the descendants of Mohammed's son-in-law Ali. This schism has led over time to doctrinal differences, and the emergence of the two major branches of Islam, the majority Sunni group (including most Arabs) and the minority Shi'ite group, found in many areas but in their strongest concentration and in the majority in Iran (Persia). This split is as great as the split between the Eastern Orthodox churches and the Roman churches in Christianity (although many Christians in Western Europe see the greatest split in Christianity to be between Protestant and Catholic, a division which underlay the recent troubles in Ireland). For Sunnis the authority of their faith is found in the consensus of the community and the texts of Islam (Quran, Sharia, Hadiths). For the Shias, authority was and is found within Muhammed's line of descent. Divine leadership was provided by the imams (literally leaders) who had the status of the divine on earth. Mohammed's grandson Husain became imam of the Shias, but was killed at Karbala (on the Euphrates and not far south of Baghdad) during an insurrection against the then caliph Yazid. The spiritual authority of the later imams became interwoven with the dynastic succession of the Safavid empire in Persia (modern Iran), where today the climax of the religious year is the mourning, on the 10th day of Muharram (the day of Ashura), of the slaughter of Husain. Shia and Sunni may both recite the same Quran in the same Arabic, they may go to Mecca together on the Haj, but both now carry with them a weight of history.

Wars of succession within the Islamic world were then as common as such wars in Europe, and, after its initial expansionary phase, the Islamic world often showed as many divisions and fractures as its contemporary Christian world, although more often than not independent princes and kings still sought the titular legitimacy of recognition by the Caliph, as during the centuries of the Holy Roman Empire many monarchs in Europe sought legitimacy from the Pope. But, independent of Baghdad or not, still the warrior princes expanded the world of Islam outwards, preaching a religion which retained its principle identity in the comprehensive dogma of the Koran.

As with Christianity, Islam also spread by another means. There was, and is, a branch of mystics known as Sufis who can be compared very approximately with ascetic missionary monks in Christianity, sometimes preaching, sometimes renouncing the world and withdrawing from it. These mystics probably reached India and settled in Sind and Punjab before the military invasions of Arabs, Afghans and Turks. In many ways they acted as a bridge between Islam and Hinduism, since the latter, too, understands the nature of mystical contemplation and experience, and the quest for absorption in God. But the path of the mystic is not one easily followed by the majority, who dream of earthly as well as heavenly empires.

Besides mystics, Islam has also spawned its own puritanical reformers. Muhammad Abd al-Wahhab (1703–1792) in the 18th century led a campaign against the veneration of holy saints and of holy sites, such as Karbala. He forged alliances with the tribes of the Najd (central Arabia) including the chief of Dariya (near Riyadh) Muhammad Ibn Saud. This was not just a puritan reform, it was also a revolt against the Turkish Caliphate. Even after Wahhab's death, the movement continued with success in much of Arabia. In 1803 Wahhabis attacked Mecca, and even set about destroying the Prophet's tomb. Much of the rest of the Islamic world was horrified. The Pasha of Egypt finally defeated the Wahhabis in Dariya in 1818, and the movement scattered. Wahhabi theology had been well argued through and well spelt out, supported by a great literature. The sixth of its seven central tenets is the obligation to wage war on the infidel, regardless of one's own suffering. The impact of the movement lived on in Arabia, and across the Islamic world, even if Wahhabis are usually a small minority. Wahhabi theology has a complex relationship with the royal family of modern Saudi Arabia, since it legitimises undemocratic 'strong' leadership. So to attack the Wahhabis is to some extent to attack the family's own authority, despite the fact that Wahhabis are more likely to be identified with the intolerant and more extreme expression of Islam.

The Submission of India

While the Arabian and Persian heartlands had been divided and embroiled in internecine struggle, the north-west passages into India had been fairly secure. Within India itself armed struggle between princes and states did not normally involve the sacking of temples – which were areas of sanctity *hors de combat*. Some of these temples had accumulated prodigious wealth, in golden idols and ornaments, and in precious stones. The temptation to invade this rich land was always present. Invasion needed sufficient power to force the north-west land routes, and perhaps too the confidence of the legitimacy of plunder.

The first attacks of Muslim Arabs on India were not motivated by such expectations, but were restricted to maritime invasions of Sind to control the pirates who attacked ships in the Arabian Sea. The goals of these invasions were limited, and the invaders' relations with neighbouring Rajput States (Rajasthan) were not uneasy. This comparatively gentle contact between two religious systems and two trading heartlands was soon to be followed by persecution and destruction on an epic scale.

A former slave[1] had founded an independent Muslim kingdom in Afghanistan, based at the town of Ghazni, in c. AD 977. One of his descendants, Mahmud, had become a skilful and powerful military leader, ruler of a large enough Afghan empire to draw on the resources necessary for the invasions of North West India. His invasions did not found a new kingdom in India, but constituted a series of deep penetrations for the purposes of plunder. In one attack in 1024 he crossed the Indian desert to attack the temple of Somnath in Kathiawar (the peninsula in modern south Gujarat) killing 50,000 'idolaters' and reaping a huge treasure of gold and jewels. The idol itself he took back to Ghazni, where it was smashed into pieces and buried beneath a mosque, so the faithful could trample it under foot. He took the temple gates too. (Nine hundred years later the British tried to persuade the Hindus of North India that they had recaptured the gates in Ghazni, and had restored them to India, avenging this slight.) Other Muslim rulers in subsequent years followed similar policies, until finally came the attempt to invade India proper. First in AD 1191 and then again in AD 1192 Muhammed Ghori tried to force his way beyond the Punjab plains into the Gangetic lowlands – and met a Rajput army at Tarain (also spelt Taraori), northwest of Delhi near Panipat. Fast-moving horse cavalry confronted slow but mighty elephants. The brotherhood of

---

1 The word 'slave' is technically correct but does not do justice to the position. Many of these were military officers who had high rank open to them. See the comments on the Delhi Sultanate which follow.

Islam confronted the fractious Rajput lineages of Kshatriya castes. Men who knew they would go straight to heaven if they died in battle, confronted a society in which the pacifism of Buddhism and Jainism had sometimes echoed, and one in which battle could be ritualised.

Mohammad Ghori won. Within two years Bihar had fallen, and shortly after that Bengal was taken too. The invasion of north India was complete. One of Ghori's successors, Iltutmish was recognised in 1226 by the Caliph of Baghdad to be Sultan of India. The seed of the independent Muslim empires of India had been sown, in rich and fertile ground.

Persecution and Resistance

The new Sultanate (Figure 3.1) was based on Delhi, and included most of north India to the fringes of the Deccan within its sway. Its source of wealth was overwhelmingly agricultural – agricultural taxes of one sort or another were the basis to maintain the military levies that tried to repress incipient rebellion. Other means of increasing the treasury included the taxes on infidels, and the plunder acquired from the destruction of some Hindu shrines.

'Empire' was founded on some contradictory bases. The new Muslim rulers kept distinct from their subjects, and were an élite who held dogmatic and 'superior' ideas and ideals, much as in later centuries the British would presume a superiority of their own 'civilisation'. But these Muslim rulers had made India their home, and so were part of India yet not of it. The opposition of ruler to ruled was one way in which the rulers would need each other and maintain a supportive collective integrity. But distinctiveness and intolerance leads to rejection, which force must contain, and force is expensive. There were indeed periods under the Sultanate when economic repression almost matched the military repression it supported. There was also accommodation, with recruitment of both Indian converts to Islam and of Hindus into state service. The social organisation of the élite was also complex and disciplined – many of the élite within the Turkish groups being 'slaves', a status which actually gave them preferment in the upper echelons of the complex of military and political power.

A second problem the rulers faced was that the source of revenue, agriculture, was as wide-spread as the people. This meant that a system of collection had to be maintained in every village of every province, and yet somehow this surplus had to be centralised or controlled from the centre to ensure the dominance of the Sultanate. Tax officers called *Muqti*s were appointed, who had the duty to transmit the assessed revenue to Delhi, deducting what was necessary for the maintenance of a levy of troops. In addition, some persons were given rights in the revenue of large or small tracts of land – *iqtar* holders – in return for their military service. None of these rights were heritable, and indeed could even be rotated during the lifetime of the officials. This was necessary to prevent the coalition of the officials' interests with local dignitaries, which would be the basis for a new landed class

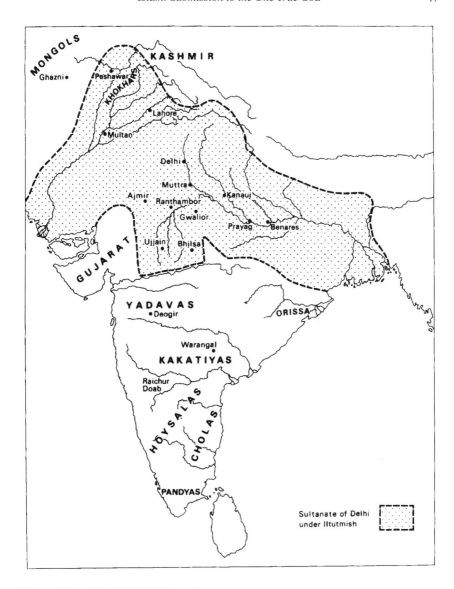

Figure 3.1　The Delhi Sultanate

Source: Davies (1959)

and the dissolution of the Sultanate – and at the end of the Sultanate this was indeed one of the processes fostering its dissolution. By contrast, it took much less time for the Norman barons of England to become a landed aristocracy with vested hereditary interests in land, and to be at war with each other.

Where movement in the northern river plains was easier, the small empire of the north more or less held together, even if wracked from time to time by wars of succession and secession. All of these were dangerous, exposing the dissenting state to the possibility of Rajput ascendancy, or further invasions from the north and west into Sind and Punjab. And indeed all of these things did happen from time to time. Mongol raids were common from the middle of the 13th century, and in 1303 a large army even invested Delhi. The border in the north-west receded beyond the Indus back towards Afghanistan whenever the Mongols themselves were divided, but would press again as far as the neck of fertile land between the Punjab and Delhi, where Panipat and Kurukshetra are located, whenever they were united. Nor were the southern borders secure. The Deccan was always beckoning, and offering that last route to cohesion and integrity – unity in pursuit of common conquest and profit, and some iconoclastic service to Islam too. The tribute exacted became an important source for financing the defence of the north-west. So at various times military expeditions penetrated into the Deccan interior, but none of them secured for the Sultanate a lasting extension in the south even though Hindu dynasties were toppled. The problem was simply that there were too many jungle hill ridges, too many empty and wild marchlands between one settled river basin and the next, to be able to maintain continuous and effective lines of communication. When permanent garrisons were attempted in the Deccan, for example by Qutb ud Din Khalji (ruled 1316–1320), the costs proved a drain on the resources of the Sultanate.

To name names, after Ghori came briefly Qutb, who built the tall stone tower, the Qutb Minar, which stands outside New Delhi, the world's highest free-standing (238 feet) stone-built tower to this day. Then came Aram Shah, and after him Iltutmish, who was succeeded, strangely, by his daughter Raziya. She was supposed to have mastered all the virtues of statecraft and had the resolve to make a great monarch: but she had the ineradicable liability of being a woman in a Muslim state. She was murdered in 1240. Ala-ud-din (1296–1316) pressed to expand the 'empire' in many directions, and during his siege of Rajput Chitor the royal women there committed mass self-immolation – *jauhar* – to avoid the potential rape and consequential pollution that defeat would bring – a ritual to be repeated in the city's history. Md. ibn Tughlag (1320–1390) was something of a megalomaniac, who moved his capital city and its population away from Delhi to the Deccan. He also built the walled city to be seen still just south of New Delhi. Muhammad (1324–1352) attempted both to develop a permanent capital in the Deccan near current Daulatabad, and also to capture both Peshawar and much of Afghanistan. To finance the costs of these enterprises, he even introduced a token currency – which prompted local revolt and dissension. Firoz-Shah was by contrast a wiser, and benign ruler, though still a purist Muslim. He began the famous irrigation canal near Delhi, later repaired both by the Mughals and then in turn the British. In 1398 Delhi was completely sacked by Timur the Lame, otherwise known as Tamerlane. He left a governor, and withdrew to his empire in Persia, Afghanistan and Central Asia (centred on Samarkand), but his governor was soon replaced by Bahlul Lodi, of the *ancien régime*. He in turn was followed by Ibrahim Lodi, the last of the Sultans, who died when the Mughals burst into India in 1526.

At different times the different provinces of the Sultanate owed more or less allegiance to the central power – and indeed independent dynasties ruled from time to time in Bengal and many of the Deccan provinces. Ala-ud-din Hasan, appointed by the Sultan to govern the north Deccan, established himself as an independent monarch in 1347. The Bahman dynasty he founded at Gulbarga lasted 150 years. The cohesion and impact of the Sultanate can thus perhaps be overstated – but for simple reasons. Iconoclasts, they destroyed much of the Hindu and Buddhist heritage, and they also took a great interest in their own history. They cultivated the arts and literature, and had their sagas recorded in poetry and prose – not particularly objective, but yet making an imprint to this day on our appreciation of their might. They built mausolea and mosques and seminaries, that still dot the plains around Delhi, some of which have now been brought within the built-up area of metropolitan Delhi. Lodi Gardens around the Lodi mausolea is one of Delhi's open parks, a kind of Hyde Park or Regent's Park. But the greatest monument is still living. It was during the Sultanate that a version of Hindustani used by the military became mixed with Persian and Arabic words, and written in the Arabic script. This is the origin of Urdu, the distinctive language of India's distinctive Islam.

Vijayanagar

Though there were embassies from North to South and vice versa, and even from time to time contracted marriages between Hindu royalty from the south and Muslim rulers from the north, much of the south retained its independence and cultural identity. It reached its climax in the two-and-a-half century-long kingdom of Vijayanagar, founded in the middle of the southern peninsula on the banks of the Tungabhadra river in 1336. The city near modern Hampi was a fortified area with outer perimeter walls variously given as 24 or 60 miles in circumference. It was a city of great wealth, and carefully developed technology in, for example, irrigation systems that still mark the landscape to this day. The society was not one that would be attractive to the non-Hindu – mass *suttee* was committed at royal funerals, and even low caste women were sometimes buried alive with their deceased husbands. When finally it was subdued by the combined forces of many northern states in 1565, the city was plundered and deserted completely. The ruined city lies to this day, its ruined marble palaces and great temples a major monument to Hindu architecture. However, it is not necessary, nor possible for reasons of space, to examine in depth in this book the Southern dynasties that were not incorporated in the northern Empires.

Second Foundation: The Mughal Empire

If ever there were a man whose own descent moulded him to be a man pivotal in history, a man who made the times as much as the times could make the man, then that man must surely have been Babur, the first Mughal Emperor of India, who was descended on his father's side from the great Timur, who had in his time sacked Delhi, and on his mother's side from Ghenghis Khan. He was only 12 years old when he succeeded to the governorship of Ferghana, north of modern Afghanistan. But he was ousted from his principality by the Usbeks, and marched south to make himself master of the kingdom of Kabul. Having already become a migrant monarch, and skilled in warfare, it is not surprising that the fabled riches of India, below the passes which he commanded from Kabul, would prove a temptation. In the cold season of 1525 he crossed the Indus with 12,000 men, a rather small contingent with which to attempt to take the Afghan Sultanate of Delhi.

Ibrahim Lodi, the Sultan, awaited him 50 miles west of Delhi at Panipat (near Kurukshetra) with 100,000 men and 1,000 war elephants, presumably not too anxious about the outcome. But Babur had with him, like Alexander the Great in his time, fast cavalry that could outflank the elephants, and even more importantly, for the first time in a major battle in India, he had artillery; and he was about to exploit his technical superiority in just the way that, in later centuries, the British would mercilessly use their field guns. At the end of the day the Sultan was dead and so were 15,000 of his troops. Babur occupied without further dispute first Delhi and then Agra.

Babur and his Turkish[2] nobility were now relatively secure: but whether they would like their new acquisition and stay was another question. The heat and desiccation of the hot season, and then the humid oppression of the monsoon, followed the early victory, and many of his army were inclined to withdraw. The principal Rajput leader was also ready to persuade them that this would be the right thing to do. Rana Sangha Sisodia saw in the defeat of the Afghans of the Sultanate his chance to establish Rajput dominion in north India, an India which even Babur seemed profoundly to dislike, lacking as it did to his mind convivial people, cold water, melons, good horses, good architecture and many other niceties. Sangha marched on Babur with a force again (so it is said) of 100,000 seasoned men, fighting in a country with which they were accustomed. It was no easy feat for Babur again to persuade his men to arms again, against an apparently superior host. But such were his powers of leadership and persuasion that he did, and at Kanhawa (Kanhua) in 1527, south of Agra, the wheeling cavalry of the Mughals ripped the Rajput army to pieces. Perhaps it had been after all a positive event, since now there was no military power left in Hindustan (North India) to dispute

---

2　The word does not mean they came from Turkey. There were several Turkish tribes in central Asia, some of whom (the Mughals) went to India, some of whom went earlier to what is now Turkey. The Mughals were thus related to the Turks of modern Turkey, but not from Turkey.

the sovereignty of the new aristocracy. Neither would it matter much to the Hindu masses of the north since the new Islamic regime was not so very different from the previous Islamic regime which it replaced, provided that it too exercised some degree of toleration of the different customs of the people.

Imperial Government under Akbar

The India bequeathed to Babur by the Afghans, who had always had a loose confederacy and who usually disputed the succession to the Sultanate, consequentially lacked established institutions of government, although periodic attempts at well-devised revenue systems had taken place. It was therefore necessary in Babur's eyes to set about creating those instruments which would translate military victory into secure dominion and lasting hegemony. Babur died too soon to achieve much in this way, and his successor Humayun, was for a time evicted from the throne, by an Afghan, Sher Shah, who is credited with some of the administrative reforms Babur knew to be necessary but never managed to complete. It was Humayun's son, Babur's grandson, who finally forged the magnificent Empire that came to represent Total Power in the minds of contemporary Europeans – and has given us the common use of 'Mughal' to mean the ultimate wielders of power. (British newspapers refer frequently to the Moguls of the film industry, or of the oil industry etc.)

This man is known as Akbar, which simply means the Great, as Alexander had once been too. In Spear's *History of India* he is one of two Indian statesmen before the twentieth century to whom he accords the rank of 'world statesman'.

He was born in 1542 in Sind while his father was a fugitive. He succeeded to the title in 1556 after his father Humayun had retaken Delhi and died in an accident: 'He fell down his library stairs, which was an appropriate end for a man of learning and culture' (Robinson 1991: 58). He then reigned until his death in 1605 – and the length of his reign was both testimony to his political abilities, and the opportunity to forge so much of what should be done. Of course, he was a skilled warrior, but he was no religious fanatic. He lifted discriminatory taxes from the unbelievers – the *jizya* tax – formed alliances by marriage with the potentially rebellious Rajputs, and recruited men of all creeds and races into his service. His power was in theory absolute, and in some ways one could almost say the same in practice. He could elevate the lowest to exalted positions, and punish with any degree of severity, including execution, any person, no matter how high, though clearly there were always political calculations involved in the exercise of such power.

Importantly, he owned all the land, and could both give and strip grants of rights in land. He could do so because the Emperor was the only embodiment of the existing concept of the State, and he could therefore use the power of the State as he personally wished, because there was no other way of expressing it. As had happened under the Sultans, in Akbar's state no official was allowed to establish hereditary rights in land. Such rights would have been the beginning of

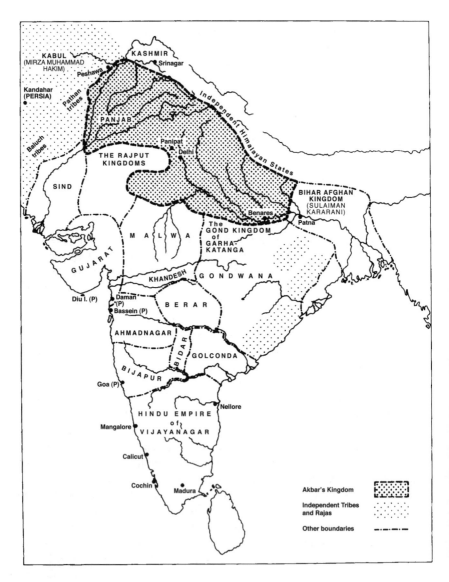

Figure 3.2   Akbar's Kingdom

Source: Davies (1959)

centrifugal and self-interested provincialism. It was the effective establishment of rights in land by the colonial aristocracy of the contemporary Spanish Empire that would ultimately lead to independence and fragmentation there. In a brief period of time after World War II the establishment of white interests in land in

Rhodesia/Zimbabwe rapidly overcame the identitive links with Britain, and led to Ian Smith's rebellion. The Mughal emperor's concern to avoid such developments can be well understood from such a perspective.

Territorially the empire was divided into provinces known as subahs, fifteen in Akbar's day, each governed by a subahdar – something akin to a viceroy. They are often referred to in English literature as satrapys, and a viceroy as a satrap, after the ancient Persian equivalent. Some of the subahdars controlled groups of provinces, which were then ruled by Nawabs. One such was the Nizam-ul-Mulk who controlled the Deccan provinces and from whom the title of the Nizam descended to the later independent princes of that area. Some – and here is a portent of the British Indian Empire – remained under the rule of indigenous princely houses, as in the case of Rajasthan under the vassal Rajputs, where principles of inheritance remained unaltered.

Below the province were the districts, forerunners of the districts of British India. At this level there was a split between the military commander, a salaried man known as the foujdar, and the revenue officer known as the dewan. The latter was very often Hindu, and therefore may have been useful in some circumstances as an informant independent of the Foujdar. Within a district a man of distinguished service might be given a grant in the revenue from a group of villages to reward him for past or present services. Such men were Jagirdars, holders of Jagirs. It is suggested that they could be a countervailing power to the Foujdar to be used by the Emperor if necessary.

All of these officers were personally appointed by the Emperor, and although clearly in many cases this would be done on advice and without a personal meeting between the two, nevertheless in theory an imperial letter of appointment, a *firman*, was necessary, such firmans being read in public at the courts of the district. At greater distances from the Imperial court itself not only was it possible but also not infrequent, that forgeries were made of firmans, to dispute local appointments. The British and French would later become drawn into such intrigues.

The principle of the division of revenue collection from the administration of civil (and military) authority was a sound one too. A foujdar who raised his own revenues would have been less likely to send the appropriate tribute to the centre. Since he was salaried and depended on centrally dispersed revenue, his ability to act independently was reduced.

The revenue was assessed and largely collected in coin. The monetisation of India, in silver, proceeded under the Mughals at an accelerating rate. Bayly (1983) even describes the empire as a great machine for the constant recycling of bullion, often moved in pack-trains under heavy military guard. Of course, it was both impractical and unnecessary to move all bullion taxes from the periphery to the centre to be disbursed again, and in practice it would be necessary only to move the net surpluses derived from imbalances in trade and the net receipts of central government. To do so meant that a system of banking and letters of credit were necessary – and in those days letters of credit would be honoured at a distance only between known and trusting bankers. Here we find another reason why the Dewans were so often Hindu. It was they, mostly castes of Kshatriya rank, who so often had the network of moneylenders that formed the basis of the banking system, and in its later years the empire rested financially on them and their services.

The central revenue and provincial revenues were dispersed for civil purposes (building roads and sometimes providing irrigation canals), but above all else were dispersed with largesse to the highly salaried military officials who maintained centrally and regionally the massive standing army of the empire. The army was colossal – Bayly estimates that at one time the livelihoods of perhaps as many as one quarter of the population were directly, or indirectly derived from military service, as dependants and camp followers. When an official was appointed, he would often select a new provincial capital, and suddenly the locus of expenditure would change. The local expenditure would raise demand and stimulate the economy locally, as farmers and artisans provided in exchange for cash, the needs of the army, sustained in turn by the cash taxes raised from the peasantry.

Military exercises and military campaigns at the borders were the essence of the lives of the ruling classes – no matter how much they were also patrons of music, literature, art and learning at the courts. An Emperor on inspection might move with an army of 100,000. The British travellers who first came to India were astounded by the sight. Unlike the British armies that moved in columns undeployed, on the plains of India the Mughal army moved in full deployment, fanning out over the countryside apparently with little regard for roads.

These peregrinations were also perceptible at the largest scales. The Emperors themselves were also inveterate migratory builders, founding and developing new capitals, moving back to re-invigorate older ones. Akbar moved his capital south of Agra to Fatehpur Sikri – a fabled and beautiful sandstone city that by fate stands ghostly and unaltered to this day. It appears that by a twist of hydrological development or, less likely, climatic change, the city ran out of water, and was abandoned. At another scale, the Emperor could and did order the wholesale migration of populations from one district to another. This might have been for military whim, but more often was because some areas went into declining production after long years of heavy use, or because the rivers and the underground waters changed on the youthful plains of Hindustan. There was therefore a constant shifting of population densities – a kind of long term and gigantic swidden system. Abandoned land might revert to jungle, only to be re-exploited decades later. This is why on the immemorial plains of India the timeless and ancient villages might appear in one sense to be just that, but in another sense might actually have been occupied or re-occupied only in recent centuries. The unbroken continuity of village site demonstrated by the Danish churches of East Anglia or the Saxon churches of Kent is not so often replicated in this more ancient cultural heart.

The Empire in Extremis and Decline

Akbar's empire was not the zenith of Mughal territorial acquisition. After him successive emperors progressively pushed their power south into the Deccan, but as the land area increased, the stability and coherence of the empire internally seemed to decrease.

Jahangir succeeded in 1605, though his son raised a revolt against him at the behest of intriguing nobles, and was blinded and then poisoned by another. He was indolent and cruel, but a great patron of the arts. The routine deadly fight for succession followed his death in 1627, with one son blinding the other. The victor, who took the title Shah Jahan, built the Taj Mahal (1632–1647) a mausoleum on the banks of the river Jumna (Yamuna) at Agra for his beloved consort Mumtaz Mahal, who bore him 14 children. It is worth dwelling on the implications of this event for a moment. Its construction and decoration involved artists and architects from France and Persia and elsewhere as well as from India, almost the best of the known world. The labour force was 20,000 – many of whom worked and died in squalor during the execution of the project. Wood was in short supply, so it was built with few derricks or other aerial hoists. Much of it was buried by 'scaffolding' of brick and earth while it was built, the final stage being, of course, the exhumation of the building. A ramp on one side, up which bullock trains and elephants could pull large stones, is reputed to have been 2 miles long. The building rests on a series of wells which are effectively a form of piling in the riverine undersoils, the whole contained within a bracework of metal clad hardwoods. This provides stability and also a degree of insulation from earthquake shock. Each of the minarets is inclined 1 degree out from the vertical, so that in an earthquake they would be more inclined to fall away from the main building than on to it. It is faced in exquisite white marble, much of which at the lower levels is inlaid with semi-precious stones in beautiful pietra dura. In earlier times many parts were said to be inlaid also with precious stones. On the opposite bank of the River Jumna Shah Jahan planned to build himself a twin mausoleum in black marble, but he was overthrown by his son before he could do so, and was finally buried alongside his wife.

This story illustrates the availability of capital in the Empire, and the abundance of labour, and its squandering in unproductive investment (at the time, though its current value to the Indian tourist business might almost make it worthwhile). It is even said that the 'misuse' of this capital took people from the land and resulted in terrible famine, though almost certainly that famine would have had several coincident causes. At a time when European traders were amassing and re-investing capital productively, here was one of the world's great powers burying it, albeit exquisitely. But in terms of world history such a use of capital must be seen as the usual one. What was happening in Europe was what was novel, and it was the beginning of the revolution that would change the world economy.

Shah Jahan's favourite son Dara Shikoh was a religious thinker in the mould of Akbar – willing to debate with Hindu sages, and to seek accommodation between different theological systems. But he was executed by his younger brother Aurangzeb in 1659 for apostasy – i.e. heretical thought that did not give Islam unquestioning acceptance. In taking the throne, Aurangzeb also had two other brothers killed. This gives something of the measure of the man – and the three deaths might not have been out of total cynicism. We cannot be sure of his beliefs in his early life, but he certainly campaigned ceaselessly throughout the rest of his life to re-impose a strict Islam, and in death left instructions for the simplest of funerals, devoid of pomp, to be paid for by money he had earned by sewing caps, and money he had made by copying out the Koran was donated to Holy men. He died in military harness campaigning in the Deccan at the age of 88 – indeed the last years of his reign from 1681 to 1707 were one long campaign. He removed the capital from Delhi to Aurangabad in Maharashtra, to support these campaigns, and it is said one third of the population died in the move. He revoked the delicate balance with the Hindus within his empire, reimposed taxes on the unbelievers, smashed temples and idols, and even built a mosque in Benares (Varanasi) Hinduism's most holy city. He fell out with the Rajputs, and broke their forts and temples too. Before his death he had extended the Mughal empire to within a few score miles of the southern tip of India (Figure 3.3), but by the time of his death he was already in retreat back to the north Indian heartland. This final explosive expansion of the Mughal empire also seemed to signal its internal collapse and decay. In alienating Hindu functionaries in particular, the land revenue on which so much depended became increasingly difficult to collect.

Why did the Empire begin to crumble? There is no 'whole truth' to be known about this period, nor is there a complete statement of the 'facts', and such facts as we have fit well with some theories, poorly with others. Let us return for a moment to Akbar. Above all others he organised the framework of the Mughal state, and achieved a degree of harmony and consent within. One of the ways in which he did this was to establish the concept of the Sovereign not just as ruler of the people, but also as their figurehead. In modern times it is accepted that there is a concept of The President (of, for example, the USA) which is different from the current incumbent. Akbar defined The Emperor in a similar way. He even went so far as to abjure Islam in its strict theological sense, to embrace aspects of Hindu thought and belief and to evolve a new religious cult around himself as a semi-divine. Partly no doubt he wanted to escape the influence of the Mullahs, the Ulema (or collective priesthood) who at times threatened to wield as much power as in contemporary Iran. What he achieved was a legitimacy for the Empire and 'the Emperor' which was Indian and acceptable to Hindus as well as Muslims in the Indian context, in which there was otherwise little sense of cohesive identity. What he achieved was of course only partial. Though he legitimised Empire, and created a clear central authority, yet in the process he so mixed the faiths and nationalities of the imperial administration that he compromised the Muslim élite, the only group to have anything near pan-imperial identitive bonds. If Islam dissolved itself in syncretic Hinduism, would there be nothing left except regional states? And if Akbar fashioned a concept of Emperor, what and whom did this

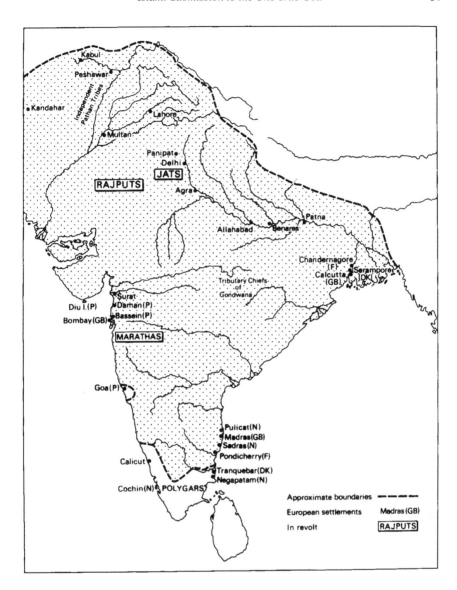

Figure 3.3   The Mughal Empire under Aurangzeb
Source: Davies (1959)

Emperor represent? An Indian nation? There was as yet no such thing, no pan-Indian identitive bonds.

If the concept is bigger than one man, and represents more than a single fief, how would the next incumbent be found? There are two separate but interrelated points here. Firstly, if there is a concept of head-of-state, Emperorship, which is something independent of the present incumbent, then there have to be some norms which define the field of candidates and the process of selection. Secondly, ascription, role assignment by birth, is common in Hindu society, but much less so within Islam. Ascription has been the basis of western kingdoms, but the rules for the selection of monarchs have survived in the remaining monarchies only because of the ever tightening circumscription of monarchical power. In India the concept of the Mughal emperor embraced phenomenal power. The field of candidates was most likely restricted to close relatives or sons of the incumbent but the rules for selection were brutal – the survival of the fittest in a period of intrigue and bloodletting. In many aristocratic families the same held true; and given the number of wives and sons a man might have, the potential rivalries were always acute. Succession was therefore accompanied by intrigue and division.

Having achieved central power, perhaps one way an Emperor had of re-uniting the empire was by waging war at the boundaries. This had several functions. The state was a military one, and expensive to maintain. The ruling aristocracy was also multiplying, and needing new positions of acceptable wealth. New conquests simultaneously united the aristocracy and offered the new wealth and power which was sought. But in addition perhaps Aurangzeb saw the possible loss of identity of the ruling class, and his return to Islamic fundamentalism, which set rulers against ruled, also re-united the rulers. Perhaps it was theologically inevitable, since, as we have seen, in the end the kind of compromise that Akbar attempted is impossible within dogmatic Islam. A religion with a revealed source of truth is not one which can freely evolve.

In the years of expansion under Aurangzeb all these features become most clearly seen with the continuing war against the Marathas, a group of Hindu castes of the high western Deccan, in Maharashtra. In the Deccan the Empire faced great problems. There was little utilitarian integration with North India, since transport was so poor and difficult. Coercion could perhaps be relied on so long as no local identitive force emerged. But in the mid-century under the leadership of a remarkable chief Sivaji, and uniting cultivator castes and Brahmins together, the Marathas emerged as a warrior Hindu force to lead successful revolts against Mughal power. They perhaps had learned something from the imperial masters, because above all they had learnt the value of the horse, of cavalry, and of mobility. They sacked the Mughal port of Surat on the west coast, and began to cut a swathe across India towards Bengal, thereby severing the communication between the North Indian heartland and the Deccan. But they never negotiated a tributary status to the Mughals as the Rajputs had done. The relentless years of guerrilla warfare turned them into a predatory and disruptive force, feared in much of India, but which ultimately left no lasting state behind them. After Aurangzeb's death, their power further expanded north, until even Delhi lay within their grasp. But the story of the resolution of their bid for power must await till we have dealt with other matters.

The Marathas were not the only new force to emerge in late Mughal India. Another, but not so significant till a century later, was the birth of a new religion in Punjab during Akbar's time – Sikhism. This had in its origins similarities with the origins of Jainism and Buddhism, in that it was a reformist revolt against inegalitarian Hinduism. But it was, of course, born into a new India that was embraced also by Islam. Sikhism has borrowed mysticism from the Hindus, and monotheism from the Muslims. It was founded by the first Guru, Nanak (1469–1538). The fifth Guru was put to death by Jahangir, partly because of events during his war of accession. Hostility between the Sikhs and Muslims grew, and reached a peak when the zealot Aurangzeb executed the ninth Guru, who refused to embrace Islam. The Sikhs, so distinguishable by their turbans, beards and swords, became at least in part a military brotherhood of the faithful, some of whom were fighting hard in the 1980s and early 1990s in independent India for that sovereign homeland (Khalistan) which briefly they enjoyed between the demise of the Mughals and the advent of the British.

The Legacy of Islam

The imprint of six centuries of Muslim rule left a legacy that would have to be considered in all future political and geographical calculations. The imprint was not so much in the physical fabric of the country. They had indeed built some roads, and done a little to improve irrigation. They had exacted as much as one third of the produce of the land as tax, but this wealth had mostly gone into the army, the palaces and tombs and the conspicuous consumption of the rulers. It has often proved difficult to re-invest surpluses cumulatively in agriculture (even in our age there are step-like shifts when yields could be raised by capital investment) and in the time of the Mughals the technology of agriculture was mostly unchanging. The possibility for capital accumulation in industry did not exist either – and given the number of artisans there was little pressure to innovate. But in the mind, the legacy was huge. Islam was implanted in India and did not die with the death of Islamic empires. There were many different kinds of Muslims in the subcontinent – all bound by the Koran, all in theory able to recite Arabic, but otherwise widely different, by mother tongue, and also still indeed by 'caste'. This sounds a contradiction: but many converts though embracing Islam had still retained the marriage groups from their original castes, as can be seen in Table 2.1 where the lowest caste of bangle seller and cotton carder etc. are recorded as Muslim. (I know of two families who were distillers and leather workers by caste until conversion two centuries ago. They are still following the same trades, and will not intermarry, despite the fact that they are the small Muslim and also economic élite of a small town. The distillers cannot, as Muslims, drink their own product.) In modern Pakistan there is a rough equivalence between what are called 'bradri' (also 'biradri'), a term used of kinship groups (it is also used in north India sometimes) – and what was once caste. After all, in Islam all men are equal; but there is nothing to say that people have to marry at random, no more than in Christian Europe upper class have to marry lower class.

Not surprisingly the society which was most Islamic from top to bottom was that of the Indus valley, from north to south, since here there had been countless invasions and settlements, by Turks, Mongols, Afghans, Persian and sea-borne Arabs. But it was not a particularly densely settled area, much of it being, before the irrigation schemes of the last 100 years, essentially semi-arid or arid and not very productive.

The most ancient and densely settled area of India was, of course, the Ganges valley – the seat of proud Hindu empires in the millennia before Islam arrived. Here was the wealth that had formed the basis of empire: this was the land ruled by the Muslim aristocracy for longest, and here by and large the Muslims were an urban élite – the obvious masters of the towns and cities, where they built their mosques and mausolea – but not so obvious in the countryside, even though there were and are Muslim cultivators in many villages. The same is true of the Deccan, under Muslim power for the least time of all in India, where the Muslims were a fort- or city-based aristocracy. Most of the Muslims of the peninsula are still urban, though there are areas of Muslim peasants, too, in the south.

In all of these regions of course there were many converts, both high and low. Islam can have a particular appeal to the lower polluted castes, since it offers, in theory, equality and dignity. But the region which matters most where such conversions were overwhelming is the eastern part of Bengal. Here the peasantry almost en masse became Muslim. The most satisfactory explanation of many put forward for why here, and not in Bihar or Uttar Pradesh, is that the trade winds and the monsoon naturally link the middle-east with both Malabar and Bengal, and that Arab traders and missionaries came to both, before the Afghans invaded from Ghazni to found the Sultanate.(The same winds would later bring the British to Bengal.) It is also true that Golden Bengal was the source of fine silks and cottons, which attracted Mughal rulers in their turn. Whatever the reasons, it was here, far from the Indus Valley in the west, that there was the second great area of Muslim population.

An imprint of the Muslim empires writ large was of course the recognition by the populace of both the probability and even the desirability of Imperial power, which provided some stability and peace, and of the fact that there would be an administration which exacted from them the surplus product of the land to maintain that empire. The Mughals had a clear conception of the unity of India, including the South, even if they never really quite integrated the subcontinent. They had a vision of unity which the populace did not have, but their vision could be transferred to the populace at large, and during the British period it would grow deeper roots. The administrative glue of the empire, the system of revenue assessment and collection, would become in turn the basis of the British system.

The churches of England have many an effigy of a worthy knight lying on his tomb, each with sword in hand. On Shah Jahan's plain tomb in the Taj Mahal is a pen-holder, symbol of the ultimate power of administration.

Hindu-Muslim Relations

The Muslims and Hindus have therefore a long history of interrelating with each other. The two religions have affected each other through their adherents, some of them great statesmen or despots, some of them holy theologians, most of them the common people. There are three levels of relationship we will consider briefly here: that of grand politics, that of theology, and that of customary daily behaviour.

There is a history of Muslim–Hindu relations at the grand political level – of Firoz Shah or Aurangzeb centuries apart but both putting unbelievers to the sword, and smashing temples – or there is the inclusive and accommodating policy of Akbar intermarrying with the Rajputs, and freeing himself from strict dogma. At this level the British would also experiment with policies in later years – according to some, dividing in order to rule.

At the theological level, although Hindus were to some extent prepared to consider many of the ideas in Islam, and certainly had no reason not to be tolerant of its right to independent thought, the reverse was rarely, if ever, true. Islam demands acceptance of its status as a revealed truth. This was bound to mean that theologians would fight periodically any relaxation in standards by their flock.

At the daily level of the ordinary populace, religious affiliation is identified not just by what the followers believe, but also by the patterns of dress and behaviour which adherents follow, to proclaim their own group identity and privileged group membership. There are rules of diet: Muslims do not eat pork, but do eat beef. Hindus of no rank will eat beef, even if the untouchable will eat a pig. Muslims go en masse to the plain and unadorned mosque free from any graven image on Friday to pray, and five times a day the Muezzin calls the faithful to prayer from the minaret (these days with electronic amplification to drown a whole city) and they bury their dead and hide their women in modesty, if not purdah. The Hindus cremate their dead in public at the burning ghats, and attend temples adorned with every kind of graven image, some of it explicitly sexual.

Though in many villages and towns Hindu and Muslim have lived peacefully as neighbours for generations, cultural differences can be exploited. There are ways which can be used in attempts to instigate riots. A moustachioed man in a turban can hurl a pig from a motorbike into a mosque. A man with a beard and a Muslim cap can slide past a wandering cow and slit its neck. Tear the veil of a Muslim woman. Smear the high-caste village well with blood. The effect of such insults is also much greater than a westerner might expect. Because individuality is much less important in India than in the west, group identity is much stronger, and vice versa. A slight against an individual which reflects religious affiliation can be felt by a whole community. There are, therefore, so many ways to start a communal riot in India. It is a potential that politicians have not always been afraid to exploit.

PART II
The British Raj

Chapter 4
The Usurpers: The Life and Death
of John Company

Preface: Changing Britain

Before we consider how the British conquered India and founded their own Indian Empire it is perhaps as well to consider in greater depth just what is meant by the 'British.' Table 4.1 provides a chronology of incidental events in three continents over a period of nearly three hundred years and is quite enough to suggest that any definition of 'British' must be a changing one. Not until 1707 was there even unity in Britain, and even then there was Bonny Prince Charlie's rebellion yet to come. Elizabeth I's England was a modernising state, and in common with other European ones, a state in which there was a growing and quite strong sense of national identity. It had a growing community of merchants and traders whose interests did not always coincide with those of the crown. In the seventeenth century conflict over religious matters and the relationship of the crown to parliament escalated to the Civil War, and for a time Cromwell's Commonwealth supplanted the monarchy. After the Restoration the balance of power within the state had changed, and so had much of the financial administration. Throughout this period the trading companies of England, such as the East India Company, founded in 1600, and of other European states continued to grow and contest with each other across the globe. As late as 1695 the Scots founded their own East India trading company, but by 1707 the Act of Union created a unified Britain and a coalition of commercial interests. British and French interests in the new world clashed, and during what was effectively a war fought across the world (the Seven Years War 1756–63) fighting was seen in both Canada and India, in the former case between government armies, in the latter between the servants of rival companies. Victory in North America and the assertion of control over French-speaking Canada gave Britain an enlarged empire for a short time. The interests of the colonialists in what became the USA and those of the home government clashed, and after the War of Independence 2,500,000 settlers were no longer subjects of his Britannic Majesty. 'New England' defeated old England. By then British subjects in India had taken a firm hold on Bengal, but they were not colonists. They were traders, who had become in effect rulers under the pressured assent of and in the name of the Mughal emperor. Other princely courts in India also had their share of Europeans – free-booting adventurers retained as mercenary officers.

Table 4.1　　A Chronology of Some Events in America, Britain, and India

	America	Britain	India
1492	Columbus discovers the New World		
1510			D'Albuquerque of Portugal takes Goa
1525			Babur wins at Panipat and founds Mughal dynasty
1558		Elizabeth I crowned	
1586		Drake singes the King of Spain's beard	
1600		East India Company founded	
1605			Death of Akbar
1607	Jamestown settlement of Virginia colony	Virginia Company of London	
1613	English settlers destroy French colonies		English receive firman from Mughals to trade – Surat founded
1620	Plymouth, MA founded		
1632			Construction of the Taj Mahal starts; Portuguese driven out of Bengal
1639			Madras founded
1642		Start of Civil War	
1659			Aurangzeb takes the throne
1660		Restoration, Charles II	
1663	Carolinas settled	Charter for Royal Africa	
1664	English annexe New Netherlands, rename New Amsterdam as New York		Sivaji, the Maratha chief, sacks Surat
1668			Bombay acquired
1690			Calcutta founded
1694		Bank of England founded	
1695		Scottish Africa and India Company	
1700	250,000 settlers in Anglo-America		
1707		Union of England and Scotland	Population of Mughal Empire 100,000,000 Death of Aurangzeb
1709		Abraham Darby smelts iron with coal in England	

1711		South Sea Company	
1713		Peace of Utrecht	
1739			Delhi sacked by Persian Nadir Shah – who takes peacock throne
1756–63	Seven Years War	Seven Years War	Seven Years War
1757			Clive takes Bengal after Battle of Plassey
1759	Wolfe defeats Montcalm at Quebec		
1761			Marathas and Afghans in mutual annihilation at Panipat
1773	Boston Tea Party	Regulating Act (of India); First Iron Bridge in Shropshire	
1775	2,500,000 settlers	British population 8,500,000	
1775	American Revolution	James Watts steam engine	
1781	Cornwallis defeated by Washington at Yorktown		
1784		India Act	
1786			Cornwallis Governor-General; population of Calcutta 250,000

Since the British merchants had had no territorial designs on India the legal status of India had not been of prime concern to the British government in the same way that the status of the American colonies had been of concern, until the accident of the acquisition of Bengal suddenly raised the stakes. So arose problems which we would call 'extra-territoriality' in our current day, the attempt by a government to assert authority over its nationals even though they be outside the national territory and perhaps even in the territory of another state. (The British for example have clashed with the USA over their attempts to legislate for subsidiary firms based in sovereign Britain.) Would the British overseas acknowledge the authority of the home government? The Portuguese sailors in many parts of the world had degenerated into privateers, and British pirates were well known in the Caribbean. A Briton would even found his own kingdom in Borneo. In controlling the British abroad would the Company or the crown devolve much power locally, and on what basis? Did the Company see India as a cohesive whole? In America there had been separate colonies. The early British saw India as a continent of different lands, and there was no necessary supposition that the affairs of the different trading factories should necessarily be linked – no more than in later centuries the different colonies of East Africa should have been one land.

Technologically the 250 years covered by Table 4.1 saw great changes in Europe, but few in India. Elizabeth's England was not technologically superior to Mughal India. But the years of national rivalry in Europe and the beginnings of the industrial revolution in England gave her a rapidly increasing technical advantage over the Indian states, which became a lethal military advantage. Retrospectively, what seems remarkable is that throughout these centuries of change a continuing cohesive sense of first English then British identity survived long enough in India for there to be a history of 'British India.' Now, as Empire fades in the memory, 'Britishness' is dissolving back into its national constituent cultures.

European Expansion

Religious intolerance and fanaticism has been as much a property of the earlier religion of revealed truth, Christianity, as it has been of equally dogmatic Islam. After the Moslems had taken the Holy Land, the Christians of West Europe struggled through the centuries of the crusades to re-establish a Christian government in Palestine. The fight was also on to push back the Moslems from Spain and Eastern Europe. Given this irreconcilable hostility, it is no surprise that Europe viewed the Islamic world of Arabia and Persia as a blockage between itself and the fabled Eldorados of the east, lands of gold, of spices and silks, and also, so it was said, of lost communities of Christians. 'The Indies' was a magnetic conception of adventure, wealth, and religious reward.

The Europe that wanted to expand East was one which, in comparison with other continents, had many coastlines and many seas – the Mediterranean, the Baltic, the North, the Irish, and one in which sea-faring and sea fighting was developing fast. The sea indeed provided the easiest and certainly most cost-effective means of transport of goods, land transport being by cart on poor rutted or muddied roads until the canal age (which in Britain did not start until the 1760s.) A 'properly drawn' transport map of the world of the 17th century would show all the trans-oceanic coasts close together (across for example the 'Atlantic River') and the landmasses huge and far-flung.

The first of the great European seafarers to seek routes to the east were Portuguese and Spanish, who took both the 'obvious' route round Africa, at all times going East when possible, and the speculative 'round-earth' route west, which resulted in Columbus mis-identifying the islands of the Caribbean as the 'Indies'. The memory of this error is retained in the contemporary term 'West Indies'. The rapid expansion of the Portuguese and Spanish Empires in the New World gave them a new wealth, much of it by plunder, but an increasing proportion by economic exploitation of new crops, principally sugar, grown by slaves on plantations. As was to be expected, trade with the new empires was a monopoly of the imperial power, and denied to the ships of other nationalities.

In 1510, before the Mughals had even come to India, the Portuguese were established at Goa on the west coast of the Deccan. Just like the Spanish in the New World, they spent much of their time and effort in missionary activity, often summarily executing or burning those who resisted. The next major European involvement in the East Indies was by the Dutch, who were much more hard-headed businessmen, and interested above all in the highly lucrative spice trade centred on 'Batavia' (Djakarta) in modern Indonesia. Europe's demand for spices was high: it seems a culinary quirk of imperial history that the Elizabethans both wanted and needed their spices in their drinks and to flavour the salt-preserved meats of the long cold winter, but by the time the British had, three centuries later, established their empire in India, many of those resident in India resolutely refused to eat any curries and preferred plain boiled vegetables.

The East India Company

The tight control by the Dutch of the trade with the Indies irked the English considerably. In 1600 the East India Company received a Royal Charter to begin English trading with the region. The significance of the charter was two-fold. First, it gave the company monopoly rights. No other English group could trade with India, and so the high level of risk the new venture faced was to some extent insured. (In actual fact there were some other smaller companies involved, but they were soon merged.) The second feature was that it was a company – a 'legal body'. This new 'fiction' which was developing in Europe was institutionally revolutionary. Akbar had to some extent separated the concept of Emperor from himself, but had not resolved the question of succession. Here was a new kind of 'body' which was separate from, but dependent on, all subscribing members, and which could recruit new subscribers on any basis. Its longevity could be indefinite. The State's interest in the Company was simple. The State taxed trade and made money from it: the more there was and the greater its value, the greater the revenues.

The company made attempts to set up a factory (trading station) in Indonesia, but the English were summarily executed by the Dutch, in a manner which was not uncommon in the rivalries of the period. To a certain extent trading with India was accepted as a second best, particularly since India had few spices except pepper. In 1613 John Hawkins received a 'firman' from the Mughal Emperor in Delhi to trade, and set up a factory at Surat (north of Bombay in Gujarat). In 1614 the first vessel laden with cottons and indigo dye reached London. The profits made were good. From then until 1657 the method of financing successive voyages was somewhat haphazard. Capital was sought for each voyage, and the original stake plus the profits divided in proportion and distributed at the completion of sales. This actually meant an investment of several years, since the round trip to India took three years, and the sales in London were also phased over a long period of time as well. The risks of losing a ship at sea were also high, but profits of 100 to 200 per cent were not unknown. The factor who was left in India had, though, a difficult task to perform, ordering textiles and dyes without knowing when and with what capital the next ship would arrive. From 1657 therefore the system was changed to permanent shares, and dividends were paid on a more regular basis, when the true costs of the operation became more apparent. The annual rate of return was usually somewhere around 10 per cent.

In 1639 Madras (now Chennai) was founded (the settlement was known as Fort St George) on the Carnatic coast, with a licence from an independent Hindu ruler whose kingdom had not yet been humbled by the Mughals. This, the south-eastern coast of India, might have been attractive to merchants, but was surely not to the sailors. Nearly the whole of Eastern India lacks a good natural harbour, and at Fort St George there was nothing but an endless beach with rolling surf. And it was also exposed to the returning November/December monsoon. (In the Fort there are contemporary cartoons of crinolined ladies being tipped by waves from the lighters which ferried goods from ship to shore.) In 1668 Bombay Island on the west coast north of Goa became the first British territory, a dowry to Charles II from his Portuguese consort which he rented out to the company. Here was an excellent harbour, but contrarily one with a very limited hinterland, hemmed in by the massive cliffs of the Western Ghats, at the top of which lay the plateau which was home to the Marathas. In 1694 work started on constructing Fort William in Calcutta – very much within the Mughal domains – and close to Danish, French and Dutch settlements. Bengal had the key features that the British required most. It is a land of rivers and, importantly, tides – so the big ships of the British could penetrate quite far upstream (Calcutta is 100 miles from the sea). This provided them with some degree of safety – since the ships carried cannon. The province provided a good trade in indigo, in cotton, and in saltpetre for making gunpowder. (Saltpetre is potassium nitrate, which can occur as a salt crust on the soil in hot countries, particularly near villages where nitrogen-rich sewage and other waste is processed by bacteria.) The ships could not get much further upstream than Bengal, since in the monsoon there are winds but strong opposing currents downstream, whereas in the dry season there is less current, but little and unpredictable wind, and a braided course. (Country boats are haled upstream by men or draught animals to this day.)

What did the Emperor and the potentates of the Deccan make of this? The Emperor enthroned in Delhi or Agra ruled a land empire, which had throughout the millennia of its history been attacked and subdued only from the North West. It was true that Arab invasions had succeeded in Sind, and that the Portuguese had made a thorough nuisance of themselves in Gujarat. But power lay with land revenues and military levies, and the fortified cities of the heartland of India. Compliant Europeans provided trade, which provided customs dues, and they had to and did pay in silver bullion. Aside from their slight usefulness, they must have appeared a sickly lot. In the long sea journeys men died of scurvy and ship's typhus, spread by lice. (It is said that in the early Portuguese voyages of a thousand men who might go, only one hundred returned.) In truth they were not a great threat: it was as much the internal collapse of the empire itself that would drag the British and the French in.

The Pattern of Trade and its Growth

In the early 17th century the initial 'fleet' of the East India Company was by any standards modest. Annually some 5 or 6 ships might sail together, not just to India but beyond to Indonesia and China, the largest of these being only 600 tons. The English had little to offer to the Indians: the aristocracy of the Mughal Empire had virtually all they needed, and aside from them there was little purchasing power. Much of the payment had therefore to be made in silver bullion – which was acceptable. But the trade threatened the English balance of payments badly, and alarmed the government. In fact, it was partly the fear of being drained of bullion wealth that prompted the European nations to found their own colonies in the New World from which they would import their own tropical and subtropical goods. Since they could not establish settlement colonies in well-populated imperial India, this option did not arise.

After the share system had been put on a permanent basis, trade began to grow quite fast. By 1675 imports into Britain are reputed to have been £860,000 – though exports were only £430,000 by comparison. The exports from England included some novel manufactures, lead, mercury, and woollens. One hundred years later the imports were £3,000,000.

The overwhelming desire for silver led the British to trade illegally with the Spanish New World. After the Peace of Utrecht in 1707 they received the Asiento, which gave them the monopoly to supply slaves from West Africa for silver. Then, following the winds they took sugar and rum to the new colonies of North America and from there tobacco and some cotton to the UK. The profits on this trade helped produce the silver: and in the far east trade with China, which held the world monopoly on the production of tea, and with the Spice Islands, produced yet other goods of value in India. Very soon after Europe's maritime expansion there was a world pattern of trade linking the great oceanic littorals – but a peasant some few miles inland in India would have known little of this unless he also indulged in some handloom production of fine textiles. Then he might have noticed that demand had increased: and that the agents were giving him strange designs in a strange artistic form from a foreign land to reproduce. (Sometimes with strange results too – the Europeans took prints of classical scenes which included perspective drawings, to India for copying. Not understanding perspective the artisans turned statues of nobles standing on flat surfaces into flying gods hovering above sloping surfaces.)

Given the appalling conditions on board ship and the rigours of the journey, it might have appeared surprising that anyone accepted the task. But there was, of course, great poverty in some sections of society in England; press-ganging was used, and as well as the stick there was a big carrot. Besides the official trade, there was also an extensive unofficial one. The Company's servants often undertook trading on their own account. Elihu Yale who gave his name to Yale University was relieved of his post as Governor of Madras in 1692, by which time he had already made 5 million dollars in 20 years of service to the Company. He stayed on a further 7 years to make yet more. Thomas Pitt (born 1653) went to India on his own account and traded in defiance of the Company's monopoly and returned home with huge wealth and a diamond worth £35,000 (now in the French Crown jewels). The Company's servants indeed appeared at times to be doing more on their own account than on the Company's.

Rivalry with the French

So long as the trade of either Britain or France remained small, neither was losing out at the expense of the other. Because of poorer commercial and political organisation in the French companies, for a long time the French posed no real threat to English economic interests in India. But after reorganisation in the latter part of the 17th century, their trade began to expand rapidly from the beginning of the 18th. From being a twentieth of the size of the East India Company's trade it became a half by 1740.

In 1744 word reached India that Britain and France were at war in Europe. The small garrisons of Europeans did have arms in their factories – to protect themselves against local disturbances. This time the disturbance was of their own (not only local) making. The French attacked Madras and took it, though it was returned to the British by treaty in 1748. Both the British and the French noted the impact of the guns they took ashore from their ships, and it also became apparent that land action on the coast could be limited by a naval blockade inhibiting reinforcements.

In the same year the Nizam of Hyderabad, overlord of the Nawab of the Carnatic, died, and the not unusual struggle for the succession took place. Dupleix, Governor of the French factory at Pondicherry some few miles south of Madras, had raised a small army of Indian troops under French officers, and used it with potent effect to intervene on one side of the dispute, thereby installing his nominee on the throne in Hyderabad. The ultimate aim was of course to use the local potentates to evict the British. It is pointless to speculate whether or not without Robert Clive the British would have been able to reverse this position, but it must be said that the position on land looked grave even if the British had assumed ascendancy at sea. Clive had been sent to Madras as 'writer' for the Company. It was a salaried job – essentially being a human Xerox-machine, copying out bills of lading and other documents of business transactions. He had been captured at Madras by the French, and taken to Pondicherry, whence he escaped south to help in the defence of the British factory, Fort St David, at Cuddalore. He had impressed his masters, was given an Ensign's Commission, and began the work of training a small army for the Company, and saw action in several minor encounters – though all of possible mortal danger. By the time of Dupleix's involvement in Hyderabad he was one of the senior officers of the small contingent of British and Indian Sepoy (enlisted mercenary regulars) troops. With his small band he reversed the French position in the campaign at Arcot, near Madras, although further engagements took place between the British protegé Muhammad Ali and his Maratha troops, and the French protegé and his troops. By 1752 the affair was over and the French bid for the Carnatic had been defeated.

The episodes there had a profound effect on Clive's understanding of Indian political and military conditions. It had appeared that troops in the employ of Indian princes, virtually all of whom were mercenaries, often in contracted bands under a contracted captain, would naturally prefer to be on the winning side, but did not necessarily assume that the winning side was the one they were currently 'fighting for'. Much of a 'battle' would be positional, and if it appeared that the position indicated that the other side might win an actual battle, should one be allowed to occur, then desertion or changing sides would gain that same result, but with less loss of human life and general destruction. In addition, it had also become apparent that Indian troops under European officers, and European troops, had been trained with a different concept of discipline (reinforced by direct action by superior officers against deserters) and with better weapons. Principal among these were the field guns, which were to the British and French what the cavalry and Turkish cannon had been to the Mughals. In short, the Europeans had developed far better the institutional and technical basis of violence, and the British foremost amongst the Europeans.

The Acquisition of Bengal

In 1739 the Persian Nadir Shah sacked Delhi, slaughtered many of its inhabitants, and returned to Persia laden with wealth and booty, including the Peacock Throne of the Mughals. After his death, Afghan chiefs achieved their own independence, and cast ambitious eyes on India. In the 1756 the Afghan leader Ahmad Shah Durrani sacked Delhi. But more than one vulture hovered around the corpse of Mughal India. Marathas seized on their opportunity to achieve empire by confronting the invaders. At first the invaders retreated beyond the Sutlej in Punjab, but then Ahmad Shah Durrani (also known as Ahmed Shah Abdali) outmanoeuvred the Marathas after they had procrastinated too long, and brought them to battle again at Panipat, in 1761. In a grizzly outcome, their principal leaders were killed along with as many as 200,000 troops and followers. Their military might was broken; so it was only the vestiges of this power that the British would have to confront later. But the Afghan army, which had been unpaid for some time, also mutinied and withdrew back beyond the Indus, leaving the heartland of Hindustan in a chaotic vacuum with a pretence of Mughal imperial rule.

In these unsettled times the Marathas had even harried Bengal, and to this day the people of Calcutta know where the ditch used to stand to protect the citizens against these predatory hordes, from whom the Nawab of Bengal offered little protection. As well as defences around the town, Fort William was also improving its protection. The improvements were built against the wishes of Alivardi Khan, Nawab of Bengal, who feared an enemy developing within as well as without. In

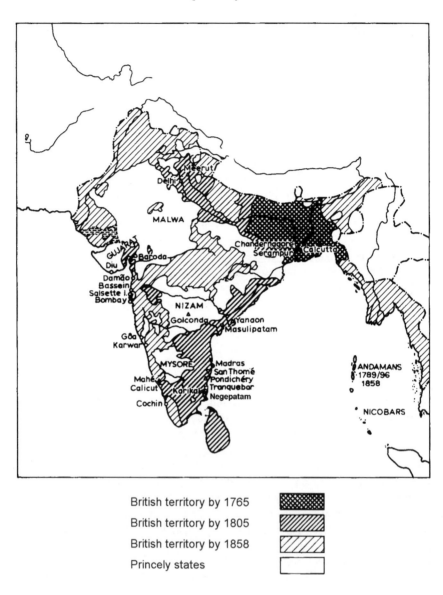

British territory by 1765

British territory by 1805

British territory by 1858

Princely states

Figure 4.1   The Expansion of British Territory in India
Source: Williams (1966)

1756 Alivardi died, and was succeeded by his grandson, Siraj-ud-Daulah, who was basically an independent monarch whether or not he wanted to be, since the empire now had few resources from the centre to offer him in time of need. He was young, inexperienced, insecure and contemptuous of his older generals and advisors. He

Table 4.2   The Inscription on a Tombstone in a Christian Cemetery in Dhaka, Bangladesh

The inscription illustrates the global nature of British imperialism, and the fear of Asian Oppressors.

Beneath this Monument lies
EZEKIEL BECK
Who departed this life 1791
Born of respectable parents in the Island of Barbados
An affectionate husband and tender father and sincere friend
Whose social disposition endeared him to all who knew him
He fell a sacrifice to power
Through the conduct of an Asiatic Oppressor
Who terminated his earthly existence
IN THE 37TH YEAR OF HIS AGE

His Disconsolate Widow Susanna
His Absence here must evermore deplore
Until like him alas (she is no more)
Who mouldering lies within this peaceful soil
And in whose name she raised this sacred pile.

moved swiftly from his seat at Murshidabad north of Calcutta to evict the British, and in 1756 he seized the town, in the process locking 146 prisoners in the 'Black Hole', a room 20 feet square, with two small windows, both shuttered. By next morning only 23 of the prisoners remained alive, the rest having suffocated, and the event passed into the folklore of British history, to symbolise the barbarity of 'the despots of the Orient' (Table 4.2). The calamity was probably unintentional – but no thought was taken to prevent it.

The Governor in Madras, though in no sense responsible directly for Bengal, took the decision to retake Calcutta, and despatched Clive with a naval squadron and troops. It was not an easy task, since his naval commander felt himself separately and directly responsible to London, and the Bengal Court of Directors were actually jealous of external interference. But with little military opposition he re-occupied Calcutta, provoking Siraj-ud-Daula to intervene. If this book had been written for an audience of young Britons eighty years ago, the Battle of Plassey in 1757, in which Clive's 3,000 defeated the Nawab's 50,000, would have been described as a stirring example of British arms and discipline. In reality it was the culmination of what he had learnt in the Carnatic. Before hostilities had been joined, he had negotiated a secret treaty with one of the Nawab's generals, Mir Jaffa, to install him as a new Nawab. The actual battle was a ritual in which some firing took place, and some men died, but whose outcome was completely prejudiced by the deliberate mis-direction of Mir Jaffa. The Nawab's hordes, true to custom, melted away as the position became clear. In current Indian demonology Mir Jaffa ranks as a traitor of the highest order: he still is to the Bengalis what Quisling is to the Norwegians. It may be acceptable for modern national demonology to disparage him; but in a way he was a token for a deeper shift that was occurring anyway. The Mughals had developed a system of government that relied on, and co-operated with, many classes – but particularly money lenders and bankers, most of whom were Hindu. These classes also prospered from the trade that the Europeans brought – and yet the Nawab had tried to eject them by force – thereby threatening one of their interests. They as a class could and did change their allegiance – mostly, but not always, succeeding in improving their position and returns by doing so.

The new Nawab, a tool of the British, assigned to Clive his own Jagir – the revenue rights of 1,200 square miles in the 24-Parganas adjacent to Calcutta. It gave him great personal wealth and roused the hostility of rivals in the Company in India and London. The situation was still not stable: the Nawab administered the province through his own officers, and collected the revenue through a deputy nawab, and another deputy was in charge of law and administration. It was by titles still a Mughal province: Persian was still the official language and all but the highest officials had kept their posts and responsibilities. In fact, since the Company had all the military power, exaction and coercion became commonplace, and Bengal began to be bled white. The inevitable tension led to further fighting, this time with assistance from the upstream provinces of Oudh and the interest of the Mughal Emperor himself. Victory for the British at Buxar (Boxer/Bakser: on the Bihar/Uttar Pradesh border) in 1764 led to a settlement in 1765 in which the Emperor recognised the right of the Company to the Diwani of Bengal – i.e. it became the agent for collecting the land revenue, which Clive estimated at £2,000,000 annually, of which only 20 per cent was to be tribute to the Emperor.

At the time that the British were transforming their toe-hold in India into a foothold, the French, too, made a major territorial gain. In return for their alliance with the Nizam of Hyderabad, they had been ceded the territory on the east coast known as the Northern Circars (Sarkars), which surrounded some British trading stations such as Masulipatam. The alliance was clearly a threat: and the simplest way to break it would be to deny the French the revenue, thereby in turn denying the Nizam the support of French-officered forces. In 1759 Clive sent Colonel Forde on a successful expedition which defeated the French, and lead to the annexation of the Circars in 1766.

The Struggle to Assert Control

A professor of economic history wrote that companies had many advantages as the forerunners of empire.

When trading with undeveloped peoples a corporation becomes a name and a power. Companies could provide relays of capital if necessary as individuals could not. They could and did take risks which no government would face, and they could, being only corporations, withdraw if necessary without loss of prestige in a way that would be impossible to a government whose national honour would at once be involved. They were, moreover, economical in working as they had not the same standard of expenditure as a government. Government servants must not make mistakes – it is the unforgivable crime – and so the tendency is naturally to play for safety. Companies on the other hand show flexibility and initiative and their officers developed a capacity for taking sudden decisions that would not have been possible to any civil servant. (Knowles, 1924: 261)

Companies were therefore admirably adapted to do pioneer work, as governments were later better qualified to organise an administration and to think in terms of future generations untroubled by questions of dividends or profit and loss of the moment.

It has been the function of the British chartered companies to go in front of the nation and to discover and organise trade. Sooner or later this trade brings the company into conflict with foreigners or with native rulers and the crown has had to intervene either in the interests of its own people or that of the natives, or to preserve order. (Knowles, 1924: 261)

Of course, the East India Company had never had any intent of being the forerunner of Empire. Yet it had become clear to Clive that British dominion was a real possibility, though, as he realised, beyond the capability of the Company alone. For more pressing practical reasons the moment had also come for some kind of role in government. At one moment Parliament was trying to limit outrageous dividends from the plunder, at another moment the Company was in ruin and seeking loans, having broken the economy of Bengal and found itself simultaneously with an ever-escalating military bill. In the midst of this uncertainty and confusion men paraded their new wealth in London – the Nabobs as the public called them in anger. The first steps were made, not to disband the Company, but to regulate it more closely. At the time the option of direct rule from Britain was not a serious possibility. Other than the Island of Bombay, the property of the Crown, the British had no territorial claim in India at all.

The struggle to assert control had several different aspects. From England the chain might have appeared clearly defined. Parliament needed control over the Court of Directors of the Company. They in turn needed control over their servants in India. And finally the servants in India needed control over events in India. The 1773 Regulating Act was designed to establish the proper lines of authority, and under its terms Hastings was appointed the first Governor-General in Bengal (1774–1785), overseen by a Council in Calcutta, and with authority over both Madras and Bombay. The power of those two Presidencies to make war or negotiate treaties independently was abolished, with the intent that the incessant and very costly interventions in the politics and warfare of the hinterlands would be halted. On Hastings' return he was impeached on charges of corruption and abuse of power brought by his political enemies. In the ensuing case, which dragged on for years and which ultimately absolved him, some major issues of principle were debated. Edmund Burke who seemed to have misjudged Hastings made several impassioned speeches for the principle of stewardship – that Parliament should be the trustee of the 'backward' people of the new Empire. The idea of 'trusteeship' was, thereafter, ever present in future relations between the British and the Indians – whether to be kept hidden under the table, waved in the open, or cynically derided.

Cornwallis replaced Hastings (1786–1793) and set about creating a fully fledged state, and a Europeanised civil service. He set about the establishment of courts and a legal system based on British custom but incorporating both Muslim and Hindu customary law – at the time the new court system seemed poorly understood in India even if in the contemporary age the Indians seem amongst the most litigious people on earth.

He also set about trying to bring order to the land revenue system, and made what came to be known as the Permanent Settlement. Under the Mughal diwans the next lower tier of tax officials were known as Zamindars – who collected the tax from an assigned group of villages – but they had no direct rights in the land. In return for fixing a final and 'fair' calculation of the revenue to be collected from each of the Zamindars' lands, the freehold of the land was effectively conferred on them, making them landlords in the European fashion. The intent was to create something like the class of landlords in England who were known as 'improving' since the majority were involved in investment in new agricultural techniques. The system, however, backfired. Many sold out to the monied classes of Calcutta, who became absentee landlords – free in effect to jack rents up far beyond the level necessary for taxation. The story of 'Zamindari' in Bengal, Orissa, Bihar and Assam from that date became one associated with a hierarchy of tenants, sub-tenants, share-croppers and landless labourers, synonymous with over-exploitation, rural poverty and stagnation. Later attempts were made to remedy the situation; but none were sufficiently radical to make drastic improvements to the fate of the peasantry.

Although he did not seek territorial aggrandisement nor war with neighbouring states, tellingly, Cornwallis was involved in an inconclusive war in the south Deccan against Tipu, Sultan of Mysore, who also had the support of the French. But, this episode aside, Cornwallis departed India 'leaving behind as he thought, an unaggressive and stable state within the complex of the Indian political system..' (Spear, 1973: 101). Yet 'Within 25 years of [his] departure ... the state had swallowed up its rivals to become the ruler of most of India and overlord of the remainder'. The reason for the failure of the policy of limited ambition and non-involvement was simply that the last condition, that the Company's servants would control events in India, was never possible. It was not to begin with a specific policy that the rest of India should be subjugated and to that extent it has often been said that Britain acquired India 'in a fit of absence of mind.'

The interior of the Deccan did not hold much commercial attraction for European traders, who were able to supply their needs by trade with the coastal zones. The lines of communication to the interior meant that trade with it was unreliable and costly. But the interior did have states like Tipu's Mysore able to launch predatory raids on the coastal settlements. And rivals of the British, such as the French, were fully committed to helping some do just that. During Lord Wellesley's Administration, in defiance of the Directors' wishes in England who saw nothing but cost in military action, British territory expanded rapidly, both directly and indirectly. He could see that years of inertia might be just as costly in terms of defence as subjugating the hinterlands to a Pax Britannica. In the south in 1799 Wellesley finally defeated Tipu and annexed half of Mysore. In 1799 Napoleon was active in Syria and Egypt – and quite clearly wondered about the possibility of a land route for the invasion of India.

Wellesley clearly saw the Maratha Confederacy as great a threat as they saw the British. In this unstable atmosphere it only needed a small wind to precipitate the storm, and it was provided by a treaty signed between one of the Maratha chiefs at Poona with the British, against the wishes of the other Maratha chiefs in the Deccan and Hindustan. When war came in 1802, the armies of all three presidencies were used. Wellesley's brother, afterwards the Duke of Wellington, was one of the leading campaigners engaged in what became a major defeat for the Marathas. By now it appeared that the three major British bases, far from being weak because each could be picked off separately by a central power, had become the reverse: bases from which the crumbling core of India could be pincered. The Marathas had been overlords of Delhi, even if they maintained a Mughal puppet on the throne. Now they were forced from the Ganges Valley, and in 1803 the British entered Delhi and took the emperor under their protection. The Marathas kept their homeland states, but were forced into tributary alliance with the British. In Hyderabad Wellesley had earlier forced on the Nizam another subsidiary alliance, and garrisoned British troops with the Nizam at the latter's expense. In 1817–18 further warfare with renegade Maratha bandit gangs, known as Pindaris, who were plunging central India into anarchy, resulted in the annexation of much of their territory by the Bombay Presidency. By 1818 the whole of India had been subjugated, from Bengal (but excluding Assam and the modern northeastern states) to the Sutlej on the south-east of the Punjab, and to the Thar desert east of the Indus. The parts that Britain administered directly became known as British India. For example, the Maratha dominions in Hindustan had been annexed to form the North West Provinces, later enlarged to form the United Provinces, and now Uttar Pradesh. Some of the empire the British ruled indirectly, acknowledging the internal autonomy of princely rulers in the Native States, also known as the Princely States, in return for control over their external affairs. Invariably, a British contingent of troops was posted in them, at the ruler's expense, 'for his own protection'. Over time many of these States were annexed on such pretexts as misrule, like Oudh (Awadh) in 1856.

The western boundary of this new empire was not so very different from the present boundary between India and Pakistan – with the exception of Kashmir and a part of East Punjab. The boundary in the east was the historic boundary – excluding the sparsely settled jungles of Assam. For the first time in South Asia an empire had grown that included the whole of the Deccan, even the extreme south, and even the rarely subjugated and remote tracts of Orissa. Now nearly the whole of India was under British hegemony, and quite clearly much of it not for the purposes of trade. What was the true purpose then? The argument that peace and stability were best assured by the British probably bears scrutiny, and indeed they did manage to re-establish law and order after the tumultuous collapse of Mughal and Maratha power, but at a cost, to the peasants. The new government reviewed the land revenue systems in operation in Madras, Bombay, and the North West Provinces, and made demands which were almost certainly excessive. History was being repeated: this geo-political region had once again passed through turmoil and fragmentation back to a single hegemony, but once again the victor sought the repayment of the costs of conquest with added profit.

In 1813 an Act of Parliament abolished the Company's monopoly rights to trade. In 1833 the trading function of the Company ceased, and the metamorphosis was almost complete: now it was in all but name simply British government in India. The new empire had almost reached its territorial limits. Events from the outside prompted only two major extensions. Firstly, expansionist Burma occupied Assam and demanded territory in Bengal. A short war resulted in the eviction of the Burmese and the annexation of Assam by British India in 1826 (and later in the century the final annexation of Burma, partly to forestall the French in Indo-China), and the creeping annexation of other frontier territory over the next century, as described in Chapter 12. Secondly, there were the struggles both territorial and political to secure the northwest frontier, adjacent to Afghanistan. This is the subject of the next Chapter.

Trusteeship and Reform

In Europe and Britain in the early 19th century social ideologies were changing fast. There were the new republican ideals in France, and in Britain there was a concerted movement to abolish first the slave trade, and then slavery in British possessions. In 1832 the Reform Bill in parliament cleansed the electoral system, another step on the evolution of democracy in Britain. India too became a subject in these debates, and reforming intentions were expressed to improve the circumstances of these new British subjects. In 1833 Indians were admitted to the officer grade of the Civil Service, though effectively on discriminatory rates of pay. Macauley in 1835 proposed a minute in parliament which was the origin of education in India on Western lines, in English, and in the same year English replaced Persian as the language of Government and the courts. Many of the reformers were quite clear that India would not remain for ever a British possession ruled by aliens, and would

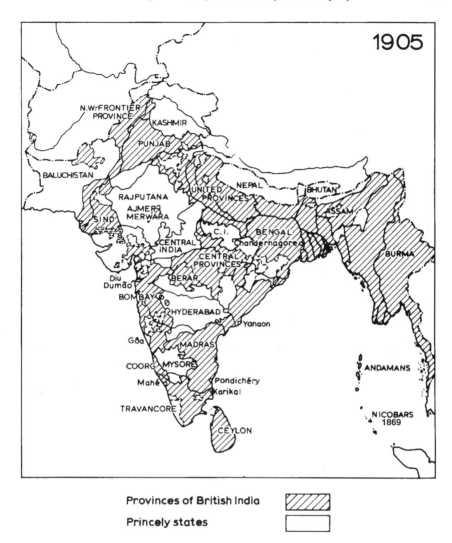

Provinces of British India

Princely states

Figure 4.2   The British Indian Empire of 1905

one day rule itself. Their intent was openly expressed as leading the Indians to a new enlightenment and the adoption of Western institutions by which, in the end, they would manage their own government. Christian missionary activity was now allowed, though neither subsidised nor encouraged by the Government as had been the case in the Spanish and Portuguese empires. Lord Bentinck, Governor-General (1828–35), is associated with many of these great reforms, some of which pressed hard on local custom. Suttee, child infanticide and sacrifice, and thuggee were all outlawed. Equality before the law was enforced, something which disturbed Indian tradition deeply, in which castes were conceived of as unequal. No Brahmin should appear before a man of a lesser caste, and the penalties enforced on Brahmins had always been far less severe than those on an equally offending but low caste culprit. Bentinck is not remembered affectionately by the current Indian public. He is thought to be something of a zealot, who, amongst other things, wanted to dismantle and sell off the Taj Mahal in pieces: but luckily for posterity the cost was prohibitive.

The Mutiny and Divorce

The events leading up to the Mutiny of 1857, which is commonly known in India as the First War of Independence, are complex. There was resentment at the contemptuous misuse by the British of many shrines, both small and great, both Hindu and Muslim. British disregard for caste taboos in the armed forces played another part, as did the much more immediate cause of pay arrears in much of the Bengal army. There had also been so many interventions in the land revenue system that a reservoir of disaffected élites had been formed. Orthodox groups in society had been offended by many of the social 'reforms' the British had attempted. In fact, much of India had been disaffected – and with good cause – for a number of years. Resentment mounted further when in 1856 the British annexed the state of Oudh, from which many army recruits had been enlisted. The final partly apocryphal flashpoint was reached when the army was issued with greased cartridges for their new Enfield breech-loading rifles: the grease was made from animal fat, both from pigs and cows – the one unclean to the Muslims, the other from the Hindu's sacred animal.

The mutineers turned on their British officers, and in many places massacred both them and their families. Delhi, Meerut, and Cawnpore (Kanpur) were taken over by complete regiments that had deserted. Lucknow was besieged, and the Residency of the British became a heroic symbol of stubborn resistance, in which the few survivors were reduced to eating rats and cockroaches. But the Mutiny was finally put down, using the armies of Madras and Bombay, and with the aid of the Sikhs who were prominent in the recapture of Delhi. The fighting was desperate and vicious – for the British were suddenly exposed as small rafts afloat in a wild sea. Reinforcements of unquestionable loyalty were summoned from England – but it was a huge distance and the time lag great. Retribution was awful – captured mutineers were regularly blown from the guns. (They were tied across the muzzle of a field gun which was charged but without a cannon ball. When fired the victim was blown into many pieces and splattered across the landscape.)

In truth, this was not a war of independence. Though many forces joined in their attacks on the British, there was no great conspiracy between them and little co-ordination. The mutiny spread by demonstration effects, not by design. Neither was the general populace particularly involved: there was no nationalism, there was no history of grass-roots state-building in India, and the troubled, ever-changing pattern of warring states of the previous hundred years had given little chance for any to emerge.

But the effects on the British were nevertheless traumatic. An alien élite ruling in an alien language, they had never become part of India as the Mughals had done: they had not cut themselves off from their roots. They had not granted themselves new estates in India. Now, after the mutiny they withdrew more, into their cantonments and clubs, but honoured pro-British Princes with styles such as knighthoods derived from Britain. British women had been taken to India before in small numbers: now as transport at sea improved (with the advent of steamships and the opening of the Suez Canal) they brought their whole families, to live in isolation separated from their Indian subjects, and during the high point of British rule it was quite common for unmarried daughters to be sent from England for the social season in India, in the hopes of finding a suitable mate. The British thus distanced themselves from groups which had previously been closer to them. In earlier days there had been a number of marriages and relationships between white men and Indian woman, the result of which was a new community of light-brown Christians, known as Eurasians.[1] These too were kept now in their proper station. The analogy is often made, falsely but usefully, with a new caste, which would, of course, require a caste-like occupation – but that was forthcoming too, on the railways.

In 1858, after the Mutiny had been suppressed the company was dissolved and the Crown took control of India through Parliament, instituting a Secretary of State with a Council of 15 members. In 1877 Victoria was proclaimed Empress of India at a Durbar in Delhi, and henceforth the Governor-General of India also added the title of Viceroy to his office. The ironies run deep. A parliament elected democratically by British voters was now ultimately the dictatorial government of India, in the name of an Empress. The British who believed they were to bring education and improvement and enlightenment to India, banned the Indians from their clubs and railway carriages. They were busy creating a new middle class of Western-oriented, liberal-minded Indians, but yet became racialists who kept the Indians at arms' length. What future could such contradictions have?

The British had to find for themselves the legitimacy of their rule, and they found it in the belief that they had something superior to give to the backward peoples. The Great Exhibition of 1851 in London was Britain's announcement to the world of the wonders of science and industry and progress. The Government in India commissioned annual reports from 1859 to 1937 into the *Moral and Material Progress of the People of India*. Look at the Table of Contents (Table 4.2) of a survey of India written in 1880 by a retired British senior office-holder in the Indian Government. This kind of legitimacy depended then on having a fountain

---

1 These people are known now as Anglo-Indians, of whom there are one hundred thousand in South Asia, mostly in the larger cities such as Calcutta and Bombay. Many of the women still wear European-style dresses, although most may now wear saris. The term Anglo-Indian was formerly used to mean British families in India, particularly those with long established connections, often over several generations.

source, from which it could be conveyed to India. And so the British in India were bound to be looking back to Britain, and bound equally to be racialist, since they saw their race to be the cradle of the new civilisation, the appointed messengers of the new truths. This is not to deny that many found in India both intellectual and moral inspiration, and that European appreciation of the ancient cultures and arts of India grew. But mostly they were separate and aloof – maintaining a cohesive identity had its uses. The bond meant that while the British were in South Asia, the unity of India in the broad sense was never in doubt. But what if they were not there? Was the pattern of their rule building an inherent unity or not? One glance at the territorial organisation would suggest not. The British Provinces were historical accidents, hotch-potch mélanges of different linguistic and religious groups within which fractious movements were stirred up. The Princely States were similarly accidental entities: and with a different pattern of administration and power-sharing were bulwarks against new grass-roots political movements spreading from province to province. These were not questions which greatly exercised many minds of the 19th century. For the majority, Indian self-government belonged to an unimaginably dim and distant future; now, it was the time to build the new economy and infrastructure of empire.

Table 4.3   The Dedication and Contents List of Sir Richard Temple's
**   *India in 1880***

Sir Richard Temple served as Governor of Bombay, Lieutenant Governor of Bengal and Finance Minister of India

a) the Dedication

TO HIS ROYAL HIGHNESS
A L B E R T   E D W A R D,
Prince of Wales
IN MEMORY OF HIS VISIT TO INDIA,
ON WHICH MOST AUSPICIOUS OCCASION
GREAT BENEFIT WAS CONFERRED
ON THE
PRINCES, NOBLES, AND PEOPLE
OF THAT EMPIRE
WITH A POTENT AND ENDURING EFFECT
ON THEIR HEARTS AND MINDS,
THIS BOOK IS,
WITH THE GRACIOUS PERMISSION OF HIS ROYAL HIGHNESS,
Dedicated,
BY HIS DUTIFUL AND LOYAL SERVANT,
THE AUTHOR

b) the Contents: Note the sub-headings are given here only for some of the chapters

CONTENTS
CHAPTER I
CLAIMS OF INDIA ON THE CONTINUOUS ATTENTION OF ENGLAND
Need of steadfast attention to Indian affairs on the part of Englishmen - Changeful character of India under British rule - Necessity of observing current events - Therefore survey undertaken of India in 1880 - Circumstances interesting to travellers and sportsmen - To those connected with art and science - To the friends of missions - To those concerned in public affairs - To economists and statesmen
CHAPTER II
OBJECTS OF BEAUTY IN NATURE
Objects of beauty in divers places - Improved facilities for travelling nowadays - Variety of climatic aspects - Phenomena of the rainy season - Different characteristics of the country on the north and on the south of hill ranges forming the backbone of the Indian continent - Pleasant mountain retreats during summer for European residents - Fine views from mountains looking towards the sea - Noble prospect from Himalayas looking towards the plains of India - River scenes - Marble rocks - Cascades - Mountains clothed with perpetual snow - Loftiest peaks yet discovered in the world - Lake region - Scenery of Cashmir valley - Features of the several Indian races - Picturesque effect of their costumes
CHAPTER III
OBJECTS OF BEAUTY IN ART
Architecture under British rule - Interesting views at Bombay, at Calcutta, and on lines of railway - European churches and cathedrals - Beauty of Native architecture - Buddhist caves, rock-cut temples, pagodas and monasteries - Ancient frescoes - Sacred mountains of the Jains - Hindu or Brahminical structures - Mention of the finest temples - Mountain fortresses - Beautiful palaces - Excellence of Muhammadan architecture - Interesting remains at many places - Largest dome in the world - City of the dead - Mosque, palace and tomb of Akber the Great - Grand mosque at Delhi - Pearl mosque at Agra - Matchless beauty of Taj Mehal mausoleum
CHAPTER IV
EUROPEAN CLASSES, OFFICIAL AND NON-OFFICIAL
CHAPTER V
NATIVE STATES
CHAPTER VI
MATERIAL PROGRESS OF THE NATIVES
Census of the population - Gradual growth of the people in numbers - Expansion of cultivation - Land can yet sustain increasing people - Cultivable waste in India itself - Emigration from India to British colonies - Sufficiency of food supply in India - Her wealth under British rule - Remittances of money, public and private, to England - Outlay of English capital in India - Wealth of former times and compared with the present - Capital largely accumulated by the Natives - Increase of cattle

- Reasons why India is inferior to Western nations in wealth - Maintenance of the poor - Public opinion regarding material improvements - Industrial employments - General condition of the Natives
CHAPTER VII
MENTAL AND MORAL PROGRESS OF THE NATIVES
Effect of British rule on the character of the several classes of the people - The peasantry - The aboriginal tribes - The temper of the Muhammadams - The Parsis - The Native nobility - The landlord-class - The traders - The priesthood - The educated classes - Their mental improvement, moral conduct and religious belief - Their loyalty - Their political aspirations - Native munificence - Culture of physical science - Vernacular press and drama - Advancement of Natives in the public service - In other professions - Good effect of Natives visiting England - State of the Native mind and disposition generally
CHAPTER VIII
NATIONAL EDUCATION
CHAPTER IX
RELIGIOUS ESTABLISHMENTS AND MISSIONS
CHAPTER X
LAW AND LEGISLATION
CHAPTER XI
CRIME, POLICE AND PRISONS
CHAPTER XII
LAND-TAX AND LANDED TENURES
CHAPTER XIII
REVENUES
CHAPTER XIV
CANALS AND IRRIGATION
CHAPTER XV
RAILWAYS AND ELECTRIC TELEGRAPHS
CHAPTER XVI
ROADS AND EMBANKMENTS
CHAPTER XVII
PRODUCTS, NATURAL, AGRICULTURAL AND INDUSTRIAL
CHAPTER XVIII
COMMERCE, EXTERNAL AND INTERNAL
CHAPTER XIX
PUBLIC HEALTH AND SANITATION
CHAPTER XX
FAMINE
CHAPTER XXI
LEARNED RESEARCH
CHAPTER XXII
PHYSICAL SCIENCE
CHAPTER XXIII

WILD ANIMALS AND SPORTS
CHAPTER XXIV
NAVY AND MARINE
CHAPTER XXV
THE ARMY
CHAPTER XXVI
FOREIGN RELATIONS
CHAPTER XXVII
FINANCE
CHAPTER XXVIII
STATISTICAL SUMMARY
CHAPTER XXIX
CONCLUSION
Recent War in Afghanistan - Its results and lessons - Prosperity of the Indian Empire, despite some inevitable drawbacks - Adverse calculations contradicted by existing facts - Satisfactory answer to question as to why England should retain possession of India - Virtues and merits in the character of the Natives - Their general contentment under British rule - Happy Prospect rising before them

Chapter 5
Securing the Empire

Geography and the North-West Frontier

The first chapter of *The Cambridge History of India: Volume I, Ancient India*, published in 1922 by Cambridge University Press, is written by the geographer Halford Mackinder. His purpose is to draw attention to those elements of the geography of South Asia which have had a significant impact on its history. He draws attention to the northwest by saying bluntly and simply:

> In all the British Empire there is but one land frontier on which war-like preparation must ever be ready. It is the north-west frontier of India. (1922: 26)

He follows this with an insightful description of the physical geography of the single plateau of Persia, Afghanistan and Baluchistan, which is reminiscent of his writing on the Eurasian heartland (considered in Chapter 14). The Persian plateau is not as lofty as that of Tibet, but it is still, as he remarks, one of the great natural features of Asia. It is bounded by escarpment to the sea in the south-west of modern Iran, and by escarpments at the border between Baluchistan and the Indus plains. It has, like Tibet, its high mountain ranges – in the north the Elburz and the mountains of the Turkish and Armenian borders, and in the east the mountains of Afghanistan, which are like splayed fingers running south-west from the western end of the Karakoram Mountains, themselves the western end of the mighty Himalayas. Near Kabul is the dividing watershed between drainage going west (the Helmand flowing to Iran), the north (the Amu-Darya), and the south (the Kabul river flowing into the Indus.) The principal mountain range, the Hindu Kush, forms a single towering ridge separating Central Asia from the Indus valley. It is crossed by a few passes, linking Chitral (in modern Pakistan) with northern Afghanistan, and by the Salang, joining southern and northern Afghanistan. From the plateau area there are two principal routes down to the Indus plains – the Khyber Pass just south of the Kabul river and, much further south, the Bolan Pass leading down from Quetta. In the British period these two were known simply as the Northern Route and the Southern Route. The first led to Punjab, and the second to Sindh. Although these two provinces are both within the Indus valley, they were separate in culture, economy, and even in the history of their most recent conquest and subsequent administration and military defence. Punjab was annexed from the Ganges valley. Sindh was annexed by forces sent by sea from Bombay, and became part of Bombay Presidency. The British Indian army was divided into a northern command, which was responsible for the defence of the northern part of the border, and a southern command, which included the forces of Madras and

Bombay, defending the borders of Sindh and, later, of Baluchistan. Contrarily, within Afghanistan the northern ends of these two routes are simply linked from north-east to south-west, by plains and valleys forming a corridor from Kabul through Ghazni to Kandahar and beyond that, north again, to Herat. There is no obvious border between Persia and Afghanistan; nor one between Iran and Baluchistan (Pakistan). Part of the border between southern Afghanistan and Baluchistan is mountainous, though there are alternative passes that may be followed.

Most of southern Afghanistan and all of Baluchistan is arid, and good grazing is rare. Cultivation in these areas is limited to small tracts where irrigation is possible. Most of the great Persian plateau has bitterly cold winters, and the mountains of Afghanistan particularly so. All of it by day, in the summer, can suffer from searing heat.

Most of these are lands which have bred fiercely independent people. Central authority is nearly always acknowledged only through some sort of titular suzerainty. Each tribe, perhaps each valley, perhaps even each village, will acknowledge only its own khan, a man whose authority stems from his skill and courage in warfare. And the replacement of an old khan by a new has often been the result of violent confrontation and death.

The origin and meaning of the word 'Afghan' is unknown, but it is used in Persia to describe the tribes of the present Afghanistan from at least c. 350 AD, long before the coming of Islam. Although most Afghans would agree to their Afghan identity, tribal division within Afghanistan is most important. A leading authority on their history and culture, Caroe (1962) distinguishes between the Afghan tribes of the plains, in both the west, around Herat, and the east, around Peshawar, and the Pathans, or highlanders, in between, who speak Pashtu or the related dialect Pakhtu. These two broad divisions mask a further division into very specific tribal groups within each. Within the Pathans, for example, two such tribes are the Afridis of the Khyber Pass and the Mahsuds.

The histories of Persia, Afghanistan and India have long been intertwined. As was noted in the first chapters, the Aryan migrations came from Central Asia through Afghanistan; and in the early centuries AD the Kushans (originally from China) established what became a Buddhist empire, based on Peshawar. They traded with China and maintained relations with the Roman Empire. During this period the massive Buddhas were carved in the rock walls of the valley of Bamiyan, in modern Afghanistan. Many later invaders came this way too, including the Afghans who founded the Delhi Sultanate. The Mughals came through the north-west, and the apotheosis of the collapse of their power was when Nadir Shah of Persia descended from the hills to sack Delhi in 1739.

The Punjab and the Seeds of the First British-Afghan War

The delineation of a separate Afghan state or kingdom in modern times dates from 1747 when Ahmed Shah Durrani, noted in Chapter 4 for his incursions into India, took the throne. Although he attempted to install a vassal Mughal ruler in Delhi, effectively he became just another part of the destruction of Delhi's authority, helping to create that vacuum which sucked in the British, and which enabled the Sikhs to establish their own state in the Punjab. In 1767 he formally ceded power in Punjab to the Sikhs. Amongst the Sikhs one young and ambitious clan head emerged as the dominant leader of all the Sikhs – Maharaja Ranjit Singh (1780 – 1839) – not only uniting the clans but arming and training a modern army, with European (many of them ex-Napoleonic French and Italian) officers and artillery. His domains included Jammu and the vale of Kashmir, but with no clear northern border. In recognition of the prowess of one of his chief generals, Gulab Singh, the Maharaja awarded him the jagir of Jammu – in effect making Gulab vassal ruler of the territory.

Early in his reign, in 1809, Lord Minto stopped Ranjit Singh's attempt to cross the Sutlej into the British sphere, and the Sutlej was confirmed as the frontier. Peshawar, on his western border, was contested by Afghanistan and subjected in 1826 to a religious jihad of the Pathans, led by a holy fanatic from India, named Sayyid Ahmad, about whom much more is said later in this chapter. Ranjit Singh occupied Peshawar in 1827, and instigated the destruction of many Muslim buildings, but he was thrown out again in 1830.

A Geographical Gazetteer published in London (Landmann, 1840) had not quite kept up with these events. Its descriptions of Afghanistan and of Peshawar are worth consideration:

AFGHANISTAN, a considerable kingdom of Asia, between Persia and Hindostan, bounded on the E. by the Nilab or Indus, N. by a range of lofty mountains, separating it from Bulkh and Budukshan, W. by Persia (Herat being its frontier town), S. by Baloochistan: it lies between Lt. 29 and 36 N., and Lg. 61 and 71 E; comprehending the ancient kingdoms of Zabulistan (Ghnizne and Kandahar) and Kabulistan. The inhabitants are esteemed hospitable and brave, but refractory and ferocious; P. about 3,000,000.

All British descriptions of Afghans from this period onwards dwell on the fractiousness and ferocity of the people, and also of their duplicity – but also on their own particular sense of honour and hospitality, and on their courage to the point of recklessness. The separateness of Bulkh (the Greek Bactria near Mazar-i-Sharif) is something to which we will return when thinking of current-day Afghanistan.

PESHAWER, Asia, a city of Afghanistan in Cabul, on the Kameh, in an extensive and fruitful plain, 142 m. E. of Cabul, the occasional residence of its sovereign; the palace is on a hill; it has many mosques, and a fine caravansary. It is an entrepôt between Persia and Hindostan, and has many wealthy inhabitants, especially of shawl dealers.

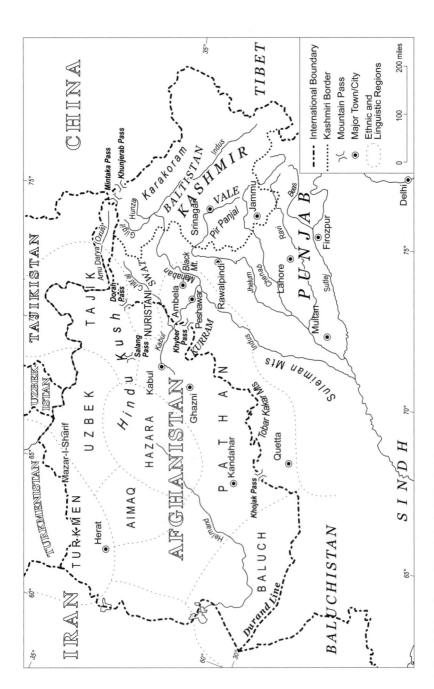

Figure 5.1   Afghanistan, Punjab and Kashmir

The Kingdom of Kabul had historically, more often than not, embraced the literally fruitful plains of Peshawar and the valuable trading city itself. Ranjit Singh was partly able to take it because of instability in Afghanistan, following the overthrow of the monarch Shuja-ul-Mulk by the new ruler Dost Mohammed. Shuja-ul-Mulk took refuge in British India. Ranjit Singh then turned his eyes south to the lower Indus and Sindh. However, William Bentinck, Governor-General, in 1835 'dissuaded' Ranjit Singh from invading and annexing the territories of the Mirs of Sindh, who then signed treaties of alliance with the British.

The last of the seeds of war concerned the geopolitical ambitions of other powers. The Russians were pushing into Central Asia, an area where the British hoped they might be able to develop new trade, and a Russian envoy appeared at the court of Dost Muhammed. Simultaneously, the Persians advanced on Herat, supposedly with Russian backing, to claim it back for the Persian crown. A British envoy, Alexander Burnes, who had been to Kabul before, was dispatched to persuade Dost Mohammed to support the British cause. But Dost Mohammed had a price – he wanted British help to regain Peshawar from the Sikhs.

The First Afghan War: The Debacle

The Governor-General, Lord Auckland, said to be an intellectual civil servant, but without the requisite field experience, backed an ill-advised plan which superficially solved everything. The British would support the reinstatement of a compliant Shuja-ul-Mulk, which would be accomplished by the army of Ranjit Singh. Once in Kabul, Shuja would help rebuff the Russians. Ranjit Singh probably never intended to risk his troops in the passes of Afghanistan, but by drawing in the British he calculated they would not be able to back out, and that they themselves would then have to commit the necessary forces, so relieving the threat to Peshawar without him having to bother too much.

With Ranjit Singh moving too slowly, if at all, on 1 October 1838 the British in Simla issued a declaration which committed them to action. They then had to consider how to move an adequate army from the east of the Sutlej, either across the five rivers of the Punjab and then the Indus, before attacking the Khyber Pass; or by taking the long southern route, using the Sutlej as protection, crossing the Indus south of its confluence with the Sutlej, then moving across the northern part of Sindh, again north to Quetta and on to Kandahar; finally, thence to Kabul. They decided on the latter, and arranged to hold a magnificent durbar for their passing army with Ranjit Singh and many of his force at Firozpur. On 27 November they met, although the regal elephants were upset by a salute from the cannons, and Ranjit Singh was at risk of being crushed to death.

The army comprised 10,000 men of the British and Indian armies, and 6,000 of Shah Shuja's men, with innumerable camp followers to provide for their wants, and 30,000 camels. During the first 450 miles down the left-bank of the Sutlej the army had sufficient water and supplies. The engineers provided excellent boats for the crossing of the Indus which was completed by 19 February 1839. But, from then on everything began to unravel. The next 450 miles to Kandahar were across un-reconnoitred, arid country. The camels could not be properly fed, and began to die, so the sepoys lost their rations too. Hostile tribesmen took advantage by attacking the straggling rear of the columns. By the 8 May 1839 the seriously depleted British force and its large number of weakened camp followers had reached Kandahar with Shuja, and installed him as monarch. They had done so with only two days of half-rations left. There was grain enough around Kandahar, but it had not yet been harvested. Money was sent by the British from India to help buy further supplies, but it disappeared somewhere in the mountains on the journey, and local merchants were reluctant to give this alien army any loans. Yet now they were supposed to move against Kabul. At Ghazni the army had a stiff fight with the garrison but, when they succeeded in blowing up the main gates, they then used superior fire power to commit substantial slaughter, followed by looting. Dost Mohammed immediately decided to confront the British on the road between Ghazni and Kabul. He is reputed to have taken the Koran in his hand to urge his army to fight against the infidels. But 'his' army refused to fight, mostly because the British had already bribed the tribal chiefs, and he was forced to flee. (Later he surrendered himself to the British and was given a pension in Calcutta – whence he in turn, later, was to find the British willing to reinstall him in Kabul.)

On the 7 August Shah Shuja re-entered Kabul and reclaimed his throne, but without any show of popular local support. A British army of 10,000, mostly comprising Indian troops, was left to occupy the capital, but their disposition did not provide them with strong defensive positions. No chiefs came from the hills to offer their allegiance, and Shah Shuja left Kabul to spend the winter in the lower and warmer city of Jalalabad. The British political and military officers spent most of the long winter of 1839/40 squabbling amongst themselves, and trying to manage some sort of government with corrupt local officials. Local leaders kept increasing the monetary demands to buy their loyalty. By 1841, Auckland could no longer afford the costs of the occupation, and he cut the level of subventions. Immediately, the Ghilzai tribes of Pathans (recorded by Caroe (1962) as Haji) revolted, blocking the passes east from Kabul back to India. All through the summer of 1841 the remaining British had to fight off attacks against them around Kabul, and gradually the numbers of dead, wounded and sick increased. In November Alexander Burnes was hacked to death by a mob, and in December the political officer McNaghten, who met with local chiefs for negotiations, was also murdered with his companions, in a manner which was to confirm British prejudice about the duplicity of the Pathans, and which rankled as a national insult at the highest levels of the British establishment. In January 1842, under poor leadership and believing they had been promised safe passage against bills of indemnity to be paid when they reached safety, the remaining 4,500 troops under British command (690 European and 3,800 Indian) and 12,000 camp-followers pulled out of Kabul. They were attacked in every defile and at every opportunity. On the 13 January the sole survivor (apart from some hostages remaining in Kabul), Dr William Brydon, reached the safety of a British garrison in Jalalabad.

Three months later Shah Shuja was murdered, and any remaining pretence of a central power in Kabul evaporated.

Reviewing the wars and skirmishes between the British and the Afghans and the frontier tribes, Caroe (1962: 397) observed wryly:

> Unlike other wars, Afghan wars become serious only when they are over; in British times at least, they were apt to produce an after-crop of tribal unrest ….

The First Afghan War: Retribution

What the British Empire lacked in political acumen, it could make up for in resources. The population of the plains of India from which an army could be raised was vastly greater than that of the arid hills from which the tribal armies could be raised in Afghanistan.

A new Governor-General, Ellenborough, took over from Auckland and instigated a policy of punishment and retreat. The British still held Kandahar and Jalalabad. They were joined by fresh forces in an Army of Retribution, which had retaken Kabul by 15 September. They then flattened as much of the city as they could, and withdrew, 'successfully' this time, to Peshawar. In October 1852 Ellenborough issued a proclamation from Simla, which read in part as follows:

> Disasters unparalleled in their extent unless by the errors in which they originated, and by the treachery by which they were completed have, in one short campaign, been avenged … The British arms now in possession of Afghanistan will now be withdrawn to the Sutlej. The Governor-General will leave it to the Afghans to create a government amidst the anarchy which is the consequence of their crimes. (Cited in Richards 1990: 55)

In January 1843, with British consent, Dost Mohammed returned to claim his throne in Kabul, where he reigned until his death in 1863. In one of his last interviews with the Governor-General he said:

> I have been struck with the magnitude of your power, of your resources, with your ships, your arsenals and your armies: but what I cannot understand is why the rulers of so vast and flourishing an empire should have gone across the Indus to deprive me of my poor and barren country. (Marshman (1867): 233; cited in Dunbar, 1943: 490)

But the experience clearly left an impression on him. In 1857 when the Mutiny broke out and the Punjab was deprived of its European and loyal Indian troops, the mullahs of Kabul, and even his sons, pleaded with him to 'bind on his head the green turban of Islam and sweep the English from the plains of India' (Aitchison

(1892): 12, cited in Dunbar, 1943: 490). But he refused, keeping his alliance with the British until his death.

The second phase of the war became nakedly imperialistic when the British provoked the Sindhis into insurrection. The British were anxious to forestall moves into Sindh by either Afghanistan or Punjab, and found a military solution to their anxieties on the open plains away from the defiles of the hills. In 1843 small contingents of British troops (mostly Irish peasants) under Sir Charles Napier did achieve what legend would like the school boy think that Clive achieved at Plassey. A force of 2,500 achieved a victory in a real bloody battle against 35,000 at Miani, although the latter's tribal loyalties made a concerted command impossible. (A larger British force had been despatched from Bombay for the purpose but had lost a significant proportion of its men through cholera.) There is a story, repeated soon after in *Punch* in London, that Napier sent a message to Bombay with but one word 'Peccavi' – the Latin meaning 'I have sinned.' (Sindh.)

The Annexation of Punjab

The Maharajah Ranjit Singh died in 1839 without any clear successor. There followed a period of intrigue and murder, from which the collective leadership of the Khalsa (here meaning Sikhism in general) passed into the hands of the army's generals. The European officers were driven out in 1841. In 1843 a five-year old son of Ranjit Singh was proclaimed Maharaja, to give a veneer of legitimacy to the true powers behind the throne. To retain power, the generals needed to retain the loyalty of the army, more than 50,000 strong. It needed to be kept occupied, and it needed to be paid. The military leadership had no clear political vision, and decided to do what most such leaderships do in such circumstances – to unite their forces in a war that promised much. There were also *causus belli* that could be cited: the British had annexed Sindh, yet their power in the north in Afghanistan had been shown not to be invincible, and they had retreated from Kabul.

McNaghten, writing from Kabul in 1841, had had his suspicions of the Sikhs, and had urged Auckland to:

> crush the Sings, macadamize the Punjab and annex the province of Peshawar to Shah Shooja. (Marshman (1867): 275, cited in Dunbar (1943): 497)

In December 1845 the Sikhs crossed the Sutlej. They were met by the British at Mudki, near Firozpur, where they were repulsed after a fierce battle that brought heavy casualties to both sides. A second battle took place at Aliwal on 28 January, again a fiercely contested affair, in which valour was recognized on both sides. At its conclusion, the British fired a salute, then played the national anthem. It is said that the Sikh band was heard playing 'God save the Queen' back from behind Sikh lines. A final decisive battle was fought at Sobraon on 10 February, and Lahore was occupied by the British shortly afterwards. One of Ranjit Singh's great soldiers, Gulab Singh, who had been awarded the jagirdar of Jammu, kept aloof from this fight, and avoided confrontation with the British.

The British did not want to annexe Punjab – as ever they thought it cheaper and more effective to allow a client state to run its own internal affairs under treaty obligations. Thus the treaty of March 1846 included some demanding clauses, but not the dissolution of the Maharaja's rule. However, the Sikh council of regency invited the British to appoint a resident British administrator, and the government became an extension of the company's rule. The Sikh army was to be reduced, the doab (interfluve) between the Beas and the Sutlej ceded to British India; and most significantly of all, the sovereignty of Hazara and Kashmir was ceded in perpetuity to the British. This was in lieu of war indemnity which the Sikhs could not afford. Within five days the British had done another deal, which lays one of the foundations for the threat of nuclear war which hangs over the subcontinent now. The British sold Kashmir to Gulab Singh for £750,000, plus an annual tribute measured in shawls and goatskins. It is almost certainly the prospect of this deal which had kept Gulab inactive during the war. Now a new dynasty, the house of Dogra (Gulab's jati name) ruled in a new state, that of Jammu and Kashmir.

In 1849 resentment against Company rule exploded into open rebellion in Multan, where the Mulraj urged a religious war against the English. Within short order the Sikhs had raised an army of between 30,000 and 40,000 men, and what might have been a small police action again became a full-scale war, forcing the British again to commit substantial troops to combat. After two fierce battles, the Sikhs again lost, and the Punjab was formally annexed. Eight short years later, Sikh support proved vital in defeating the Mutiny in Delhi.

Interim Conclusions

There are few places in the world which have not at some time in their history been colonized by the powers of Western Europe. Those that have not are mostly those that the maritime powers cannot penetrate, as pointed out famously by Mackinder in 1904. They include the massive Eurasian heartland colonized by Russia (it is the opening subject of Chapter 14); and the Persian Plateau, including Afghanistan. There are exceptions; China's population in the main lives by large navigable rivers, but, arguably, China was forced by the maritime powers to open itself to a colonial trading pattern; and Thailand – a true exception.

The story of the northwest frontier told so far includes all the ingredients of the continuing story from then until now. The geography of isolation, the wild and rugged terrain, the harsh climate, and the lack of major resources, mean that the mountains are sparsely settled. Neither have they ever attracted outside powers for their own sake. But they have a strategic geopolitical position. They have been crossed time and time again by invaders on their way to the population and wealth of the Indian plains. They have been the route for a small trade, mostly in high-value fabrics like silk and Kashmir shawls. Today they could be the route for a massive trade between the oilfields of central Asia and the oil-deficient populations of Pakistan and India.

The isolated mountain valleys are home to fiercely independent villages and clans. Islam alone does not breed war-like fanatics. In the Middle Ages the main export of the Swiss mountains was mercenary soldiers – hence the Swiss guards of the Pope in Rome. Even today all men of military age in Switzerland have a rifle at home, and annually have to join in training. Where now you see civilians, tomorrow you could see an army. In a less organized and less formal way the same has always been the case in the mountains of the Northwest Frontier. Men are armed, with the best weapons they can buy or steal, prepared to fight, sometimes in age-long feuds between families, sometimes between valleys, often behind a tribal chief, between tribes. Or depending on money and politics, perhaps the tribes unite to form an army – a lashkar. Such an army does not fight as a unitary whole: tribes may join or leave the fray according to their opinion of its worth and progress. At any time a fixed formation can dissolve back into the hills, to resort to guerrilla war and raiding; or simply to become farmers and herdsmen again. Mountains also attract mystics and holy men. Communities in the mountains accept wandering mullahs and Sufis.

Religion also comes into the story. Although not much written into the first part of this chapter so far, it is always there in the subtext. European memories included the Moorish occupation of Spain and the Ottoman attacks up to the gates of Vienna. Islamic memories include the Crusades (literally the followers of the cross) of the Christians attacking their holy city of al-Khuds, or al-Quds (simply meaning The Holy, and the third shrine of Islam after Mecca and Medina), which most English-speaking readers would know by the name of Jerusalem. One way of uniting a fractious people is to appeal, as the Sikhs did, as the Amir's advisers did, to take on war as a holy duty; in the case of Moslems, to proclaim it a jihad. The Christian British saw their crusade in terms of being a civilizing force in India, and also as the even hand between the 'inferior' religions of India – be it Islam or Hinduism, or indeed Sikhism. When the Army of Retribution returned from Afghanistan, Ellenborough ordered General Nott to bring back from Ghazni the gate stolen from the temple of Somnath. At an elaborate reception at Firozpur he proclaimed their return to 'avenge the insult of eight hundred years' (of Muslim domination of Hindu India). As a propaganda coup it was a failure: the gates had not been made in India (they had in fact been ordered by Sabuktigin, AD 977–997, an early Muslim monarch of Ghazni) and, therefore, did not come from Somnath; and they were subsequently left in a lumber room of the Muslim Red Fort at Agra.

To the extent that the British had 'won', their victory had depended firstly on the resources of a large empire, and on technology – since usually they managed to keep their weaponry one step ahead of the tribal fighters, and on superior sources of wealth which sometimes, but unpredictably, enabled them to bribe the tribes.

Finally, it is clear that events in one part of the northwest can easily be linked with events elsewhere, from Afghanistan to Kashmir.

'Muslim Fanatics' and the Revolt of 1863

In the late 18th century, a reformed brigand Zamin Shah from Buner (also spelt Bonair) established himself as a recluse at a place called Sittana on the southern flanks of the Mahaban (Great Forest) Mountain west of the Indus near Amb. Two of his grandsons became lieutenants of the Muslim Crescentader[1] Sayyid Ahmad,[2] known at the time as the Prophet, who had united some of the Pathan tribes against Ranjit Singh in 1824 and had taken Peshawar by 1826. There he struck coins with the legend 'Ahmad the Just, Defender of the Faith; the glitter of whose scimitar scatters destruction among the Infidels' (Hunter,1871). Sayyid Ahmad was finally killed by the Sikh army in 1831, but the two brothers escaped back to Sittana with some of the remnants of Ahmad's forces. One remained there, the other was enthroned as the King of Swat (who died in 1857).

The surviving brother, Sayyid Umar Shah, continued with the crescentade that Sayyid Ahmad had begun. Now, Sayyid Ahmad himself had been born in Bareli District of modern Uttar Pradesh, on the Ganges Plains. He had been on the pilgrimage to Mecca, where he had attracted the authorities for his fundamentalist pro-Wahhabi leanings. On his return to India he preached across the country with passion, became convinced of his own mission, and set up a religious foundation at Patna, run by his appointed Caliphs. They taxed the local faithful, so that the institution was permanently funded, and it spawned other local centres from Bengal across northern India supported by dedicated preachers. Sayyid Umar Shah used the chain to find recruits, money and supplies, from right under the noses of the British.

Until 1849 Sayyid Umar Shah's forces descended from the hills in sporadic attacks on Ranjit Singh's Punjab – so it appeared to be no great matter to the British anyway. However, in that year the Punjab was annexed, and the northern border did become of interest to the British, who were disquietened by the raiding and

---

1 This term is currently unusual, but was used by the British to express the Muslim equivalent of a Crusader. Much of this account is based on Hunter (1871) who used the term throughout his book.

2 Sayyid Ahmad's supporters circulated a manifesto across north India which goes as follows:

'The Sikh nation have long held sway in Lahore and other places. Their oppressions have exceeded all bounds. Thousands of Muhammadans have they unjustly killed, and on thousands they have heaped disgrace. No longer do they allow the call to prayer from the mosques, and the killing of cows they have entirely prohibited. When at last their insulting tyranny could no longer be endured, Hazrat Sayyid Ahmad (may his fortunes and blessings ever abide), having for his single object the protection of the faith, took with a few Musalmans, and, going in the direction of Cabul and Peshawar, succeeded in rousing Muhammadans from their slumber of indifference, and nerving their courage for action. Praise be to God, some thousands of believers became ready at his call to tread the path of God's service; and on 21 December 1826 (in the original: 20th Tamadi-ul-Sani, 1242 Hijra) the Jihad against the Infidel Sikhs begins'.

hostage-taking, which continually increased. The general approach to controlling them was to exert pressure on the local tribes, by threatening their villages or by fining them, so that they in turn would control the crescentaders in the hills. Between 1850 and 1857 sixteen expeditions totalling 33,000 regular troops were committed to these exercises. In 1857 the crescentaders sensed they could form a coalition with the tribes to attack the British while the Mutiny threatened Delhi. General Sir Sidney Cotton[3] commanded 5,000 troops in an expedition which:

> burned the villages of the rebel allies, razed or blew up the two most important forts, and destroyed the Traitor Settlement at Sittana. (Hunter, 1871: 24)

The next sentence continues

> The fanatics, however, merely fell back into the fastnesses of the Mahaban mountain; and so little was their power shaken, that a new Settlement at Mulka was immediately granted them by a neighbouring tribe.

By 1861 they were back in Sittana.

> So long as we left it alone, it steadily sent forth bands to kidnap and murder our allies: when we tried to extirpate it by our arms, it baffled our leaders, inflicted severe losses on our troops, and for a time defied the whole Frontier Force of British India. (28)

In 1863 a major British Force of 7,000 men with 4,000 pack animals and supplies under General Sir Neville Chamberlain, was sent across the Indus, beyond Amb, and into the Ambela Pass ('La' simply means 'pass'). The idea was to go round the Mahaban mountain so the rebels could not escape north, but would be driven east and south towards the Indus and other British forces. Reports had suggested the tribes would not be hostile. On 19 October the first troops reached the narrow wooded defiles at the head of the pass, and were immediately attacked. Though the top of the pass was successfully occupied, further passage proved impossible. Seeing the British in difficulties, more and more of the local tribes sided with the rebels, until some 60,000 were pitched against Chamberlain's army. Fighting to take, to hold, to repulse, to re-take the ridges of the mountain crests raged night and day for weeks, while cold and sickness also took its toll. Word spread southwest into the Pathan areas of Afghanistan, where further insurrections threatened. By late November disaster clearly threatened. Then the tide turned, when the Commissioner from Peshawar managed to buy-off some of the Buner clans, backing up the carrot with some violence in which Ambela village was destroyed and 200 clansmen left dead or wounded. Other tribes began to defect, the British forced the pass, and succeeded in destroying the settlement at Mulka,

---

3 A brother of General Sir Arthur Cotton, the irrigation pioneer quoted in the next chapter.

hunted down the fanatics, and returned to the plains by the end of December. Some peace and stability was brought to the area for some years, but at a high price: British losses had been heavy.

It is the aftermath that proves so interesting. Hunter's book was written while the trials of the conspirators were still being conducted in India. His book begins as follows:

Chapter 1

The Rebel Camp on our Frontier

The Bengal Muhammadans are again in a strange state. For years a Rebel Colony has threatened our Frontier; from time to time sending forth fanatic swarms, who have attacked our camps, murdered our subjects, and involved our troops in three costly Wars. Month by month, this hostile Settlement across the border has systematically recruited from the heart of Bengal. Successive State trials prove that a network of conspiracy has spread itself over our Provinces, and that the bleak mountains which rise beyond the Panjab are united by an unbroken chain of treason-depots with the tropical swamps through which the Ganges merges into the sea. They disclose an organisation which systematically levies money and men in the Delta, and forwards them by regular stages along our high-roads to the rebel camp two thousand miles off. Men of keen intelligence and ample fortune have embarked in the plot, and a skilful system of remittances has reduced one of the most perilous enterprises of treason to a safe banking operation.' (9)

How the power of Empire could long be resisted was a question needing to be addressed.

It is easy to understand how a Settlement of traitors and refugees, backed by the seditious and fanatical masses within our Empire, could, in an excess of bigoted hatred, throw down the gauntlet. But it is difficult to comprehend how they could, even for a time, withstand the combined strategy and weight of a civilised army. (28)

The allure of the extremists was felt in all sections of Muslim society.

While the more fanatical of the Musalmans have thus engaged in overt sedition, the whole Muhammadan community has been openly deliberating on their obligation to rebel. (10)

Somehow or other, every Musalman seems to have found himself called upon to declare his faith; to state, in the face of his co-religionists, whether he will or will not contribute to the Traitor's Camp on our Frontier; and to elect, once and for all, whether he shall play the part of a devoted follower of Islam or of a peaceable subject of the Queen. (10)

Hunter's book has the delightful title: *Our Indian Musalmans: are they bound in conscience to rebel against the Queen?* Victoria had only been proclaimed Queen in 1858, after the Mutiny, and why many, if any, Moslems should feel loyalty to a distant, absent woman, is unclear. But his point was real enough. He knew that such a wide-spread treason could only have worked if there had been sufficient numbers of people sufficiently aggrieved, and he took Muslim grievances seriously. They had lost power and office under the British and had not adapted as well as Hindus to the new education on offer. Hunter discusses in great detail the exact question of his title. In Islamic law there are two kinds of political states: Dar-ul-Islam, countries of submission (to the faith), and Dar-ul-Harb, countries of the enemies. There are definitions given for these terms. If a Muslim is in Dar-ul-Harb, he should rebel (it is lawful to make jihad), or else leave for Dar-ul-Islam. The Muslim lawyers of Northern India, the three most senior lawyers in Mecca, and the Calcutta Muhammadan Society were all asked to rule on the matter of whether India under Christian British rule was or was not Dar-ul-Harb. In all three cases the ruling (fatwah) was that because the British afforded protection and because they permitted Islamic worship to continue, India was not Dar-ul-Harb, and for these reasons and the further reason that there was no realistic probability of overthrowing them by force, jihad was not lawful.

In case, which I very much doubt, the reader has missed the parallels: for Wahhabi read Taleban, for Sayyid Ahmad read Osama Bin Laden, for Sittana read Tora Bora; for Sayyid Umar Shah read Mullah Omar; for Treason-depot read terrorist cell; for banking operation – well, read banking operation. In June 2002 during the same month as the jubilee celebrations in Britain for the fifty-year reign of the Queen, *The Guardian* newspaper ran a series of articles on the role of moderate Moslems in Britain, and asked whether or not they could be both Muslim and English or British. In India it is all too easy in time of tension with Pakistan for *agents provocateurs* to suggest the disloyalty of Indian Moslems, so that non-Moslems turn on them.

The Search for a Frontier, 1860s to the Second World War

Some of the British wanted to establish a 'scientific border', following a 'forward policy', in which the British would take control of the high passes in the Hindu Kush, necessarily annexing part of Afghanistan. It would not only have been 'scientific'; it would have been supremely historical as well. This was the frontier of the Indian empire of the great Mauryan monarch Chandragupta in 305 BC. (The name of the mountains is supposed to denote the deaths of Hindus guarding the border passes.) The British experiences of direct involvement were that this would be a painful and costly policy to adopt. It is also the case that it is not necessarily rational to use watersheds as borders. In historical and cultural terms different societies evolve and settle different altitudinal zones, so that at higher levels the people on both sides of a 'natural border' are often closely related. The Ladakhis of

India are essentially Tibetans, and in the British period Baltistan was subtitled on maps 'Little Tibet'. In Afghanistan, Hazaras and Tajiks straddle mountain divides, and so do most of the Pathan tribes.

If the forward policy could not be adopted, the second best option was to keep Afghanistan as a client state, and a buffer between the British and the Russians. In 1877 a British mission was sent to Kabul which achieved nothing, and in 1878 a Russian mission arrived. When a British ultimatum was ignored, again they invaded in force, at the beginning of what is known as the Second Afghan War. Again they had major debacles, and again there was heroic retribution; it was a costly and bloody affair, which created new heroes for the Victorian public in London. After a series of coups and different attempts at finding a new ruler, Abdur Rahman was installed as Amir in 1880. He signed a treaty with the British, which gave them control over his foreign affairs, but which also committed them to paying him a handsome subvention. They then proceeded to negotiate with the Russians the definition of Afghanistan's northern border – which they managed to do more to Afghanistan's favour than Russia's.

But they had not yet negotiated formally a border between Afghanistan and British India. By the 1890s they were avidly building railway lines to supply the garrisons beyond Peshawar in the Khyber Pass, and beyond Quetta to the Khojak Pass commanding Kandahar. They dug a tunnel under the Khojak Pass. At the Amir's request a conference was held with the British to demarcate the border. The senior British negotiator was Sir Mortimer Durand, whose name is now attached to the border between Afghanistan and Pakistan. (He was assisted by Henry McMahon, who later was central to attempts to demarcate the northeastern border with Tibet and China.) The agreement was signed in 1893, and then for two years a Joint Commission worked to demarcate it on the ground – with small pillars erected every few miles. It has frequently been attacked as the freezing of a temporary and illogical demarcation – which splits ethnic groups and tribes, and which produces strategic nightmares on both sides. Many local people neither knew then, nor know or care now, where exactly it is. Later Afghan governments have on occasion taken to repudiating it, claiming that it was a treaty signed under duress, but there have never been serious negotiations to change it.

In 1901 Amir Abdur Rahman died and was succeeded by Habibullah Khan. During the First World War the latter was approached by a Turco-German mission, who promised him a treaty recognising Afghanistan as an independent state. Although here was a major opportunity to create trouble for the British, who had deployed much of the Indian army in the Middle East, Habibullah maintained his neutrality, to the annoyance of many in his higher counsels. In 1919 he was murdered, and after the usual struggles, a new Amir Amanullah Khan emerged. He set about organising an invasion of British India. It is also often charged that he preceded this by fomenting trouble for the British in Punjab by supporting the nationalists there, in part being implicated in the events which led up to the massacre by the British under General Dyer in Amritsar in the same year. The British had to fight the Afghans both in the Khyber and in the Kurram Valley to

Table 5.1   A Record of 'Pacification': Table of Expeditions against Frontier Tribes

Year	Tribe	Commander	Numbers	Casualties	
				Killed	Wounded
1847	Baizais	Lt.Col. Bradshaw, C.B.	Mixed brigade	1	13
1849....			2,300	7	40
1850	Afridis, Kohat pass	Brig. Sir C. Campbell, K.C.B.	3,200	19	74
1851	Miranzai Tribes	Capt. J. Coke	2,050, including levies	2	3
1851–2	Mohmands	Brig. Sir C. Campbell, K.C.B.	1,597	4	5
1852.....			600	2	8
..	Ranizais ..		3,270	11	29
..	Utman Khels ..		2,200	3	15
..	Wazirs, Darwesh Khel	Major J.Nicholson	1,500	24	7
1852–3	Hassanzais	Lt.Col.Mackeson, C.B.	3,800, incl. levies	5	10
1853	Hindustani Fanatics ..		2,400	nil	nil
..	Afridis, Ada Khel	Col. Boileau	1,740	8	29
1854	Mohmands	Col. Cotton	1,782	1	16
1855	Afridis, Akra Khel	Lt.Col. Craigie, C.B.	1,500	9	25
..	Miranzai Tribes	Brig. Chamberlain	3,766	14..	
..	Orakzais, Rabia Khel ..		2,547	11	4
1856	Turis ..		4,896, incl. levies	5	3
1857	Yusafzais	Major Vaughan	400	2	3
			990	5	21
			1,625, inc. levies	1	8

Year	Tribe	Commander	Strength		
1858	Khudu Khels M.	Gen.Sir S. Cotton, K.C.B.	4,877	6	29
1859–60,	Wazirs	Brig.Gen. Chamberlain, C.B.	5,372,	1	19
	Darwesh Khels		incl. levies		
	Wazirs, Mahsuds ..		6,796,incl.levies	100	261
			900	238	670
1863 Ambela Expedition ..	(Hindustani Fanatics, etc)	(later M.Gen. Garvock)			
1864	Mohmand	Col. McDonell, C.B.	1,801	2	17
1868	Orakzais	Major Jones	970, incl. levies	11	44
	Black Mountain Tribes	M.Gen. Wilde, C.B., C.S.I.	12,544	5	29
1869	Orakzais	Lt.Col. Keyes, C.B.	2,080,incl. levies	3	33
1872	Dawaris	Brig. Gen. Keyes, C.B.	1,826	-	6
1877	Afridis, Jowaki	Col. Mocatta	1,750	1	10
1877–8 ..		Brig. Gens. Keyes and Ross	7,400	11	51
1878	Utman Khels	Capt. Battye	280	-	8
..	Ranizais	Major Campbell	860	nil	
..	Utman Khels	Lt.Col. Jenkins	875	1	
..	Afridis, Zakha Khel	Lt.Gen.Maude, V.C., C.B.	2,500	2	9
1879			3,750	5	13
..	Mohmands	Capt. Creagh and Major Dycce	600	6	18

Year	Tribe	Commander			
..	Zaimukhts	Brig. Gen. Tytler, V.C., C.B.	3,226	2	2
1880	Mohmands	Brig. Gen. Doran, C.B.	2,300	2	3
..	Batannis	Lt.Col. Rynd	721	5	
..	Wazirs, Darwesh Khels	Brig.Gen. Gordon, C.B.	800	nil	
1881	Wazirs, Mahsuds	Brig.Gens. Gordon and Kennedy	8,531	8	24
1887	Bunerwals	Col. Broome	460	3	2
1888	Black Mountain Tribes	B.Gen. J.McQueen, C.B., A.D.C.	9,416	25	57
1891 ..		M.-Gen.W.K. Elles, C.B.	7,289	9	39
..	Orakzais	Brig.Gen.Sir W.Lockhart, K.C.B.	4,600	nil	
.. ...			8,000	28	73
1894	Wazirs	Brig.Gen. Turner and Lt. Gen.Sir W. Lockhart, K.C.B / M.Gen. Sir R.Low, K.C.B.	11,150	45	75
1895	Chitralis	M.Gen. Sir R. Low, K.C.B.	15,249	21	101
....		Col. Kelly	1,400	165	88
1897	Wazirs, Darwesh Khel	M.Gen. Corrie-Bird, C.B.	8,000	29	61
	Akozais (Swat)	Col. Meikeljohn, C.B. C.M.G.	12,650	97	386

Year	Tribe	Commander	Troops		
	Mohmands	and M.Gen. Sir B. Blood, K.C.B. Brig. Gen. E. Elles, C.B.	8,500	12	96
	Akozais and Tarkanris (Dir and Bajaur)	M.Gen. Sir B. Blood	12,200	61	218
	Utman Khels	Col. A. Reid, C.B.	2,900	nil	
	Orakzais	M.Gen. Yeatman-Biggs, C.B.	9,500	26	54
	Afridis	Lt.Gen. Sir W. Lockhart, K.C.B.	34,550	287	853
	Chamkannis	Lt.Gen. Gaselee, C.B. and Col. Hill, C.B.	9,700	26	35
1898	Bunerwals	M.Gen. Sir B. Blood, K.C.B.	8,800	1	
1899	Chamkannis	Capt. Roos-Keppel	1,200	1	
1900–1	Wazirs, Mahsuds	Brig. Gen. Dening	small column	32	114
1901–2	Wazirs, Darwesh Khels	M.Gen. Egerton, C.B.		4	15
1908	Afridis, Zakka Khel	M.Gen. Sir J. Willcocks, K.C.M.G.	14,000	3	37
..	Mohmands ..		12,000	38	184

Source: Wylly (1912)

Note: This does not include the very much more substantial number of troops committed in the Afghan Wars; nor does it indicate the names of tribal leaders, and the casualties they suffered.

Table 5.2   British 'Control' of the Tribes

Dep. Commissioner, Hazara	Cis-Indus Swatis – Allai, Tikari, Deshi, Nadihar and Thakot. Yusafzais – Trans-Indus Utmanzai, Mada Khel, Amazai, Hassanzai, Akazai, and Cis-Indus Chagarzai.
Political Agent, Dir, Swat and Chitral	Yusafzais – Trans-frontier Akozai. Sam Ranizais Bajauris. Chitralis.
Dep.Commissioner, Peshawar	Yusafzais – Trans-Indus Chgarzai, Kudu Khel, Chamlawals, Sam Baizai and Cis-Indus Utmanzai. Utman Khel. Mohmands. Gaduns. Bunerwals. Afridis – Adam Khel of Janakor and Kandar.
Political Agent, Khyber	Afridis – except Adam Khel. Mullagoris. Mohmands – Shilmani. Shinwaris.
Dep. Commissioner, Kohat	Orakzais – except Massuzai. Afridis – Adam Khel. Bangash.
Policial Agent, Kurram	Zaimuhkts. Turis. Orakzai-Massuzai. Chamkannis.
Dep. Commissioner, Bannu	Bannuchis.
Political Agent, Tochi	Darwaris. Wazirs – Darwesh Khel.
Political Agent, Wana	Wazirs – Mahsuds.
Dep. Commissioner, Dera Ismail Khan	Batannis.

Note: The extent to which the word 'controlled' actually applied is debatable
Source: Wylly (1912) Appendix C

the south. For the first time airplanes were used, aging biplanes from the War, though to begin with the rules of engagement were very restrictive. To protect women and children, pilots had to drop leaflets 24 hours before a bombing raid. This caution was not popular with the troops since women, and indeed children, were also happily capable of torturing prisoners in grotesque ways. This, the Third Afghan War, ended in failure for the Afghans, though only after costly fighting and the commitment by the British of huge resources. It also left the frontier again in a troubled and unsettled condition, and with the British laying waste to the country where local tribals had supported the invasion.

The 'fanatical revolts' also continued. In 1897 a holy man the British nicknamed the Mad Mullah proclaimed himself Zia-ul-Millat wa ud-Din (the light of the nation and religion). He preached to the tribes in Swat, published tracts justifying jihad, and then led an uprising, which grew in scale until major military forces had to be committed against it. The revolt also spread among most of the tribes along the frontier – perhaps because of their resentment at the demarcation of the Durand line. In the late 1930s Mirza Ali Khan, known as the Fakir of Ipi, fanned revolt by the Mahsuds and Wazirs, with a mixture of religious and tribal fervour. In the Second World War Hitler also used agents to try stir the border, but without success. British policy to control 'their side' of the frontier involved the hiving off, under Curzon, of a North-West Frontier Province from Punjab in 1901. This was administered by a Commissioner in Peshawar, answering to Delhi. It meant that the laws and courts of Punjab were no longer relevant, and that the internal autonomy of the tribes could be recognised and used. Frontier Forces were then raised locally, under British officers, in a further attempt at pacification and integration – but though they were as capable of fighting with their usual ferocity for the British, they also proved equally capable of defection with their weaponry in circumstances they favoured. The system was discontinued. Armed marches of 'pacification' by British and Indian troops continued right through to Independence.

Afghanistan from 1947 to the Taleban and Al-Qaeda

The British might have left South Asia, and Russia's Czar been overthrown by a communist revolution and the establishment of the Soviet Union, but Afghanistan stayed in the same strategic place. More is said in chapters 14 and 15 of this book about how and why it has been swept up in the web of relations between the USA, the USSR and Pakistan. Here I wish to draw a few swift parallels between the historical events just described in this chapter, and later developments. After World War II the Americans became as determined to contain the USSR as the British had been determined to contain Russia. Pakistan became an ally in CENTO on the USSR's southern flank, and the base for U2 spy planes that could overfly the USSR in the days before satellite surveillance. To begin with, Afghanistan was left as a poor and backward monarchy under king Zahir Shah, in which all areas and tribes exercised considerable if not complete autonomy. The USSR began aid projects, which included roads that would be of strategic use in their later invasion. The US was also involved in projects, including in irrigation. But Afghanistan's major foreign policy issue was something the USSR could meddle in, to discomfort America's ally. Since the British withdrawal, the Afghans pursued the issue of a Greater Pashtunistan (incorporating the Northwest Frontier province, thereby a recreating a greater Afghanistan), in the process coming close to major wars with Pakistan. In 1973 the king was overthrown in a coup by Muhammad Daud, who had the support of the USSR, but having gained power and declared a republic, Daud then distanced himself from the USSR, only to be overthrown by a pro-communist coup. When that showed signs of faltering, the Soviets invaded in 1979. The Americans then used Pakistan as the conduit for arming the Islamic Mujahideen resistance, including international volunteers such as Osama Bin Laden and his Al Qaeda fighters. During the long Afghan War, a third of the population was displaced, with millions of refuges fleeing to Pakistan and Iran. In 1989 the resistance won, and the Soviet Union withdrew, later to collapse. The West then turned its back on Afghanistan, and walked away, leaving these 'refractory and ferocious' inhabitants to fight for power among themselves. These were anarchic years of the worst sort, brought to a sort of conclusion by the emergence of the Wahabist Taleban, who almost succeeded in taking over the whole country before the attacks on New York by Al Qaida in September 2001.

The counter-attack by US led NATO forces on Afghanistan, starting in October the same year, mostly used bought-off northern tribal leaders to evict the Taleban from Kabul, helping the leaders avenge their defeats, and reinforcing irredentist feudalism. The Taleban may have been evicted from Kabul but their resistance against NATO's intrusion continues from Greater Pashtunistan. Of course, on paper this does not exist: it is split between the Pashtun areas of Afghanistan, and the Northwest Frontier District of Pakistan. But the Pakistanis, despite huge subsidies from the USA, have made few inroads into controlling the area, and the Americans and other NATO troops in Afghanistan have to be careful of how far they cross the Durand line in hot pursuit, to avoid antagonising their essential ally, Pakistan. And as ever, the trouble spreads. Radical leaders have again taken over Swat, further undermining Pakistani authority in the mountains, and spreading insurrection into Punjab.

Conclusions

The long and difficult search for a secure frontier in the northwest of India produced a multi-layered result: a border (which was not the best that Russia might have achieved) between Russia and Afghanistan; a buffer and client state (Afghanistan itself), and finally a border Province of British India which never pretended to the same civilian law and administration as in the other provinces. The instability of the region depends in part on the fact that there are no obvious borders. In the Northwest Frontier Province and Afghanistan, spatial continuities and discontinuities in ethnicity, language, religion and culture are as confusing as the patterns of mountains and passes. Given the difficulties of transport, and the primitive near self-sufficiency of the fiercely independent tribes, there is no utilitarian integration. Exactly the same conditions mean that coercive integration is extremely difficult. This leaves identitive integration, which is possible, but which rarely holds for long, given the fractious and feuding nature of the society. The higher levels of identity that seem to work best are ethnicity, as for example when the Pathan tribes unite, or religion – the call to jihad against the infidels. The two working together can be momentarily quite explosive, the most recent example being that of the Taleban.

In the British period the frontier became a mesmerising obsession, where the virility of Empire could be tested against a doughty foe. In every year for the military officers there was a sporting chance of a jolly good scrap, which would test them out and keep them on their toes. A whole literature grew around this manly way of life and the confrontation with the idealised Pathan, perhaps best summed up in the way the unrelenting brutality of one characterisation is sanitised in the title 'Exploits of Asaf Khan' ('Afghan'(1922), with an introduction by Sir George Younghusband). In 2008 Prince Harry Windsor, third in line to the British throne, as a young officer in the British Army, has had his chance to play his part in the 'shoot first, develop later' policy that has characterised outside intervention in Afghanistan. The other side, the Pathans in the Taleban, might be less literate, but they in their turn have their folk memories of heroes fighting the British, and seem just as happy for history to be rejoined.

Only one state outside of Sri Lanka has directly meddled in its civil war. Sri Lanka is of very limited geopolitical importance. By contrast, it seems that any number of states have meddled in Afghanistan and the northwest – certainly since Alexander the Great's invasions we have a continuous record of Persian, Indian, Central Asian, Turkish, Russian, British, German and now American (and other NATO) intervention. At every stage of history the mountains have been awash with arms and intrigue. The culture of instability and bigotry of this area is not simply the fault of its own population. What the armed Congo has become, Afghanistan has always been. But it has no diamonds. It has one thing only: its position as the greatest mountain cross-roads in the world, and a redoubt for fugitives, who can strike north, east, south and west, and in these days at ever greater distances.

Chapter 6
A New Geography: A New Economy

The Railroading of Empire

In the region which was the most favoured for transport in India, the Ganges valley, in 1812 a fast 500-mile journey by boat from Calcutta to Cawnpore/Kanpur could be accomplished in about 11 weeks. In 1832 that new-fangled contraption the steam boat, could reach Allahabad (short of Cawnpore/Kanpur, but not by a large fraction) even when the wind did not blow, in three weeks. In 1852 it took Lord Roberts three months to ride on horseback from Calcutta to Peshawar near the Afghan border (note he did not go up the Indus Valley), which was twice as fast as it took a regiment to march the same distance. The electric telegraph reached India in 1864 linking the small British system with the smallish Indian system already developing, and was used *inter alia* to report jute and cotton prices to manufacturers in Britain. In 1869 the Suez Canal was opened – and cut the journey from England to India and vice-versa, from a fastest possible time of 100 days to 25, and, in addition, progress could now be reported en route. It also proved, against expectations, that steam could compete in cargo trade profitably against sail, since sailing ships had no wind to blow them through the Red Sea, and still had to use the Cape Route. And it further meant that agricultural produce could be exported much more safely, since evil weevils now had far less time in which to multiply and sabotage the cargo. By 1880 when Assam was still awaiting its railway, the journey from England to Calcutta was quicker than the journey from Calcutta to Upper Assam.

The biggest change that occurred in the new Indian Empire in the 19th century was clearly the rapid transformation of communications: on which nearly all other major changes in the geography and economy hinged. At the beginning of the 19th century even the roads which had been delineated and guarded by the Mughals had broken down, and in many parts not even a bullock cart could pass. In 1839 the Company's Engineers embarked on the complete reconstruction of the Grand Trunk Road, now to run from Calcutta to Delhi and, later to Peshawar, a length of 1,400 miles. Before it was half-finished a count at one point in 1846 estimated 390,000 tons of goods were being transported annually by bullock and buffalo carts, and camels and mule trains. The start was made in 1840 on a road from Bombay to Agra, and another from Calcutta to Bombay. As the new communications structure was being laid down, it occurred to several men that the fastest and most capacious communications could be had by using the latest and best, the steam railways. There were doubts expressed: that the white ants would eat the sleepers; that the monsoon torrents would wash away the tracks;

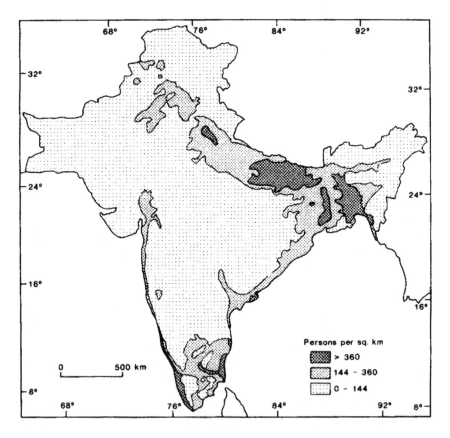

Figure 6.1   Rural Population Density

Source: Spate and Learmonth (1967)

that, indeed, the cyclonic conditions which occurred would simply blow the trains over – and indeed the doubts later proved to have been founded at least in part on some realistic assessments. Neither was there in India the industry to produce the capital goods, nor yet the men trained in surveying and civil construction. In short, there were doubts about the appropriateness of the technology.

Besides, it was hugely expensive. In 1845 Mr MacDonald Stephenson recommended to the Directors the construction of a line from Calcutta to Delhi at a cost of £15,000,000, a figure which can be put in perspective by comparing it with the figure for India's Imports of £14,250,000 in 1856. The cost was clearly beyond the Company's means, and the project might have stagnated indefinitely, except that some small experimental railways in Bombay and Calcutta opened successfully in 1852, and Lord Dalhousie, Governor-General, was a complete enthusiast. He wrote a minute to the Court of Directors in 1853 as follows:

A single glance cast upon the map recalling to mind the vast extent of the empire we hold, the various classes and interests it includes, the wide distances which separate the several points at which hostile attack may at any time be expected; the perpetual risk of such hostility appearing in quarters where it is least expected; the expenditure of time, of treasure[1] and of life that are involved in even the ordinary routine of military movements over such a tract ... will suffice to show how immeasurable are the political advantages to be derived from a system of internal communication which would admit of full intelligence of every event being transmitted to the Government under all circumstances, at a speed exceeding five-fold its present rate; and would enable the government to bring the main bulk of its military strength to bear upon any given point, in as many days as it would now require months, and to an extent which at present is physically impossible ... The commercial and social advantages which India would derive from their establishment are, I believe, beyond all present calculation. (Cited in Griffiths, 1952: 423)

I will continue his minute, which deals with his other expectations, below. But his priorities were clearly laid down in this first part. The railways would be of military value first and foremost, and his plans were for 'Presidency Lines', i.e. those that would link the Presidency towns of Calcutta, Bombay and Madras, and also the new North West Province centred on Delhi. A cursory glance at the population map of India would show that the greatest part of this mileage would make little economic sense at all – the wide distances across the Deccan crossed dry tracts of sparse use. The map of rural population (Figure 6.1) is a fair indication of the relative weight of population and trade in the different regions. In seeking investment capital in England it appeared that because of suspicions about the environmental problems and about political stability, potential backers were dubious. The impasse was broken by the Railway Guarantee Scheme, by which the Company would guarantee a return of 5 per cent (a good return at the time) on capital invested by railway companies. This attractive offer resulted in sufficient investment, 99 per cent from England, although there was nothing to stop Indian investors subscribing if they so wished. Several companies began building the first trunk lines. Then the Mutiny of 1857 provided dramatic proof of Dalhousie's prophetic words about the railways' military utility, but it also rattled potential investors even more. The Guarantee Scheme had to be continued at considerable cost. By 1868 eight companies had between them opened 4008 miles of trunk line. This was a prodigious feat in many ways – rails, engines, wagons, and the engineers and craftsman, all were brought (such things as locomotives in knock-down form of course) from England, and off-loaded at small jetties in Calcutta or Bombay, or onto the lighters that struggled through Madras' rolling surf. It was also prodigious in standards. The gauge adopted, to stop trains being blown over, was a broad gauge of 5'6"; and all the lines were double lines, with very restricted minimum curvature and gentle maximum gradients. The result of this policy was

---

1 The word means silver – the bullion in which the rupee was denominated.

of course expensive cuttings and earthworks throughout the lines, which employed vast armies of local castes and tribes as coolies. As such groups migrated with the new lines, the railways had their first major social impact on India, as the canals had moved Irish 'navvies' around England.

On all of this the East India Company had to pay 5 per cent (even, as it was wryly observed, if the Railway Companies chose to invest by throwing bricks into the River Hooghly at Calcutta). The revenues were not enough to make the railways pay, and indeed the railways in India did not pay until into the twentieth century. So the people who subsidised the investors were of course the peasants whose land taxes were the basis for the government's revenue with which it could honour its Guarantee. The people who made the most money were the railway manufacturers in Britain, who had a powerful lobby in Parliament.

The railways were proving though to have other virtues. In 1865 there was a severe famine in Orissa; in 1868–70 in United Provinces (Uttar Pradesh). For the first time the Government had the means to help move major stocks of food, and to alleviate some of the suffering in those regions they could reach. Realising the excesses of the original companies and finding itself able to borrow money at a lower rate than the guarantee, the post-mutiny Government of India began building the second phase in 1867 directly as State Railways, much of it metre and standard gauge. But in the 1870s and 1880s the Rupee, which was denominated in silver, began fast to lose its value against gold-backed currencies such as Sterling. The Government again had to borrow capital by reverting to guarantees for private companies.

By 1905 the system was one of the world's largest, incorporating 28,054 miles of track, and often viewed with pride by the British as possibly their greatest and most beneficial achievement in India. But it was also chaotic: 15,000 miles of State Railways; 7,000 miles of Guarantee Scheme; 3,500 miles in Native States; 1,400 in Assisted Non-Guarantee Schemes; and 1,300 in Non-Assisted Companies. There was a lot of 5'6" track, a little standard gauge track, a lot of metre gauge track, and a large number of narrow gauge railways (in hill regions and Princely States mostly) of 2'6" and 2'0" and any other gauge someone had cared to build. To this day, transhipment from one gauge to another is a not infrequent, costly and slow business.

Dalhousie's minute of 1853 went on;

Great tracts are teeming with produce they cannot dispose of. Others are scantily bearing what they would raise in abundance, if only it could be conveyed whither it is needed. England is calling aloud for the cotton which India does already produce in some degree, and would produce sufficient in quality and plentiful in quantity, if only there were provided the fitting means of conveyance for it, from distant plains to the several ports adapted for its shipment ... Ships from every part of the world crowd our ports in search of produce which we have, or could obtain, in the interior but which at present we cannot profitably fetch to them, and new markets are opening to us on this side of the globe under circumstances which defy the foresight of the wisest to estimate their probable value, or calculate their future extent ... A system of railways, judiciously selected and formed would surely and rapidly give rise within this empire to the same encouragement of enterprise, the same multiplication of produce. the same discovery of latent resource, to the same increase of natural wealth, and to some similar progress in social improvement, that have marked the introduction and improvement of and extended communication in various kingdoms of the western world. (Cited in Griffiths, 1952: 423)

The prophetic remarks about the commercial impact were also true, and we will deal with these in a moment. What no-one had anticipated was the social impact: it had been believed that the Indians were too poor to create much demand for passenger traffic, but the railways suddenly opened the possibility of making pilgrimages to holy shrines, reducing perilous and lengthy journeys to quick and comparatively safe ones. The safety of the railways was indeed accredited with depriving Thugs of travelling victims, and playing no small part in the end of Thuggee.

In 1908 it was said:

... the passenger traffic contributes to the business of railways to a very much larger extent than was anticipated. The development has been in all classes; but the principal increase has been in the third class passengers, of whom nearly 200,000,000 were carried in 1904. (Government of India, 1908: 386)

This meant that in crude per capita terms the whole population of India took at least one railway trip. The same source mused:

It is less easy to gauge the moral influence which railways have exercised on the habits and customs of the people. It is often said that they are helping to break down caste; but it is doubted by many, whose opinions are entitled to respect, whether there has been any weakening of caste prejudices amongst the orthodox. There can however be little doubt that increased travel and the mixing of castes in carriages which railway travel necessitates, must produce greater tolerance, if it does no more. (Government of India, 1908: 388)

Such mixing and tolerance was of course not required of the British, who reserved carriages and waiting rooms for Europeans. Neither can it be said that in the staffing of the railways was caste broken down. The 500,000 employees (in regular pay: this excludes piece workers, casual workers, and contractors' labourers) were neatly differentiated. Moslems were drivers and workshop men; Eurasians were predominantly clerical and lower executive, the British higher executive and administrative, the Hindus the porters and incidental labourers.

The commercial impact was two-fold – on the relationships between markets inside India, and the relationships between India and the outside world. Since time immemorial there could be wide fluctuations in the price and availability of basic

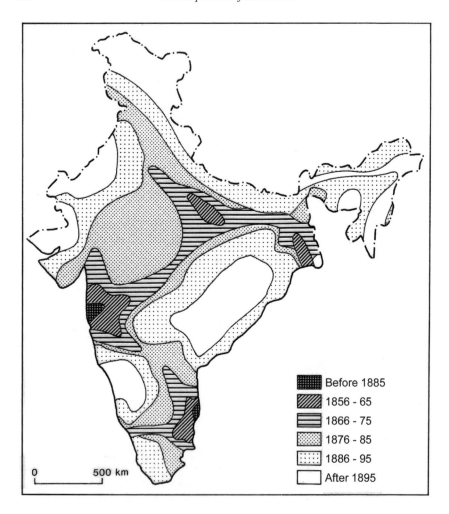

Figure 6.2   The Dates of Railway Expansion in India

commodities between different districts, even on occasion districts which were very close to each other. Without transport for bulk goods, a surplus in one area could not be moved to a deficit area. The British had from time to time to suspend the collection of land revenue from famine districts: but they also had to suspend revenue payments in districts which sometimes produced too much. In the late 18th century they had occasion to relieve the collection in Sylhet District, since the rice harvest was so bountiful that the price was almost zero, and not even enough to cover the cost of transport to market. Times of surplus were usually celebrated with great festivals and feasts: there was nothing else to be done. Clearly, the inability to turn surplus into capital was one of the many reasons why Indian

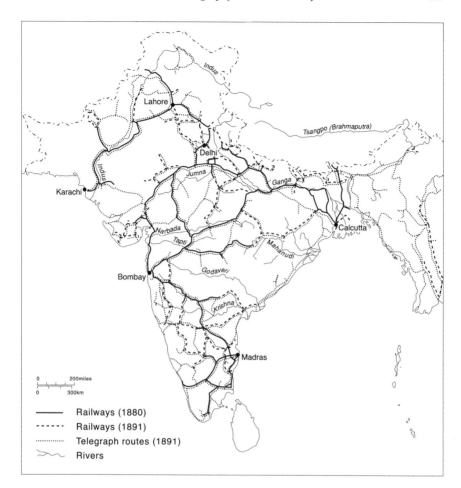

Figure 6.3   Railways and Telegraphs in India 1880

agriculture stagnated. But the country – sub-continent – is so large that in every year there are areas of surplus and areas of deficit. Linking these would enable those in need to acquire food, those in surplus to turn it into expendable capital, and would also lead to increasing stability in food prices. There is overwhelming evidence that this is exactly what did happen.

Both Mughal and British Governments had sought to monopolise the trade in common salt, using it as a source of tax revenue. The main sources of salt in North India were the salt pans in the Bay of Bengal, and the salt mines of Rajasthan. From the latter trains of pack animals moved laboriously down the Ganges valley heavily guarded and moving along a garrisoned trail. A locked and sealed carriage made thousands redundant, enhanced the monopoly, and reduced the price of salt at a stroke. Good quality coal in large deposits is primarily available in one region in India – the Bengal and Bihar coalfield, some two hundred miles west of Calcutta. Before the railways, in cost terms this was further from Bombay than England. Indeed the west coast settlements were supplied with coal by ship from England – although this was also partly because India never imported as much from Britain as Britain did from India, and freight rates on the outward journey were therefore lower. To some extent coal was half a ballast item. But the railways enabled the Indian coalfields to develop, and indeed they became one of the principal industrial employers before independence.

The other impact was the linking of the interior with the external world economy, in the way that the littoral regions had been for some centuries. British and other overseas trade with India had either stagnated or grown only slowly in the seventy-five years before the Mutiny. Indeed, the lack of opportunity in India was one of the reasons the Company would always produce to defend its monopoly. Some theorised that it was because India was too poor to import enough from Europe. Per capita incomes in India in the nineteenth century were undoubtedly lower than in Europe, but the number of capita was of course very large, and obviously the inability to penetrate much beyond the coast, and the anarchic political conditions of the interior until 1820, were also significant factors. After the advent of the railways, imports from Britain grew fast in volume, and to some extent exposed the 'market poverty' argument. Exports grew apace too: the railways began to haul food grain surpluses for sale on the international market. When the first canal colonies were developed north of Delhi in the 1840s, the aim was to use irrigation to secure local food supply. By the 1880s the purpose of developing irrigation in the Punjab was to produce export crops of cotton. After more than 200 years of European trade, suddenly, there was an impact that could be felt in many districts right down to the grass-roots of the economy.

There are two schools of thought about this impact: the immiserationists, who claim that the railways commercialised agriculture, producing industrial crops for profit rather than food crops, and raised land values, all in all creating a new class of rural landless poor. There are the ameliorationists, who claim that new wealth and prosperity trickled through the economy to everyone's gain. The truth must surely be a composite – depending on the region and the time. There is no doubt that there was a rising tide of prosperity in Punjab. There seems evidence that local artisan production in small towns in Bihar and Bengal and elsewhere suffered badly from new competition. We will return to some of these issues below.

Beside the economic impact, the railways had a particular physical impact. In the riverine lowlands of India, for mile after mile the railways had to be raised on embankments above the maximum known flood levels. This meant that huge new barriers were drawn across the landscape which impeded drainage. And the embankments took earth from near at hand as well – creating what are known as 'borrow-pits' which remained as shallow swamps afterwards. The combined effect was to create literally hundreds of thousands of acres of stagnant pools in the plains, with the consequent risk of an increase in malaria. It does seem likely

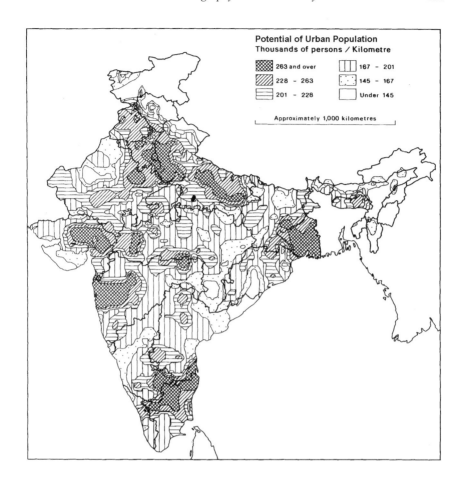

Figure 6.4   Urban Fields of Influence in India 1961

Source: Chapman and Wanmali (1981)

that the railways were indeed one of the causes of a resurgence of the disease in the second half of the nineteenth century.

The railways were also the cause of other ecological pressures. Already by the 18th century the British demand for tropical hardwoods for shipping and wharfage (ships for His Majesty's navy and the merchant marine were built in Bombay and Chittagong from teak) had started extraction from the forests, but this was little compared with the voracious wood demands of the railways. Each mile of track required 860 sleepers – and each sleeper lasted only 10–12 years. More than 1 million were required annually – and from superior sources such as teak and sal. When these seemed to be in short supply, experiments were conducted with the Himalayan cedar, the deodar, and that was judged adequate too, so the cutting of

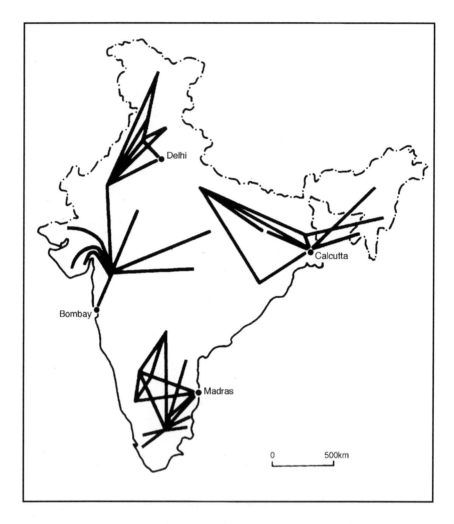

Figure 6.5   Commodity Flows in the Indian Economy 1960s
Source: Berry (1968)

the forests of the middle Himalayas started. Then there was the question of fuel. In the early decades this was also wood – not until the last quarter of the 19th century were the railways themselves and the mines of Bihar developed enough for the whole network to become coal-fired.

It was the demands of the railways more than anything else that led to the foundation of the Forest Department in 1864 and the passing of the first Forest Act in 1865. It was the beginning of a process of the extinction of local customary rights, and the commodification of the forests, that has marked India down to the present day.

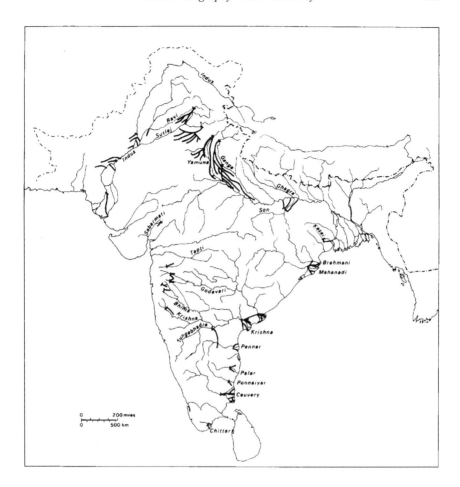

Figure 6.6   Canal Irrigation in India 1870
Source: Schwartzberg (1992)

Irrigation

Given the extreme seasonality of India's rainfall, it is inevitable that irrigation has
had a long history. Indeed, more than four thousand years ago in the Indus Valley,
Mohenjo Daro was an 'hydraulic' city based on irrigation systems. There are small
wells, sometimes with water wheels, unchanged in design for thousands of years;
there are small 'tanks', or local reservoirs, that govern most of the paddy lands of
peninsular India, and for millennia there have been quite large barrages and canals
in some of the lowlands – in the Kaveri (Cauvery) delta in Madras, for example.
Most of the larger schemes had fallen into disuse during the latter half of the 18th

century because of the political turmoil and not infrequent war. When the smoke began to clear in the early 19th century, the Military Engineers of the East India Company began to take a strong interest in the restoration of some of these older works. This was not as incongruous as it seems: much that was taught to military engineers of the period concerned fortifications, with massive stone abutments and revetments; in short, building skills closely allied to those now required. Starting in the 1820s, the Western and Eastern Jumna in the North West Province near Delhi canals were repaired and renovated. In the 1830s in Madras Major Arthur Cotton redeveloped the Grand Anicut (Barrage) on the Kaveri (Cauvery) and its distributaries, and then later the Krishna-Godavari delta as well.

The income from these works was collected in different ways in different provinces. In the south it was customary to charge an enhanced land revenue for wet land as opposed to dry land, regardless of what crops were grown. In some schemes in the north, records were kept of the crops grown, and farmers were charged a water revenue accordingly, in addition to land revenue. Early calculations seemed to indicate that the rates of return were very high; in the first fourteen years the Cauvery Anicut scheme repaid 70 per cent of its original capital. In the north, the West Jumna rehabilitation scheme, started in 1820, had by 1845 accumulated a surplus of 20 per cent of revenue over expenditure, and by then the annual income stream was little encumbered by further capital cost. But these high rates of return were to some extent a fiction, since these schemes benefited greatly from the existing capital works, even if in disrepair, and in the case of income calculated on land revenue, this should on occasion have been attributed to factors other than the canals. Yet success for such schemes nevertheless seemed assured. And the administration wanted to prove their success: there would be nothing less hostile to the British than a richer and more contented countryside, which happily also swelled government coffers. Further, control of the headworks of a canal several hundred miles long (the West Jumna canal had 445 miles of main canals and distributaries, its irrigated 900,000 acres covered a quarter of four revenue districts) was a form of political domination. There has been considerable discussion over whether more small wells would have been better: certainly they would not have given rise to the problems of water-logging associated with the canals. The water-logging was mostly the result of building unlined canals – i.e. canals were dug with simple earthen banks and beds, and not lined with an impermeable layer such as (in the current age) concrete. As in a natural river, clay and silt would form some kind of natural lining, but water transmission losses were still high, often as much as 60 per cent of the water at the head. The other way in which canals caused water-logging was by impeding natural lines of drainage. When a canal left the headworks on a river and traversed 'inland', it would meet small tributaries. These would have variable discharges, and often considerable silt loads, so they could not be admitted to the canal. Aqueducts over such tributaries, and siphons beneath them, were expensive, and so were restricted to larger streams. Water-logging itself had two effects: ultimately an increase in salinity which first depressed crop yields and then made soil sterile, and, secondly, an increase in breeding grounds for anopheles mosquitoes. The canals, like the railways, brought malaria in their train.

But canals also saved the populace from famine in drought years. And on average they boosted production – perhaps by 20 to 40 per cent in their command areas. In 1840 Major Probey Cautley began the Ganges Canal, from Haridwar, to irrigate the whole of the upper Doab.[2] Cautley was a pioneering zealot of an engineer. This, the first of the wholly new large canal schemes of the British, was built on an epic scale, and at an epic cost. The first twenty miles of the canal contained a string of astounding engineering features, aqueducts, siphons, level revetments against erosion, all built in a superbly engineered manner. Cotton believed it unnecessary, and thought the headworks could have been located lower down the Ganges. In the present day the Irrigation Department have to pay twice as much to modify Cautley's masonry and brickwork as they pay to modify equivalent later structures, because of the toughness of what has to be demolished.

The costs of Cautley's dream mounted. Periodically he would have to ride the two or three months journey to Calcutta, to seek further financial support, which was not easily forthcoming. He had to struggle with other problems too: as happened with the railways later, he had a shortage of skilled surveyors. So, at Roorkee near the headworks he built an engineering college, the origin of Roorkee University, still the premier centre for irrigation engineers in India. In the University there is a workshop built by Cautley before the Mutiny for the making of sluice and lock gates, with a tall chimney rising above it. The chimney is not what it seems. It is in fact a ventilation shaft for a secret underground room in which the British hid their families in time of trouble (It is reputed that one of Cautley's children was born there). They knew their canal was part of a process of pacification of a very troubled land.

Cautley's Ganga canal scheme did not give the kinds of returns government expected: in fact it lost money. The costs had been too high. It was also deliberately designed on uneconomic principles, in that it had a command area more than twice the size of the area it could irrigate. Therefore the capital cost was twice as high on that account alone as it need have been. The logic of this, pursued also in the Punjab, was that the greatest number of cultivators were reached that way, all of whom had some water, all of whom, therefore, had some protection against famine. The principle was to give protection to the greatest number possible. In the Punjab it became fairly standard in the early schemes to provide enough water so that each farmer could irrigate only one third of his fields. Given that the Ganga canal was built before the railways had reached the north-west, the idea that it was for local relief and not economic development for export seems reasonable.

Arthur Cotton, who had rebuilt the Grand Anicut on the Cauvery near Madras, continued his distinguished career, and became a knighted general. In Britain he

---

2 The word 'Doab' means literally 'Two Rivers', and refers to the interfluve between two rivers. These are many Doabs in Punjab, all with descriptive names, such as Rechna Doab. When people talk of The Doab, they mean the land between the Ganges and the Jumna.

campaigned ceaselessly for more support and investment in irrigation – he saw it as the government's Christian duty to ward off famine amongst its Indian subjects. In 1866 he gave a lecture in Manchester at the Social Science Congress held that year.

> The first question is, What is the nature of these famines – as to their occurrence, extent, effect

> &c.?...They occur in some parts of India perhaps every ten years, and we may therefore consider their return in future a certain event, and at least as much to be provided for as war, or any other of the matters for which Government is instituted. (1866: 4)

Famine in Orissa in the 1870s and in Madras persuaded the government in India that unremunerative canals were actually justifiable in some circumstances. Therefore the Public Works Departments (created in the 1850s to take over from the Military Engineers) classified works into Productive (revenue raising) and Protective (loss-making within reasonable limits, but improving agricultural yields), and they were not unsympathetic to the latter when finances could allow and the proven need was urgent.

Cotton continued his campaigning. In 1877 he issued a pamphlet on famine in the north east Deccan, which reiterated his line that not only was the control of water the answer to India's problems (both for irrigation and transport), but it had perforce to be large-scale canal irrigation – that in which engineers like himself were superior to any indigenous talent.

> It will be well perhaps to remark on some mistakes which are almost universal on this subject. The first is, that if a tract has plenty of rain, there is no necessity for irrigation.

> One plain answer to this is, that the Famine in Orissa occurred after a Monsoon of 60 inches. The question is not how much rain falls, but how it falls. In Orissa 30 inches fell in June and July. There was then a pause of six weeks during which the whole crop perished, and the 30 inches that fell in September could not restore it. ...

> Again, when we say "irrigation", we always mean the complete regulation of the water, that is including draining; and so there is never a season when there is not at some moment excess of rain, which requires to be carried off by a system of drains. It is this regulation of the water that is needed, and which so abundantly repays the cost of works. God gives us the rain, but as in everything else, he leaves something for us to do, which if we are too indolent to do, we must suffer for it.

> The second is that water is water, but this is also a great mistake: there are three kinds of water in agriculture. That from rain, water that has been stored in tanks, and water led direct from the rivers to the field. The first has been filtered, and does little more than afford moisture; the second has deposited most of its rich contents which have

been held in suspension, though it conveys to the field what was in solution; but the third comes to the land loaded with everything that the plant can want. With this, the land is perfectly renewed. Lands that have been watered for hundreds of years from rivers, continue to afford white crops without diminution, though without manure. The district of Tanjore which has lands in it that were watered from the Grand Anicut in the second century, and ever since, continues as fertile as ever. No application of well or rain water, can make up for the want of river water. Thus the Midnapore ryots begin to understand this, so that after a fall of rain they empty their fields as quickly as they can, and fill them again from the canals. A third is that irrigation as a rule produces fever. (1877: 33)

Fever did indeed accompany the march of irrigation just as it did the march of the railway lines, but these correlations, although believed by many, were dismissed by Cotton, peremptorily observing that he never suffered from any, and he certainly lived to a vigorous and active old age, hardly stopping until he died at the age of 96.

Cotton was a committed Christian who saw his life's work tied up with Christian duty, and argued the case for irrigation from that stance:

But what shall we say to the loss of our character as a Christian Government from our having so neglected both to execute those works that can alone prevent these famines, and also to prepare for each famine when it was imminent? (1866: 6)

and indeed he actively understood the value of recruiting the Great and Good of his day to support his cause. In the 1866 pamphlet there is a supporting letter appended, written by Florence Nightingale, which shows the true nature of Christian mercy:

If all England could set their face against the Suez Canal, we must not be surprised if there are other people almost as stolid. Another Nation had to cut the Canal for us and thus force upon us an incalculable benefit. In England and Bengal you must take people as you find them and force blessings upon them. So we thank God and take courage. We are really gaining ground. (Nightingale in Cotton, 1866: 30)

By the 1870s the irrigation works in India were achieving a world-wide reputation for their size and performance, and they were set to become something of a focal point for the international traffic in engineers and politicians interested in them. Visitors and ideas flowed between the Americas, the Mediterranean, Egypt, India and Australia. One such rover was Alfred Deakin, formerly Chief Secretary and Minister of Water Supply, Victoria, Australia, who visited the United States in 1885, Egypt and Italy in 1887, and India in 1890–91 – publishing a valuable and detailed survey entitled *Irrigated India. An Australian view of India and Ceylon. Their Irrigation and Agriculture* in 1892/3. He was quite clearly a very Australian and very independently-minded individual, who admired the technology, but wondered about its purpose. He had a kind of Malthusian anxiety over the teeming Asian hordes.

The main purpose of this book is to sketch the superb systems of water supply, by means of which many millions maintain existence upon tracts that without would only support a fraction of their number. At this stage it is but just to indicate that there is a point of view from which the great schemes appear less admirable in their net results.

As the real notion of irrigation in India is to maintain life, and its success lies in minimising famine, it brings those who would sum up the case for and against it fairly face to face with an old problem of history, pertaining in some degree to all races, but especially under Asiatic conditions. Progress in numbers is readily measured, and at each census the totals of the Indian Empire are enlarged. ... The prospect of a country doubling its population in five or ten years may appear at first sight matter for congratulations. It means peace and plenty, to some extent health and morality, increased production, increased consumption, increased trade, and increased wealth. All these can be predicted of India, whose total population for British and Feudatory states alike was 256,000,000 in 1881, and was 286,000,000 in 1891. In the same period Australia has added 1,000,000 million as against this 30,000,000; and though the latter total has been swollen by annexation and improved methods of enumeration, the broad fact remains that the gain in 10 years exceeds the population of Italy or Prussia. Among the most potent means of this rapid growth in the population is unquestionably the irrigation, which not only makes agricultural settlement closer wherever it obtains, but provides the vegetable food of the Hindus for countless thousands beyond the schemes. It may be held to have saved the lives of millions who would otherwise have perished, and to have enabled them to beget millions more, whom it now assists to maintain.

Is this a real gain? Does it deserve the name of progress? Does it benefit either the individual or the race? Many will reply without hesitation in the affirmative; but surely in so doing they confuse the size of a nation with its eminence – they mistake quantity for quality? (1893: 53)

The struggle of the British government to raise the masses, like that of the daughters of Danaus, seems fruitless as well as endless; the courage, energy, self – sacrifice, ability, and benevolence of its rule, idle and without avail. The history of its superb conquest of the elements, like that of its conquest of the country, when viewed from the standpoint of philosophic history, concludes, not with a paean, but with a melancholy question – Cui bono? (1893: 55)

Little irrigation was developed in Bengal, for a number of reasons. The drought risk was less, to be sure, but in Bengal there was also under Zamindari a more tangled hierarchy of vested interests in land, greater population density, and greater fragmentation of holdings. All of this made the acquisition of land more difficult and much more costly, and the later revenue-raising more difficult. (Land acquisition costs in the poor Doab had been low for Cautley. He bought enough land to allow for the many changes and improvements, which are still being made even now.) In addition, in the majority of years rainfall would be adequate, so that there was not quite the same degree of urgency as in the drier districts to the north-west.

The Land of the Five Rivers

The Province in which the development of irrigation under the British most concerns us in this book is Punjab, which means the Five Rivers, the great tributaries of the Indus. They flow down out of the Himalayas onto what is naturally a semi-arid land. There had been little cultivation away from the banks of the rivers and the rim line of the hills themselves, and so here the great Doabs were mostly scrublands used by wandering pastoralists but not formally owned by them in any British legalistic sense. In 1849, after the Second Sikh War, the Punjab was formally annexed and the Sikh army disbanded. As with the independence fighters of Mugabe's Zimbabwe, it was imperative to find occupations and incomes for the former warriors before they became lawless gangs of bandits, and in a society based on agriculture or husbandry, that meant giving them land. The wastelands of the Doabs were formally claimed Crown Lands, available to the new administration for use as it saw fit. A solution to these various problems was, of course, then sought in developing irrigation on the Doabs and founding new settlements.

In 1850 the first plans were laid for the Upper Bari Doab[3] Canal, which was planned to be 247 miles long from its headworks at Madhopur. Like the Ganga canal it would have to start right where the rivers debouched from the hills onto the plains, since the rivers were incised, leaving the Doabs as slightly raised tablelands. The planners were in buoyant mood: '... striking into the deeps of the wildest wastes of the lower doab, and running past the ruined cities, tanks, temples, and canals, all of which it is to vivify and regenerate, it will join the Ravee 56 miles above Multan.' (General Report on the Administration of the Punjab, cited in Michel, 1967: 59). With no clear idea of the discharge of the Ravi, Colonel Napier set the capacity of the canal to be 3,000 cusecs (cubic feet per second.) For 180 miles it would provide irrigation: for the last 80 it would reconnect with the Ravi so that it could be used for navigation. In practice it never reached its goal, so navigation did not become an important source of revenue. It was also built with too steep a gradient, and the bed eroded. The engineers were, in fact, conducting what amounted to massive experiments in hydrology, and were building themselves a flume the envy of any modern geomorphologist. The canal was not a great success financially, but it did again have a beneficial impact on agriculture. (It also became blamed for water-logging and malaria, and part of its

3 The Doab names are easily remembered as they combine the names of their two rivers. The Bari Doab is that which lies between the Beas and the Ravi.

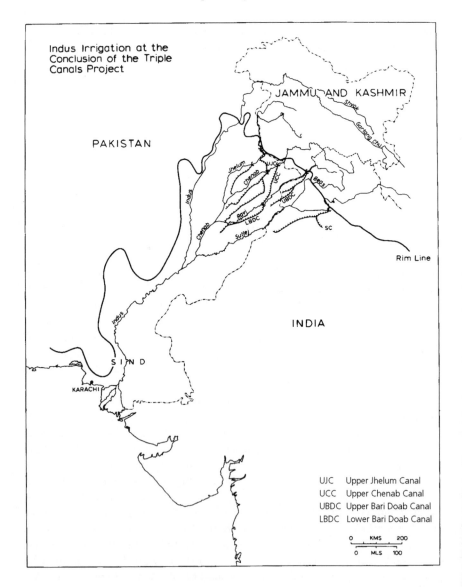

Figure 6.7   Canal Irrigation in the Upper Indus at the Conclusion of the Triple Canals Project

Source: Michel (1967)

command in the early twentieth century became the first to be treated with diesel and then electric tube wells to reduce local water tables.)

The next canal attempted was the Sirhind on the Sutlej, sanctioned in 1869. This had a gradient set at 30 per cent of the Upper Bari Doab Canal's gradient, which proved too shallow as siltation began to choke it. It also employed a barrage across the river for the first time (the UBDC had only had a side weir: the engineers had not thought they could build a barrage to withstand destruction by peak floods). In 1895 the barrage was redesigned to include a silt trap, and its performance again improved. But, as with Cautley's Ganges Canal, the costs had been too high in relation to revenue, and the Government felt disinclined to try further investment. Famine in 1879 proved the spur to overcome caution, and further schemes were sanctioned for the lower Doabs, each being built with an increasing competence and confidence, and, finally, with growing financial success. The Chenab Canal of 8,000 cusecs on the lower Rechna Doab was sanctioned in 1891 to irrigate 1 million acres, and so was born Lyallpur (now Faisalabad), named after the Governor of the Punjab at the time. The settlement patterns changed: the Sikhs came down from the hills and became the dominant farming and commercial group everywhere, though outnumbered by the Muslims in nearly every district.

All of these canals were limited by the fact that they were run-of-the-river systems: they used barrages to head up water, but not to store it. In those days large storage dams were a dream, not yet an engineering commonplace. This meant that when a river had been tapped, further expansion could only be achieved by bringing water from elsewhere. And by a quirk of geography, the eastern rivers of the Punjab were those with the smallest discharge, despite the fact it is the west that is driest. The biggest discharges were down the Jhelum, and of course the Indus itself. To overcome these imbalances in 1905 the Triple Canals Project (Figure 6.7) was sanctioned to bring water from the west to the east. The Upper Jhelum Canal was to run from Mangla to the Chenab above Khanki. The Upper Chenab Canal ran from Marala on the Chenab and crossed the Ravi at Balloki, and from Balloki ran the Lower Bari Doab Canal. Nearly another 2,000,000 acres were brought under cultivation. Since by now the railways were an established fact of life, the schemes could be designed from the outset to maximise the production of cotton for export, to the hungry mills of Lancashire.

The important point that comes from this is that the British had now created a system-wide single scheme, each part dependent on the others. No part could be operated without consideration of the others. It irrigated land both directly under British control, and in the territory of princes, since the British had decided that the overriding goal must be the irrigation of the greatest area regardless of political considerations. In the early twentieth century it even came to mean that agreements had to be negotiated between the Government of British Punjab and British Sind, since the whole Indus Valley became interdependent.

Any irrigation scheme, no matter how small, is a political statement first and foremost. It is also of course an economic statement of costs and revenues, and a technological artefact of good or bad design. The acquisition of land, the methods of revenue raising, the distribution of water, all take place *given* the nature of the political system. The political system here was British imperial hegemony. It created the largest integrated irrigation schemes ever seen on earth. What would happen if the hegemony ever ended? That is a theme to which we must return later in the book.

International Trade in the 19th Century and the Balance of Payments

Before 1757 the British were almost unable to export anything to India to pay for their imports because they produced very little that was in demand, and perhaps as much as 80–90 per cent of British trade was settled by payment in silver, the drain which the British Government so disliked. By about 1800 British manufactures exported to India were worth about a third of the value of imports from India, which were still dominated by cotton piece goods. Instability in Europe during the Napoleonic Wars and the turmoil in India itself combined to produce erratic trading conditions for the next twenty years, with no discernible signs of growth. But, from somewhere about 1825, a long-term secular change starts, with trade growing at an ever-increasing rate in the second half of the 19th century as the railways expanded. Imports grew fast, but so too did exports, although of a changing nature. Throughout the whole century (except for a brief period 1855–63) India remained in trading surplus, although in the latter part of the century she was in trading deficit with her principal partner, Britain.

From 1828 to 1840 the export of cotton piece goods fell by 48 per cent, as British cotton manufacturers supplied their own home market, and indeed as they increasingly supplied manufactured cotton to India. In the same period India's imports of yarn rose 80 per cent and of cotton cloth 50 per cent. After 1860 imports of cloth rose dramatically, and of course imports, too, of railway goods. A major shift was occurring: India that had once exported manufactures, was now increasingly importing them. To pay for these, her external trade began to be more and more in primary produce, including by the latter part of the century, raw cotton, jute, food grains (rice and wheat), indigo, tea, and opium. This new relationship is depicted as the stereo-typical colonial dependency: the colony was the source of raw materials and ultimate captive market for the Mother industrial economy.

There were obvious consequential internal adjustments. The distressed plight of weavers, particularly of the finest muslins, in Bengal was noticed even by the hard-nosed British administration, though ideas of protection had no support in an era ideologically committed to free trade – and indeed the disturbances of the Maratha raids and the blockades of the Napoleonic Wars had also been to blame for the loss in trade. There was good as well as bad: the imports of British cotton were predominantly 'grey cloth' i.e. cloth that had yet to be bleached, dyed and, of course, tailored. The consistent quality and the comparative cheapness of British cloth gave a stimulus to these sectors, and prices for the Indian consumer fell.

The continual surplus of India's balance of trade meant that the hunt for bullion (treasure) never ceased. One estimate suggests that India imported 1/3 of the world's silver production in the last quarter of the 19th century (and although not a producer, it is currently thought to have more than half of world stocks). It was here that opium was so useful. This was grown in Bengal under government supervision and control, and in the native states near Bombay under private control, but ultimately exported under government control from the port itself. Its greatest market was China, which the British insisted on supplying, despite the hostility of the Chinese Government. Indeed the British fought two Opium Wars in 1839 and 1855 to keep the trade open. It was also very heavily taxed, and in the late 19th century opium was the second largest source of revenue to the Government, only being exceeded by land taxes.

Sir Richard Temple, the Contents of whose 1880 survey of India was listed in Table 4.2, in the same volume compared the taxation of opium with the taxation of spirits, and said of the trade in general:

> There cannot ... be any objection to the taxation as it falls upon an article which is a luxury and which, if it be useful in extreme moderation, is more deleterious if used in excess ... inasmuch as the culture is profitable to thousands of cultivators, and as the exportation is still more profitable to traders and capitalists, any attempt on the part of the State at suppression would be futile, and would only lead to serious abuses ... It is manifest, however, that if the Chinese Government shall ever attempt to prevent the importation of an article of luxury much desired by the people, it will be essaying a task which has rarely been undertaken by any Government with success anywhere and which the Government in most civilised countries does not attempt. (Temple 1880: 240 and 241)

The absolutely addictive nature of the drug was not well understood at the time, and its use was not uncommon in British society itself – as witnessed by the central confusion in Wilkie Collins' novel *The Moonstone*. Indeed Robert Clive was a frequent user of laudanum, a liquid potion containing opium and alcohol, used as a painkiller and sedative. Only with hindsight does Temple's justification of the trade seem quite so appalling.

Opium was thus an important part of the triangle of trade between Britain, India and China. When the trade collapsed after the turn of the century, it seemed to do so for economic reasons as much as a result of tightening government policy. Other competing sources developed abroad, and in consequence the accumulated profits were taken out and invested in new textile mills in India.

Exports of raw cotton were slowly increasing in the first half of the nineteenth century, but did not really take-off until the latter half. At that time Britain's favoured supply was from the Southern States of the USA. India's cotton had proved inferior for the new machinery, being of shorter staple. But the stimulus provided by the American Civil War (1861–65) which badly disrupted supplies, was followed by the successful development of long-staple hybrids suitable to Indian conditions, and then the development of the Punjab and Sind Canal colonies in conditions which were also suitable for long staple production. By the end of the century cotton was exported to China in great amounts as well, and then also to Japan.

An agricultural industry that was new to India was opened up in the second half of the nineteenth century – the cultivation of tea. It seems impossible now to realise that only eighty years ago British writers noted that people in Indian towns were beginning to take up the new custom of drinking tea – its is now such a deeply ingrained part of Indian life. Indeed, it is hard now to remember that there ever was an India without railways and the chai-wallas who pitch on the platforms. Tea had been exported originally from China to Britain, being apparently unknown in India. After the takeover of Assam in the 1830s Britain was in possession of wet sub-tropical Himalayan areas similar to those in China – in fact the western flanks of the mountains that divided China's tea area of Yunnan from India and Burma. British investors could cultivate the crop, and export it down the Brahmaputra valley to Calcutta, saving distance and cost. Indeed tea was produced, by 1900 being 1/3 the value of cotton's exports. The story of exploitation and money is taken up again in Chapter 12.

The New Geography

The geography of India was indelibly changed by the pattern of infrastructural development in the nineteenth century. Figures 6.2–6.4 show the way in which the railways entrenched the port cities of Bombay, Calcutta, and Madras, and produced an urban system in India that was at variance with the previous urban structure, and with the distribution of the bulk of the populace, which was, of course, rural. Discriminatory freight rate policies underlined the supremacy of each of these cities in their own regional economies. The pattern was still visible in the freight flow maps produced by Berry in the 1960s (Figure 6.5), and persisted in the period of India's autarkic development since Independence. These economic regions paid little regard to cultural and linguistic regions, a fact which was to prove of significance. Population redistribution had also occurred in specific areas, such as the canal colonies of Punjab and Sind, and in the export-oriented tea gardens of Assam, and in new farming areas in the Valley of Assam. Areas of new industry grew – though one must stress that industry singularly failed to achieve the general take-off widely hoped for and predicted, but never actually encouraged, by the British. Such industry as did develop formed small islands in a sea of rurality. North of Calcutta there were the extensive jute mills, mostly in British ownership, and nearby the expansive railway works at Kharagpur (which built far too few locomotives compared with the numbers imported from Britain: before independence 12,000 locomotives were imported from Britain, and 700 built in India). Around Bombay and in the Madras Presidency, there were new cotton mills, most of which were built with Indian capital and management. Some of these, emulating Lancashire, began to achieve export success in China, until early in the 20th century Japanese competition defeated them. In the Damodar Valley of Bengal and Bihar were the new coal mines, sunk with British capital. At Jamshedpur the Indian capitalist Jamshedji Tata built India's first, and at that time one of the world's biggest, iron and steel works, much to the surprise of the British.

    At a smaller scale there was a new urban form imprinted on the landscape to one side of most cities and towns. Mortality rates amongst the early British traders and soldiers from diseases such as malaria (literally 'bad air') and cholera had often exceeded those of the native population. The cause was nearly always put down to impure air and water, with the obvious conclusions that cleaner air and water was desirable. New cantonments for the British soldiers and civilians were built in prodigiously spacious settings, segregated from and upwind of the dense and noxious native settlements.

> Impure air and water may not be the only source of cholera, dysentery, malaria ... but when the source of these impurities is the exhalations from the human body, they are the most powerful exciting causes of those diseases. ... I believe that the part which the excrementitious matters of the affected play in the dissemination of epidemic diseases is chiefly due to the gases given off by them immediately or soon after they are voided, which gases enter the system through the lungs. (Clerk 1864: 103, cited in King, 1976: 109)

This led to the conclusion that airiness was necessary. Buildings for the colonialists needed more space:

> European inmates in tropical climates to have 1,000 to 1,500 cubic feet of air. ... For natives, these numbers admit of reduction (to) ... 600 cubic feet of air. (King 1875: 220, cited in King, 1976: 110)

To be sure the spacious nature of cantonments and civil lines had other purposes as well, providing fields of fire for artillery, and minimising the hiding places for rebellious infantry. The doctrine of space was pursued to its climax in the building of New Delhi, a city seemingly built with the kind of personal mobility in mind which is available in modern Los Angeles, although in New Delhi's case that mobility could be had at a slower pace by horse-drawn tonga or the messenger's bicycle. Motor transport was for the very few.

The Language of Empire

Depending on the definitions used and the extent to which a dialect becomes another language, there are between 200 and 1,600 different languages spoken in India. There are at least fifteen different scripts, and although for some of the northern languages derived from Sanskrit the scripts are very similar to the Devanagari used by Sanskrit and Hindi, there are at least ten different calligraphic systems. Administrative systems would work better if there were some major official language, and indeed English had replaced Persian at official levels in the early 19th century. Table 6.1 documents a stage in this ascendancy. English also became the favoured language of new institutes of higher education. Ancient India was noted for its scientific achievements, and in particular its contribution to mathematics, but since the Renaissance scientific advance had been fastest in

Europe, foremost perhaps in England. English became a prominent language for the communication of the new knowledge.

English was not adopted in India as the language of higher education without debate. But the British rejected the use of vernacular languages because

> the vernaculars of the country do not as yet afford the materials for conveying instruction of the comparatively high order (Naik, 1963 quoting government documents from 1860, cited in Khubchandani, 1983: 20).

Table 6.1   A Letter from the Governor-General to the Nawab of Bengal, 1833, in the Hazaduari Palace at Murshidabad, West Bengal

His Highness
Newab Shuja ul Mulk Ehtesham ud
Dowlah Humayoon Jah Syyed Mobaruck
Ali Khan (Behadur Feroze Jung)
My Honored and Valued Friend,

Your Highness has probably heard that in the early part of the cold weather I left Simla and Sabatheo and proceeded on my journey to Calcutta and I have now the pleasure of informing your Highness that I arrived at the seat of Government on the 2nd February last. As I am sure that a knowledge of this event will be gratifying to your Highness I have hastened to communicate with your Highness on the subject. I am always conscious to receive accounts of your Highness' health and welfare and I shall therefore hope to be favored occasionally with your Highness' friendly letters.

I am not myself conversant with the Persian language and it is of course satisfactory to me to communicate my sentiments to my faithful Allies in my own words unaltered by any foreign mediation. With this view I propose to carry on my correspondence with you in the English language. I presume you can have no difficulty in commanding the services of persons acquainted with that language and in the meantime I shall direct my letters to be accompanied with Persian translations.

In conclusion I beg to express the high consideration I entertain for your Highness and to subscribe myself.

Your Highness' sincere friend
W. Bentinck
Fort William
16th March 1833

The situation has never really changed, since scientific knowledge has grown apace throughout the 20th century and English has acquired new ideas and terms faster than they can be translated or assimilated by the vernaculars. Thus, English became the new lingua franca of India. Major newspaper publication in began in English – *The Times of India*, in Bombay 1838; *Amrita Bazar Patrika* and *The Statesman*, both in Calcutta in 1868 and 1875 respectively; *The Hindu* in Madras in 1878. The new nationalist leaders of India used this same foreign lingua franca not for local affairs and not for daily business, but they did use it to communicate between East and West, North and South. Indeed, perhaps the trains, the post office, the use of English and the press were essential preconditions for an independent and united India. There was also a vernacular press – which was accused from time to time of being too anti-government, so that an Act was passed permitting its muzzling if necessary. Many of the vernacular papers were not founded until into the 20th century. But English literally ruled:

> The imposition of English as the language of politics transfigured Indian public life in at least two ways: it obviously divided the British rulers from their Indian subjects; and it also divided Indians themselves, between those who could speak English, who knew their Dicey from their Dickens, and those who did not. These often immaculately anglicised élites were also, it is essential to remember, fully bicultural, entirely comfortable in their own Marathi, Bengali or Hindustani milieux: it was, after all, exactly this amphibious quality which made them useful porters to the British Raj. The slow extraction of power from the society and its concentration in the state was in India's case crucially a matter of language. The social power that Sanskrit – and Persian – had once held was replaced by a new, still more mysterious, more potent language of state: English. (Khilnani, 1997: 23)

Lord Curzon ordered the erection of an obelisk on the site of Clive's command post at the Battle of Plassey. It had an inscription commemorating the valiant soldiers and their victory. Recently it was refurbished, and re-opened in 1998 by Sri R. Bandopadhyay, IAS, Commissioner of Presidency Division. The plaque now reads:

Battlefield of Plassey June 23rd 1757

In memory of the valiant warriors of Siraj-ud-Daula

Plassey – Nadia

This re-interpretation of history has been written in the language of conquest.

A Necessary Understatement

In the 1920s when India's population was about 275 million, enrolment in secondary education was 0.6 per cent of the population, and in Universities 0.03 per cent. Official British thinking suggested that these figures were good and stood comparison with more highly developed economies in Europe. The figures reflected in particular the schools and colleges of urban India, which really were producing a new small, but influential, Indian urban middle class. But India was only 10 per cent urban, and the figures for primary education, which should have embraced the countryside as well, were actually extremely bad. Less than 4 per cent of the population were in any form of instruction, and the literacy rate was about 10 per cent. There is a suggestion that the syllabi in schools (where there were any) was so completely irrelevant to farming needs that it was rejected even by those who could have had it. Education then formed small islands of privilege in a sea of ignorance, just as industry and other aspects of modernisation were similarly restricted. It also formed disciplinary islands: one curious feature of higher education was the disproportion between graduates in popular subjects such as law, and scientific subjects which were not widely available.

It is therefore possible to claim far too much for the transformations that were undoubtedly taking place in the 19th century. In 1901 Digby wrote:

> There are two countries: Anglostan, the land especially ruled by the English: in which English investments have been made, and Hindustan, practically all India fifty miles from each side of the railway lines. (Digby,1901: 291 cited in Bose, 1973: 51)

There was, despite the size of the network, a huge amount of India that lay more than 50 miles from a railway line. And overall, more than 70 per cent of the populace still relied on agriculture for its income, and 90 per cent were rural inhabitants. They were for the most part still interested in self-sufficiency: cropping for export was restricted to specific areas like Punjab and the cotton-growing parts of the Deccan, or perhaps to many peasants in a region, such as the jute growers of Bengal, but even then each farmer would usually put only a lesser part of his holdings to jute. Perhaps the influence that reached furthest through the economy was the linking of internal prices from region to region, which had a beneficial dampening effect, and to world markets, which was not always so beneficial. At a time of local plenty, the poor could still be outpriced in their quest for food if prices quoted in Chicago's grain markets had risen high. And the major traders at the turn of the century knew by telegraph any price changes within 24 hours.

The contributions that the British like to be remembered for are in the fields of administration (the iron frame of the Indian Civil Service which at one time commanded greater respect than the home civil service), the army, the posts, telegraphs and banking systems, the irrigation schemes, and above all the railways. They had forged a new imperial state which was in one aspect much stronger than the Mughal state. It had developed aspects of utilitarian integration which the former had lacked, based on this new infrastructure. But it was not an Indian State, in that the British never made their home in India and were never automatically committed first to India's interest before Britain's imperial interest, although it *was* an Indian State in that most of the personnel who ran it were Indian. The few British who were there represented just the top of the pyramid of government, whose employees as police constables and railways guards at the base were all Indians – with Indian fingers wrapped round the handles of the lathis, and poised on the triggers of rifles. There were in the 1930s only 4,000 British gazetted officers in the civil service, the police, the railways and in forestry. There were the traders, managing agents, tea-planters, bankers as well, perhaps only a maximum of 30,000 civilian Britons in India, and in addition units of the British Army garrison which peaked at 70,000.

Given this fragile hold, the British were obsessed with how to maintain law and order and to retain power, for as long as practically possible. What their imperial state failed to do in fields such as agriculture, education, public health, and industry, is a story that should be written alongside the march of the railways lines. But since they did not do very much, it is a hard story to write. Above all; their imperious, timid and reactionary attitudes meant that they were never in much haste to develop new political infrastructure, and it is perhaps no surprise that the new middle classes would become impatient for change in that too; and that ultimately the loyalty of the government servants could be stretched to the limits.

Concluding Remarks

The empire the British created surpassed the Mughal empire in size. It included the furthest south, which the Mughals had never dominated, and similarly included Assam and Burma temporarily. It did not however encompass Afghanistan, and at best that only came under British 'influence'. Although the precursor to empire had been a trading company, the basis of government finance became land revenue – just as in Mughal times – for all the areas of direct British government. Perhaps more than the Mughals had done, the British also relied on indirect control via tributary chiefs – the Princes of the Princely States. The prime bond maintaining the integrity of the area was undoubtedly the identitive bond of the British themselves – a bond which showed less sign of dissolving into the local territories of India than it had for the Mughals, probably mostly because the vast majority of British never actually settled themselves as colonisers. In that sense they did not make India their home. They also, like the Mughals, devised a system of administration – the Iron Frame of the Indian Civil Service – which has to be included in the fourth category of bonds – the bonds forged by an able bureaucracy. Towards the end of empire more Indians were incorporated into this administration, and given rank and security which brought quite a strong degree of loyalty and co-operation with the Government, although the British incorporated the local élites in Government less than the Mughals had done. But in the building of the transport networks they provided the first strong elements of utilitarian integration in South Asia – even if on a regional basis. The imposition of English in government and higher education together with the communications networks, permitted the beginnings of an articulated Indian sense of 'Indian' identity – something that was new to South Asia. The British had inadvertently stimulated it, and they were always ambivalent in their reaction to it.

This account is, however, rather one-sided – it is written from the viewpoint of the British impact on India, as though that was what stimulated change. The impact is exaggerated because this book is written in English, and from sources written in English. There are many of them – the British in the 19th century accumulated facts and data and wrote about India with an extensiveness that that is astounding.

There would have been change in India anyway – although the speed and kind of such counterfactual change cannot of course be proved, but only suggested as a kind of thought experiment. It is possible, though, to write accounts which stress movements and change at the level of the peasant masses much more than older writing, and recent studies such as that by Sarkar (1989) and the series of 'subaltern' (non-élite) studies by Guha and his associates do adopt this wider perspective. It is also possible to quote the outsider, Deakin, observing the British not just as irrigation engineers, but as imperialist administrators.

> Finally, then, the British government of India is a compound of contradictions, for, while practically absolute in authority and vested in two or three men entirely, it is supposed by many to be controlled by a popular assembly; military in spirit, it is bureaucratic in method, and pacific in end; conservative in practice, it adopts many radical principles; and, committed wholly at first, and often still, to the energy, judgement, and initiative of individuals, it has created for them a complete system of written regulations embracing the whole field of possible activity. No public service is so constrained by the pen, and yet even the civil members of it may be said to live in the shadow of the sword. Separated by immense distances which forbid frequent personal association, all business is conducted by correspondence; the affairs of the country, from the most momentous foreign relations to the pettiest details, being set out upon papers which are passed from hand to hand. It is a Government of minutes based upon memos. (Deakin, 1893: 28)

> British India, in short, is British neither in race, religion, language, policy, sentiment, nor aspiration. Garrisoned by a few Britons, and governed by still fewer, it not only retains its Asiatic complexion, but impresses its character to a large extent upon its conquerors. The British in India have themselves ceased to be British in many respects. They have developed castes and curious creeds, walk with troops of retainers, live like Persian satraps or Roman proconsuls, coming at last to think and speak in the phrase of the Orient, and with its vivid colouring. It is they who have adapted themselves to the Hindu, and not the Hindu who has taken their imprint. It was not to strengthen her hold upon her British subjects that the time-honoured title of the Queen was altered to that of Empress of India. Bearing in mind how few are the whites in proportion to the hordes of varied hue who swarm from Cape Comorin to the Himalayas, and the extent to which they have required to stoop to the conditions of life in the tropics in order to conquer, it is not too much to say that the first fact requiring to be fixed in the mind of the inquirer is, that India today is altogether Asiatic in the spirit and form of its life and institutions, and British only in flag, in fame, and in name. (Deakin, 1893: 29)

Chapter 7
The New Nationalisms and the Politics of Reaction

Contesting Dynamics

The new geography of India also had a new cultural dimension, in particular there now existed a new lingua franca that enabled educated people to communicate with one another from furthest north to the deepest south. It is possible to exaggerate the importance of this – it is said that in the 1930s less than half a percent spoke English. But equally one should not underestimate the significance of that half percent, the new intelligentsia and middle class, nor forget that it was the language of the courts, of much of business and commerce, and it was also the most important access to the new institutions of higher learning. And in addition, the railways and the posts conveyed the speakers and the pamphlets, newspapers and private messages from one region to another.

Here were the means to promote new senses of identity across India. The new communications also connected India increasingly with political ideas derived from other parts of the world (initially mostly from parliamentary Britain) and also led to an increasing awareness of the great events elsewhere – such as early in the twentieth century the defeat of great imperialist Russia at the hands of the Japanese. That was followed by the First World War, in which omnipotent Europe was exposed as a divided and destructive civilisation, and in which Britain in particular came perilously close to defeat. Out of the ashes emerged the post-isolationist anti-colonialist democratic America and the anti-imperialist revolutionary Russia of the masses.

Here were new models of power and independence.

But there was an unevenness with which the two largest communities of India – the Hindus and the Muslims – took up the new ideas and the new opportunities. After the Mutiny the Muslims were to an extent kept out by the British (Persian was no longer the language of Government), but also to at least as great an extent, opted out. They had their traditional patterns of Islamic education, and had during their time of imperial ascendancy largely left commerce in the hands of the Hindu merchant groups. It was mostly, therefore, the Hindus who filled the ranks of the emerging urban middle classes, and who accepted much of the higher education that was becoming available. In 1878 there were 3,200 graduates from the new universities – nearly 3,150 were Hindu. Of course, this is only a generalisation; but even the exceptions to it demonstrated the way in which thoughtful Muslims perceived the trend in history. There were indeed great Muslim families who accepted much that the West had to offer, and some, for example, became eminent barristers. On of the greatest of these was Sir Sayyid Ahmad Khan, a member of the Vice-roy's Executive Council and the founder in 1875 of the Muslim University at Aligarh, in the heartland of the former Mughal Empire. He was concerned that when the British arbiters left, the Muslims' interests would be subservient to the Hindus'.

Here was the anxiety of minority groups wondering what a future majority rule would mean.

The British knew that they would not rule India for ever. Lord Hastings wrote in India in 1818:

> a time not very remote will arrive when England will, on sound principles of policy, wish to relinquish the domination which she has gradually and unintentionally assumed over this country and from which she cannot at present recede. (cited in Coupland, 1942 Pt 1: 18)

In 1844 Henry Lawrence wrote:

> We cannot expect to hold India for ever. Let us so conduct ourselves ... as, when the connexion ceases, it may do so not with convulsions but with mutual esteem and affection, and that England may then have in India a noble ally, enlightened and brought into the scale of nations under her guidance and fostering care. (cited in Coupland, 1942 Pt 1: 18)

There seemed in these dimly perceived futures the tacit acceptance that the India left behind would be a single India, because that was what the British had to some degree created, and were both intentionally and unintentionally cementing. But the policy by which such a goal could be achieved was not seen, and indeed the goal seemed so remote that there seemed little point in trying to define prematurely the form or forms of the successor state. The parliamentarian John Bright in 1877 mused in prescient form:

> Thus, if the time should come – and it will come ... – when the power of England, from some cause or other, is withdrawn from India, then each one of those States would be able to sustain itself as a compact, as a self-governing community. You would have five or six great States there, as you have five or six great States in Europe: but that would be a thousand times better than our being withdrawn from it now when there is no coherence amongst those twenty nations, and when we should find the whole country, in all probability, lapse into chaos and anarchy and into sanguinary and interminable warfare. (Cited in Coupland, 1942 Pt 1: 50–51)

The British had founded an Empire and seen most of it become a Commonwealth, an imperial grouping of states that evolved through a 'colonial model' to (in the later nineteenth and early twentieth centuries) Dominion Status, that is to say self-governing Canada, Australia, New Zealand, South Africa, all linked to the crown, and all contributors to a global defence network. India, of course, was different – a thing unto itself, *sui generis* – by accident part of this Empire, but not settled by white English-speaking people and with such an ancient and strong culture of its own, and yet such a valuable part of the Empire, particularly in terms of global defence strategies, to say nothing of its economic and psychological value. Here was the parent with a wayward adopted child, anxious that he should grow up to be a partner in the family firm.

Looked at with hindsight it is almost as if from the moment that British power was consolidated the race was on to determine the form of the successor state(s) and the division of the spoils between the inheritors. Why, given the ineluctable progress of neutral time, call this a race? Because, the different forces we have just outlined developed at different speeds; but all had a bearing on the outcome, while they were developing with momenta that sometimes rebounded off each other, sometimes developed their own internal dynamic, and sometimes were randomly spurred by the emergence of new charismatic leaders.

When Lord Stockton was asked what had been the greatest difficulties with which his government had had to contend when, as Harold Macmillan, he was Prime Minister in the UK in the 1960s, he replied 'Events.' Perhaps above all it was the events that happened outside of the forces I have named that determined the outcome of the race, for these would change the backdrop against which the players played, and would indeed change the personalities of the players too. Above all it was the British who failed to see that 'Events' would not provide them with the time to give ground inch by inch until a Dominion of brown Englishmen became a fully-fledged partner of the family firm. 'Events' could shake the sands of time a little faster through the hour glass.

The Structure of Government in British India and the Problem of an Evolutionary Transfer of Power

At the time of the Mutiny two thirds of India was directly governed by the British, the other third being governed by the Princes of the Native States, who were internally autocrats, but externally bound by treaty as vassals of the British. There were 562 of these, 100 or so being of major significance or some degree of importance, another 100 of some note, and the remainder minor estates, sometimes not much more than a village or two altogether. Virtually none of the significant states, with the possible exception of Travancore (in modern Kerala), approximated any geographical or cultural homogeneity, and very often not only did they comprise a mixture of races and languages, but quite often the ruler was of a different religion from the majority of his subjects – a Muslim Nizam in Hyderabad ruling over Hindus, or a Hindu Maharaja in Kashmir ruling over Muslims.

It could be argued in hindsight that these anachronisms should have been tidied up, and that, as the paramount power, the British could have forced new treaties on the Princes. And, indeed, independent India and Pakistan have basically done just that, and finally eliminated them (except there is no agreed solution on Kashmir). But the British were aware, in post-Mutiny India, that not only was their annexation of the State of Oudh (modern Awadh) one of the many resentments that had fuelled it, they had also only survived it with the help of States that had remained loyal. There was also the profound British reverence for heredity and aristocracy. The Viceroys were drawn from the aristocracy (or recruited to it). British democracy's Upper Chamber was filled with hereditary peers – 'a stabilising influence' – and both houses acted in the name of the Monarch. The princes not only saw themselves as stabilising influences in India, the British accepted too that they had that role. With Durbars and twenty-one-gun salutes (or less, depending on keenly contested status), both sides could use each other as excuses for mutually satisfying displays of pomp, privilege and power. One way to have incorporated the princes totally within the British establishment would have been to establish a British peerage system for them, and Lord Lytton (Viceroy 1876–80) even contemplated just such a move.

British India was entirely different. Its Government was a unitary executive – that is to say a single executive structure with responsibility for all of the British Provinces, with the Governor-General/Viceroy at the apex. Although responsible for the governance of British India, the Governor-General was responsible, by Act of Parliament, to the Secretary of State for India, a Cabinet Minister, who acted with the assistance of the India Council in London. The Council's membership was appointed by the Secretary, and usually comprised senior British India hands. As a Cabinet Minister the Secretary of State was obviously finally accountable to Parliament. Thus, we have the Government of India ultimately being accountable to the elected members and unelected peers of the British Houses of Parliament. Mostly by historical accident, the British had assumed extra-territorial power. The East India Company had been granted the governance of Bengal by the Mughal Emperor, but the British Parliament had exerted closer control over the Company as its power grew. When, finally, the last pretence and symbol of Mughal power had been abolished, and the Company wound up too, the only power in British India was of this extra-territorial kind. In what ways could power be repatriated? In what ways could responsibility be repatriated? And how would the repatriation of the two rebound on each other?

Government operates through institutions, is staffed by personnel, has responsibilities it defines for itself, and applies them to the given territory. It needs money to finance its own establishment, and to discharge the other functions it has defined. We can examine the problems posed by each of these in turn. Put simply, it was impossible for the Secretary of State in London to take day-to-day decisions of government and, therefore, much of his actual power was vested in the Governor-General in India. He too could not assimilate all functions to himself – indeed could do so less than the Mughal Emperors had done, because each Governor-General served for a much shorter period of time, and because the complexity of the modernising economy was greater. He, therefore, like the Secretary of State in

London also had an Executive Council, to which he appointed members, known as official members, who guided his conduct of government. In many spheres he needed their majority consent for action: but the principle of accountability was not violated, since the members were appointed by him, and there were also emergency provisions. There was nothing to prevent the appointment of Indians to the Council, and after the Act of 1861 indeed Indians were appointed; but, as one might imagine, these were senior and conservative supporters of the British Government, whose criticisms were circumspect.

The staff of the Indian Civil Service were also ultimately appointed by the Secretary of State, were salaried by him, and he was responsible for the terms and conditions of their appointment. There was nothing to prevent the Secretary appointing as many Indians as he felt fit; but in practice since the examinations for recruitment were held in Britain, the senior ranks were in the majority British. In fact, one of the bones of contention of the new nationalists in India in the late nineteenth century was that at a time when it was hoped the service would be Indianised, and when Indians were prepared to travel to England for the examinations, the age of candidates was lowered to a point which effectively precluded all but those who were schooled in Britain.

The functions which the Government had assumed for itself pre-eminently included defence and internal law and order, but also of necessity a growing number of 'development' fields such as communications, large-scale irrigation, and the regulation of commerce from the local scale (through regulated markets and official weights and measures) to the international scale (through the determination of tariffs etc.).

In the same way that the complexity of Government meant in practice that the Secretary of State in London had to operate through the Governor-General in India, so he in his turn could not administratively cope with the whole of British India as a single unit, and hence the Province again duplicated the same institutional ideas. They had Governors (some lesser ones had only Lieutenant-Governors etc) who also in the main operated with their Executive Councils, to whom again they appointed members. In their day the Mughals had also had their provinces, but in those days required more because of the difficulties in communication than the complexity of government, whereas in the late nineteenth century although the complexity of government increased, the growing communications meant that the Secretary of State in London became more closely involved, and the government in Calcutta more strongly felt in the Provinces.

The principle of responsibility which lead to such centralisation was obviously reflected in the financial structure as well. The finances raised in the Provinces were all part and parcel of the 'Revenues of the Government of India'. It would not have been easily possible for the Governor-General to discharge his responsibility to the Secretary of State, if the Provincial Governors could raise and disburse money as they felt fit.

What Indian national sentiment would want in due time, of course, was for their government to be *responsible* to them; in the broadest sense simply to be Indian.

It was apparently impossible for a government to be responsible simultaneously to two different polities, one British and the other Indian, and hence the obvious demand would be for the immediate transfer of all powers to a native government. This immediately raised the question of what sort of government, and there was a broad consensus that it would be in some manner *representative*. But here was a 'Catch-22'. Until such time as the mechanisms and institutions for representation existed, the only possibility would have been to transfer power arbitrarily to some autocratic person or institution. Before such a transfer of power, representative institutions could be created, but to whom the government would not be responsible. This was not a scenario for a friction-free evolutionary process: and indeed during the heat of the friction of the next decades Congress and Gandhi in the 1930s did demand that the British simply go, and leave Congress as the sole authority in India, to devise whatever constitution it thought best.

The Process of Constitutional Concession

The British did attempt over the decades from 1857 to 1947 to guide and constrain the different forces at work within a changing constitutional framework. They made changes in 1861, 1892 and 1909, all of which increased Indian representation but without conceding self-government as a long term goal. In 1917 that goal was conceded, and followed by the Act of 1919 which made some degree of responsibility possible, and the Act of 1935 which continued to work towards self-government in a Federal framework, and which remained in force during the Second World War and to Independence in 1947. It remained the Constitution for India until 1950 and for Pakistan until 1956. The success of these constitutional changes in constraining the forces at work was limited but important. The Indian National Congress basically refused to cooperate except between 1934 and 1939, and thereby provoked some of the tension for change: the Muslim League in the end refused to cooperate, and demanded Partition with Independence. Perhaps, no matter what the constitutional changes, the end result would have been very nearly the same: but there is no denying that the expectations of the outcome of the dialogues were greatly influenced by these institutional developments.

In 1861 the Indian Councils Act provided for the expansion of the Executive Councils to become Legislative Councils: and at the same time new members were added in a 'Non-official Capacity' – at least half of whom had to be from outside the civil service, and were in practice senior figures in Indian life (as perceived by the British). But the official members of the Councils outnumbered them – if they had not done so it might have been possible for the 'unofficial members' to thwart the concept of responsibility to the British Parliament. In 1883 Lord Ripon instituted the principle of elected membership of local Municipal Councils and Rural District Boards, although not with the intention that it represented any abdication of the powers of central government. But, ominously, the leader of the Muslim community, Syed Ahmad Khan observed:

... in borrowing from England the system of representative institutions, it is of the greatest importance to remember those socio-political matters in which India is distinguishable from England. The system of representation by election means the representation of the views and interests of the majority of the population, and, in countries where the population is composed of one race and one creed, it is no doubt the best system that can be adopted. But ... in a country such as India ... I am convinced that the introduction of the principle of election, pure and simple, for representation of various interests on the local boards and district councils, would be attended with evils of greater significance than purely economic considerations ....Government, in reserving to itself the power of appointing one third of the members of the local boards and district councils, is adopting the only measure which can be adopted to guarantee the success of local self-government, by securing and maintaining that due balance in the representation of the various sections of the Indian population which the system of election, pure and simple, would fail to achieve. (Cited in Coupland, 1942 Pt 1: 155)

In 1885 the Indian National Congress held its inaugural meeting in Bombay, attended by 72 delegates from most parts of India. Its foundation was accepted with some mild degree of enthusiasm by the Viceroy Lord Dufferin, who felt that here would be a useful body of expression of the feelings of the subjects. The delegates hoped that its foundation would prove the germ of a native parliament, but nevertheless avowed their loyalty to the Empress. The membership included retired British officials from the Civil Service as well as the majority of Indian delegates. Naturally, the body wished not only to express opinions, but to push for more representative government, and shortly after its foundation had asked for the principle of election to be conceded for the Provincial and Central Councils. The Act of 1892 compromised over the issue: recommendations were sought for some of the non-official members – and, in effect, those who were recommended were chosen by unofficial election.

Congress had Muslim members from the beginning – and always has had. Several times Muslims were elected to the party Presidency. But they were not many – in fact at all times proportionately less than their share of the population. Some Muslim leaders were not slow to realise that the Act of 1892 might well be followed by others widening the extent of representative government and, therefore, began a campaign to urge the government to protect the rights of minority groups. In 1906, as the next move came closer, they held the inaugural meeting of the All-India Muslim League in Dacca (now Dhaka, capital of Bangladesh).

The tempo of change did not satisfy the growing body of nationalist opinion, who became increasingly concerned to exercise some influence over, for example, the budget, which was as closely controlled by the Governors as ever. Much of the strongest sentiment was felt in Bengal, where extremists, although not with a wide popular base, began terrorist attacks on government officials. The new nationalism was in fact pushing not just against the principle of accountability, but also against the conservatism and rigidity of mind of the bureaucracy which the British had created. In the end it was bound to lay bare the essential imperialism that gripped

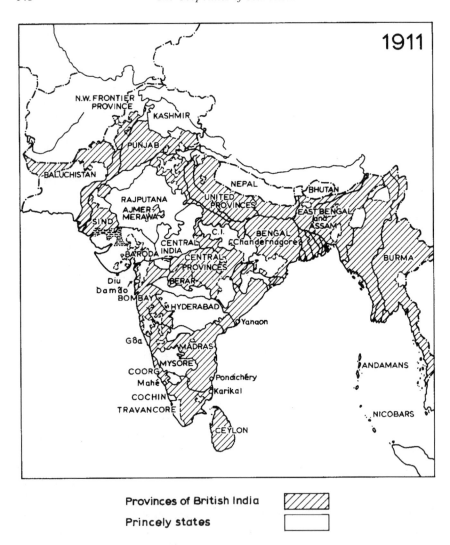

Provinces of British India

Princely states

Figure 7.1   The Indian Empire 1911

the highest levels of the administration, exemplified at its apogee by Lord Curzon (Viceroy 1899–1905), noted for his establishment of the North-West Frontier Province, his enthusiastic founding of the Archaeological Survey, some beneficial measures in land reform and rural debt, scientific progress in agriculture, but also a number of arrogant follies culminating in the partition of Bengal in 1905. This was a significant stimulus to the foundation of the Muslim League.

Representation had a territorial basis. Provincial Councils were Councils of Provinces with defined borders, which of course defined the persons affected by

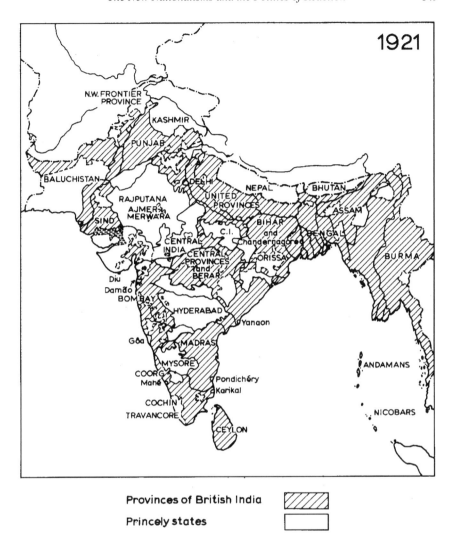

Provinces of British India

Princely states

Figure 7.2   The Indian Empire 1921

those councils, and the arena within which debate over provincial issues would occur. The Bengal that Curzon inherited was huge, far too big for effective administration, and clearly it had to be subdivided. Later, a 'sensible' solution was derived – to divide it into three (Bihar with Orissa, Bengal, and Assam) – but Curzon's solution lacked such sensitivity. With scant consultation and apparent august impulse he divided it into two – Bengal, encompassing modern Bihar and Orissa, and East Bengal, approximating modern Bangladesh and Assam (Figure 7.1). In the western part, the Bengalis felt outnumbered by the Biharis.

The eastern part was predominantly Muslim, though with a Hindu landlord class, many of whom were absentees living in Calcutta. The outburst of agitation against the evisceration of the identity of Bengal was widespread and deeply felt – and culminated in more terrorism and a boycott against British goods. Violence was now a real part of the political agenda, and it would not be monopolised by one side alone. In 1909 one of the leaders of Congress, Bal Gangadhar Tilak, was sentenced to six years imprisonment for incitement to violence. The Partition of Bengal was repealed in 1911, but by then Congress had become a less emollient organisation, but still meeting and debating legally and openly.

In 1909 constitutional reform took another momentous step forward, that is to say momentous in its consequences, even if from the Indian viewpoint the advance seemed no more than a minimal concession. Lord Minto, who had assumed the Viceroyalty in 1905, agreed with the Secretary of State Morley the measures which became known as the Morley-Minto Reforms. These were a considered attempt to win back moderate Indian nationalist opinion, by providing for the election of Indian members to the Legislative Council at the Centre, although power still rested with the British since officially appointed members were still in a majority, and because the Executive Council still remained the Governor-General's own preserve. The momentous part of this grass-hopper's leap in the jungle was that the principle of Separate Electorates was accepted – that is to say, in order to preserve the rights of the minority, Muslims and non-Muslims were registered on separate lists each of which was associated with a block of reserved seats. The numbers of these reserved seats were also given weightage; those representing the minority were more than proportional to the fraction of the minority in the population at large in the different Provinces. The number of Indians in the Provincial Legislative Councils was also increased, and in Bengal it was now theoretically possible for the elected members to form a majority over the official members. This did not quite violate the principle of responsibility to London – since the Executive Councils still had greater power, and the Centre in any case had a veto over the Provinces. In other words, Government was still not responsible to these electorates, even though there was this wider degree of representation.

In retrospect, it can be seen that these were the opening moves in the process that would lead South Asia to representative and responsible government – in two parts: but, precisely because at the time they were not seen as that, the different players were not so careful about the consequences of their attitudes and demands towards each other. The British averred privately, and to some extent publicly, that these were not the first moves towards parliamentary government: India simply was not ready for that. Indeed, at this very time at the Delhi Durbar of 1911, the King Emperor himself announced the move of the Capital from Calcutta to Delhi (this was the beginning of the New Delhi of today). The move was highly symbolic as well as perhaps having a practical element. The pre-British empires and urban structures were all land-based – stretching from the upper Ganges plains down into the central Deccan. The British had added the new coastal super-port cities to this pattern. Now the maritime power had assumed the role of land power, and proclaimed itself as such in the heartland of the Mughals.

The Indian National Congress in 1910 minimised the Muslim-Hindu differences, and claimed that separate electorates were unnecessary and that the Muslim-Hindu differences would subside when the British had gone – but to all participants the power of the state must have seemed such that they would take a long time agoing. In fact, except for the brief period following the Lucknow Pact of 1916 (discussed below), Congress always opposed separate electorates, as writing a constitutional division and barrier into the Indian public.

Muslim notables accepted the changes precisely because they were only an improved and more favourable form of representation, and not an advance towards responsibility – something which they feared because no matter what weightage was given to them, they could never be a majority. Yet, these fears were only the articulations of these senior men: they did not necessarily reflect a widespread mass political anxiety. By and large the masses were politically apathetic and uninvolved.

Then the 'Event' of the First World War shook the Empire to its foundations. The British attitude to the Indian nationalists during the First, as during the Second, World War, can be summarised: 'please do not cause any trouble while this Great Event passes. Let us freeze matters where they are, and resume negotiation afterwards'. The initial, and to some extent the main, response of India throughout the long ordeal, was to close ranks behind the British and fight German militarism. Public donations for the war effort flooded in, and many of the Princes contributed troops from their own armies. 500,000 Indian troops took an active part in the Middle East and Europe, and suffered 96,000 casualties. But as the war dragged on without victory, British power and credibility waned. As economic shortages made themselves felt, murmuring of discontent increased. The nationalists became less inclined to wait – this was after all an opportunity while their masters had one hand twisted behind their backs. And sensing that they would have more power if undivided, they concluded a pact at Lucknow in 1916 between the Indian National Congress and the All-India Muslim league. This required major concessions on both sides, but perhaps the greatest were by Congress, which accepted the principle of Separate Electorates – although it was not a concession it could make binding on all future policy resolutions. The Pact also proposed much greater devolution to the Provinces – something which was also acceptable to the Muslims, since although in a minority overall, and therefore in a minority in centre government, they would be more powerful in their own majority areas. The other declared policy objectives of the Pact included the power after an interval of a year to reject a Governor's veto, and the repatriation to India of the power to direct foreign affairs and defence. The former was a beginning to make the Executive responsible to the Legislature, but it was sufficiently muted that the Muslims acquiesced.

To the British, who could always claim that they must remain as arbiters between conflicting groups, the Pact might have taken some of the wind from their sails. It was, at least, a small part of the events of those years which finally led to the policy declaration in 1917 in the House of Commons by Secretary of State Montagu, that the policy of the British Government was that of 'the increasing association of Indians in every branch of the administration and the gradual development of self-governing institutions with a view to the progressive realisation of responsible government in India as an integral part of the British Empire'. The statement was the first major public admission by the British that responsible government was the aim of policy: but it was simultaneously contradictory because this was to be circumscribed by imperial constraint. What would happen if the new responsible government wanted complete independence? This was an issue to be faced at a future stage – since the declaration contained caveats about 'successive stages' and 'the British Government and the Government of India ... must be the judges of the time and measure of each advance ...'

Montagu followed this announcement by his visit to India and the production with the Viceroy Chelmsford of the Montagu-Chelmsford Report of 1918 and the subsequent new constitution of 1919 which came into force in 1921, with the explicit intention that its working should be reviewed after 10 years. The Report noted the Lucknow Pact as a sign of the growing national feeling in India, stated plainly that separate electorates were a hindrance to the growth of national consciousness, but agreed with the pact in retaining the principle of separate electorates, in its view for the time being. In other ways it was a complicated document: it had to show that the first of the 'successive stages' was being achieved, while ultimate responsibility still rested in the UK. It achieved this firstly by devolving most power at the Provincial not the Central level, and secondly by distinguishing between the functions of government transferred to a representative Legislative Council, and those retained by the Governor's Executive Council. This idea survives in the present Indian Constitution, in which responsibilities are divided between the States List, the Central List and the Concurrent List. As it was conceived in 1918 it was known as 'dyarchy' – the duality being between the untransferred powers of the executive and the transferred powers of the legislature. But the power of veto by a Governor was still very strong, and in particular with regard to finance.

There was another way in which the Constitution seemed far from representative and responsible government. Because the franchise was so hedged by property and other qualifications, the electorate was actually very small – in 1920, 7.4 millions out of a population in British India of 240 millions.

The subjects of the Princes were of course almost all completely disenfranchised by their autocracies – but the Report was aware that they could not stand aloof from external change forever. The year 1921 therefore also saw a major development in the conduct of their affairs too. The Princes hitherto had had bilateral relationships with the Government of India. Now it was proposed that a Council of Princes be established, so that they could take part in future dialogue with each other and jointly with British India, clearly with the idea of accommodating them in some way to the changes taking place around them.

The framing of government was one thing. The actions it took were another. In 1919, as a result of a committee of enquiry chaired by Justice Rowlatt into terrorist violence, mostly in Bengal, Bills were passed which allowed trial before two judges without jury, and also internment. Public resentment was acute. In the same year a protest rally at Jallianwalla Bagh in Amritsar, Punjab, was brutally terminated by police firing. On the orders of General Dyer, they fired .303 rifles into the crowd in a confined urban park: 379 people were killed, and 1,208 injured, women and children being amongst both groups. It was the kind of incident which, like Sharpeville years later in South Africa, or Tiananmen Square in Beijing, simultaneously demonstrated the power of the State and yet demolished its legitimacy.

Gandhi and the Nationalist Response

The new Constitution did not come near making the kinds of concessions which the nationalists had hoped for. The Congress leadership became more strident in its demands, and also in 1921 adopted a new party constitution of its own, which demanded Swaraj (self rule) and which no longer made reference to continued membership of the Empire. But the independence movement had a quality to it which would now seem unlikely – it was committed to the attainment of independence by peaceful and legitimate means. The effective leader of Congress was now Mohandas Karamchand Gandhi – known as the Mahatma (the Great Soul) – who was almost a one-man Great Event of the kind which the British would have preferred to have wished away. Gandhi launched a non-cooperation movement which, amongst other things, enjoined people not to pay their taxes. This was legitimate in his view, since the British Government in India was illegitimate. The struggle revealed two things – firstly that Congress was not yet sufficiently grounded in mass, as opposed to middle class, support that it could really threaten the State and, secondly, that such campaigns could easily get out of hand and degenerate into violence, something which Gandhi abhorred, because the ends could never justify the means, and could themselves become distorted. Twenty policemen were murdered in a violent riot in the United Provinces. If it happened that Hindus murdered Muslim policemen, such acts would precipitate communal violence; and so Gandhi called the movement off.

Gandhi's leadership of Congress (often officially uninvolved, but in fact always the spiritual and main strategic mastermind) has been described as a 'peculiar blend of bold advances followed by sudden and capricious halts, challenges succeeded by unwarranted compromises ...' (Desai, 1966: 372) His subtle complexity at times seemed more like perplexity. But there are very strong and consistent threads in the pattern he was trying to weave for India, but which in the end failed in one signal respect. He attempted more than ever was realistically possible, yet, perhaps because of that, achieved much more than otherwise would have been achieved. He was a profound revolutionary – yet would be prepared to stop his revolutions in their tracks if he felt they were going astray. He wanted immediate independence from the British, but was prepared to wait indefinitely if in his view Indians had not yet proved worthy of his trust in them. Gandhi rejected the new industrialism and materialism of the West, and wanted a return to village self-sufficiency. His spinning wheel, at which he toiled each day, was incorporated on the Congress flag as a demonstration of his commitment to this ideal. He abhorred the injustices of the caste system, and championed the cause of the polluted untouchables, giving them a new name – Harijans, the people of God. He was devout, but in the tradition of many synthesising Hindu reformers over the centuries, he was theologically eclectic and non-dogmatic. He had his favourite passages in the Koran and the Bible as much as in the Mahabharata. He gave Congress the contact they needed with the masses of India. But he also asked of the masses and the leaders what they were unable in the end to give: the eradication of communal distrust and discord. He knew his rapid strides must be slow because he was flying not just in the face of the British, but also in the face of Hindu orthodoxy and Muslim dogma.

Not all the Congress high command agreed with the total rejection of the 1921 Constitution, and some who were dissatisfied with the concessions made, but who nevertheless felt that an evolutionary approach might work best, left the Congress and founded the Liberal Party – in a sense the Conservative Part of Congress. They contested the 1920 elections which the rest of Congress boycotted. In the 1923 Elections the Swarajists of Congress contested the elections in order to paralyse dyarchy from within, and in Bengal and Central Provinces they managed to block the payment of Ministers salaries, with the result that the Governors had to suspend expenditure on the transferred subjects.

At the end of the First World War one of the concluding treaties, signed at Sèvres, redounded on all sides of the Indian struggle in an unexpected manner. Turkey's Ottoman Empire had been defeated, and the Treaty stripped the Caliph of Istanbul of any remaining political power. Muslim opinion in India was mobilised in the Khilafat (Khalifat, Caliphate) Movement, which led to general unrest and disturbance. Gandhi welcomed it and tried to use it as a common cause between Hindus and the Muslims against the British; but its peculiar involvement with non-Indian affairs was not attractive to many Hindus, and the disturbances were partly implicated in violence by an extreme Muslim group in South India, the Moplahs (Mopillas), which turned in 1921 from resentment against the government to the wholesale butchery of Hindus. Indeed, while the Muslim League and the Congress appeared to be making common cause, the level of communal violence escalated throughout the 1920s. Much of this was of the spontaneous discord noted in Chapter 3 above – the result of clashes on religious feast days, the playing of music during processions which passed mosques at prayer time, or local anger against some alleged beef-eaters. The new Constitution did not necessarily cause many such riots: but certainly the realisation that there was, in fact, a small but tangible devolution of power to provincial politicians did nothing to extinguish them. Instead it forced home the realisation that the use of that power in the India of the future could not be guaranteed to be 'neutral'. Here in fact was one of

Gandhi's threads which, sustained in a different form in Nehru, ultimately led to catastrophe. Since Hindu and Muslim must sink their differences, Congress had to be the sole representative of Indian nationalism. It would embrace Hindu and Muslim alike, and need make no further concessions to leaders of Muslim parties. It was a noble if arrogant dream.

As the Constitution creaked through the 1920s, the promised review was brought forward, but in a move of crass stupidity reminiscent of some of Curzon's greater insensitivities, the Simon Commission of 1927, nominated by Baldwin's Government, included two British peers and four members of the House of Commons,[1] but no Indians. When the Commission went to India to take evidence from Provincial Committees formed for the purpose, Congress boycotted them. Congress had by now adopted a demand for total independence (purna swaraj), meaning complete severance with Britain and the Commonwealth.

Nor was the reaction of the nationalists solely negative and destructive. In 1928 an all-party conference was called in Delhi, to discuss the next steps to be taken as the nationalists saw them, independently of British manoeuvrings. The hard wings of the two communities, that is to say the Muslim League and the Hindu Mahasabha, obstructed progress, with the result that a committee was formed with representatives of the different communities charged with determining the principles of a new constitution. It produced the bones and in some parts the flesh of a constitution. The committee was chaired by Motilal Nehru, General Secretary of Congress (father of Jawaharlal Nehru, first Prime Minister of independent India). The Nehru Report it drafted was not binding on any of the political parties, and certainly not on the Muslim League, which was not formally represented on the committee. It spent much of its effort on the communal problem. It resolved that the future constitution might have aspects of federalism in it, but basically the legacy of a strong centre and unitary government remained. The whiff of federalism was really an inducement aimed at the Princes, although it was clear they would have to allow representative government in their states too. The Report also opposed the retention of separate electorates, but was not completely immune to the need to assuage communalism.

There were two ways in which communal anxieties were accommodated. Firstly, it suggested some improvements in the territorial division of India. Sind, a Muslim area, should be elevated as a Province in its own right, separate from Bombay, and Orissa should be separated from Bihar. Secondly, it proposed that reserved seats be instituted at the Centre. This needs a brief explanation, since it sounds akin to separate electorates, but is in fact different.

With separate electorates, Muslims on one register were alone allowed to vote for candidates for the Muslim seats. The effect of this was to make Muslim politicians responsible to the Muslim electorate alone, and naturally led to a narrow communal viewpoint. In the alternative system there may be reserved

---

1 One of these was Clement Attlee, who, as Prime Minister after the Second World War, would grant India its independence.

Table 7.1　　The Act of 1919: Government Functions

Central subjects
Defence, foreign affairs, relations with the Indian States, communications, customs, commerce and banking, criminal law, census and surveys, All-India services (e.g. PTT, railways)
Provincial Subjects
A) Untransferred
Justice, police, jails, land revenue, forests (in all but one Province)
B) Transferred
education, agriculture, public health, local government

seats to ensure that there were Muslims in the Assemblies, but all voters could vote for candidates both for Muslim seats, and also for the general seats. Then the elected Muslims would know they were responsible to the whole electorate, and the representatives in the general seats would be too. The policies they would campaign on could not then be sectarian. The Nehru committee expressed the view that debate in free India would in fact create political parties around economic and social groups, but not around cultural and religious groups.

The Report was never ratified by Congress, although neither was it rejected. The more radical wing, led by Jawaharlal Nehru opposed acquiescence to Dominion Status, and favoured total independence. The Muslim League noted that separate electorates were the law, and should be retained. The nationalists' independent attempt to close the communal schism had failed. The report of the Simon Commission duly appeared in 1930, and incorporated recommendations for the promotion of Sind and Orissa to provincehood. It transpired that this was not completely a dialogue of the deaf. The government proposed a series of Round Table Conferences at which representatives of the British Parliament debated the issues with representatives of Indian opinion. Although the Simon Report provided a background to these meetings, it did not constrain the agenda or the discussions. They were constrained from the British side mostly by the declared intention of giving India Dominion Status – but not 'purna swaraj', or the total independence that the radicals of Congress wanted.

A large number of Congress members might have been happy with just such a conference, one that was close to all they had demanded. The issue over whether to attend or not, could have split Congress. To Gandhi its unity mattered most – because in that alone did he see hope for the unity of India. So he demanded immediate, complete and total independence (1930) by transferring power to Congress, and he backed this with a civil disobedience campaign and a boycott of the conference. The campaign of Civil Disobedience and non-payment of taxes included Gandhi's celebrated march to the sea from his ashram in Gujarat. He declared his intention was to make some salt, an act which was illegal since the government had a revenue-raising monopoly on the salt trade. It took Gandhi 24 days to walk through the villages to the sea, pursued by the world's press. When there he both humbled and crumbled the British Raj as he picked a grain of salt from the sand.

Gandhi was persuaded to attend the Second Round Table Conference as the sole representative of Congress, and claiming simultaneously that Congress was the sole representative of India. He resolutely objected to some of the common ground that had been agreed thus far – in particular he objected vehemently to separate electorates, not only for the Muslims, but as now had been proposed, for the untouchables too. Here was division being written in, not just between Muslim and Hindu, but, as Gandhi saw it, within Hinduism too, for he saw the Hindu manifold as one, and the incorporation of the Harijans, as one of his life's goals. He was not averse to reserved seats – but the electorate had to be general to avoid sectarian campaigning.

Since the Round Table Conferences failed to find agreement with Congress, civil disobedience intensified, and a boycott of British goods was also organised. Riots and instability followed: the Congress leadership was imprisoned. The riots fizzled out, and the British imposed their new constitution in 1935, which retained and indeed strengthened separate electorates. It also widened the franchise, from 7 million to 30 million voters.

As noted earlier, the Act of 1935 remained the basic framework for the Government of India until Independence in 1947, and the basis of the constitution of India until it adopted its new constitution in 1950 and of Pakistan until 1956. Broadly, the Act transferred more power to the Provinces where dyarchy waned as few powers were left reserved to the Governors, and now introduced dyarchy at the centre, reserving to the Governor-General/Viceroy several responsibilities, including, importantly, defence and foreign affairs. It also drew up the framework for a federal government for India, in which the members of the upper chamber the Federal Assembly would be elected indirectly by the provinces (or be appointed by federated princely states), and the lower chamber the Federal Legislature by direct elections, but on the basis of separate electorates in the provinces. The Federation would only come into being when a sufficient number (approximately one half) of the princely states had signed deeds of accession. The Federal part was never enacted, because the Event of the Second World War changed the backcloth against which the final denouement would be resolved. Until such time as it was enacted, the central government continued in its 1919 form – that is to say with an elected Legislative Assembly, but a Viceroy's Executive Council ultimately responsible to Parliament in London. There was therefore no cabinet of elected ministers at the centre for the time being.

At this time the Muslim League was not strong. It was given most support by the Muslim community where that community was small – in the United Provinces and Bihar – but it was not strong where the Muslims were in a clear majority, as in Punjab and Sind. The elections of 1936–37, therefore, produced some impressive gains for Congress – they had majorities in all the Provinces (including the Muslim North-West Frontier Province) except Sind, Punjab and

Bengal, and in those latter three the non-Congress administrations were coalitions and not Muslim League parties on their own. Congress, after a few necessary objections and bridlings, accepted the power given to it, and used it. It used it in a way that the constitution makers had not really expected. The party organisation remained strong nationally, and kept the provincial governments under the control of the party high command. Here was the antithesis of federalism, one of the routes by which minority fears could be allayed. In the provinces where Congress held strong majorities, nevertheless the Muslim League hoped that it would offer coalition governments, and incorporate League Ministers. But why should it? Congress had Muslims within its ranks, claimed to represent all India, and saw no need to share. Indeed, in a mass contacts movement it even strove to recruit more and more Muslim peasants since politics were secular and not matters of religion, something promptly repudiated by local mullahs who saw (and see) the Islamic way of life as incorporating and defining permissible politics. These two moves and many others began to alarm Muslim opinion. Congress underestimated the strength of Muslim feeling, almost flew in the face of Muslim sensitivities, and far from using strength to be generous, used strength in an attempt to destroy separate Muslim politics for good. Nor were the princes encouraged by the attitude of Congress, and their enthusiasm for federation began to wane.

At the outbreak of the Second World War the Viceroy, Lord Linlithgow, declared war on behalf of India against Germany and its ally, Italy. He did so, as he was constitutionally able to do, without consultation with Indian opinion – something which was difficult to some extent because the centre did not have that kind of responsible government. But the manner of doing so did not go down well. In protest against the lack of consultation Congress ordered the resignation of all Congress ministries. The Muslim League reacted with a day of thanksgiving for deliverance, and set about the strengthening of mass Muslim participation in their aims. In 1940 at Lahore, capital of Punjab, Mohammed Ali Jinnah proclaimed the independent homeland of Pakistan the object of Muslim League policy. The Viceroy hoped to freeze all political development during the war, promising Dominion status and constituent assemblies at the end. He put negotiations with the princes on ice – and thereby killed for ever the proposed federation. Congress responded by demanding immediate, total, independence.

The war did not go well for Britain in the early years. France fell, the U-boat war started, Egypt nearly fell, and above all from the Indian point of view, the Japanese seemed clearly in the ascendant. Malaya fell, Singapore followed, and then after that Burma. The Japanese stood at the gates of India. In 1942 Sir Stafford Cripps was despatched to India and gave a guaranteed offer, so he thought, of Dominion status and a constituent assembly at the end of the war, and incorporation there and then of Indian party political leaders in what would effectively be a central war-time cabinet, although not formally established as such under the 1935 Act. The Congress leadership almost accepted this; but Gandhi held on to the demands for an unconditional British withdrawal. This had a huge air of unreality about it, stemming in part from his pacifism. He might have been gambling that if the

British gave so much now, what would they do when the Japanese were even stronger? But he did not realistically calculate how a suddenly independent India would resist the Japanese, other than to speculate that the British presence was a provocation to the Japanese, with the implication that they would withdraw if the British did as well. Indeed Gandhi not only characterised the offer as a 'post-dated cheque drawn on a failing bank', he also instigated another round of civil disobedience, the Quit India movement, which resulted in sabotage, arson and riots in many parts of the Ganges Valley. For some months this did disrupt the war effort, particularly communications, but the leadership was again imprisoned, and the Government used force to quell the disturbances. The Cripps offer had incorporated what seemed to be an admission of the right of the Muslims to self-determination in some form – perhaps that was partly why Gandhi rejected the offer. The Muslims therefore reacted in their turn to 'Quit India' with 'Divide and Quit'. As the war began to turn more favourably for the British and their allies, political stagnation seemed to grip the confrontation within India – stagnant that is, except for the continued and dramatic growth of the Muslim League, which seized the opportunity to campaign while the Congress leaders remained interned.

By the end of the war the genie of communal separatism was well and truly out of the bottle and could not be wished away. And the British were by now as anxious to withdraw themselves as the Indians had been desperate for them to go. Churchill, the arch imperialist and great war leader, lost the general election in Britain to a Labour Government committed to building a welfare state in Britain, anxious to shed international responsibilities, and committed to leave India. On 20 February 1947 Britain declared it would leave India no later than June 1948. Now, instead of dragging its feet, it was imposing a deadline on all the negotiating parties. The independence of the two successor states, India and Pakistan, was achieved in fact by August 1947 – with a rapidity the effects of which are still the cause of endless speculation.

The Two Nations

An American political geographer writing in the early 1920s identified Hindu-Muslim relationships as the most pressing issue facing the government of India, and there is a hint in his analysis of the partition to come.

> Mohamedanism ... is growing with terrific speed ... (it) represents a fanatical religion whose political power will try the tact, and it may be the military strength, of the Western Powers. (Bowman, 1921: 54)

> Disorder in India is a particularly grave matter since it affects not only the control of the country but also the distribution of food and the whole modern system of trade that has become established there. India now has 112,000 square miles of irrigated land, and irrigation works require cooperative control and an orderly government. Were the

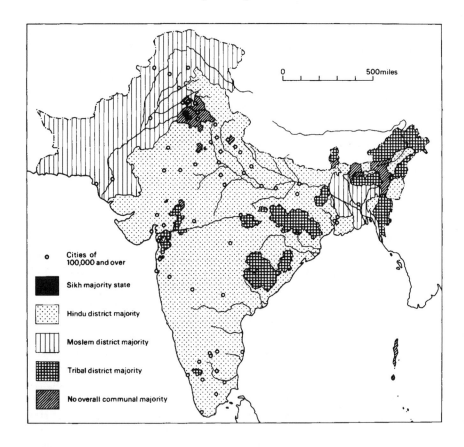

Figure 7.3   The Distribution of Religions 1947

Source: Spate and Learmonth (1967)

railroads and the irrigation works, the ports, and the whole machinery of commercial life to be disorganised, India would be ripe for a great disaster. (Bowman, 1921: 45)

But this kind of foresight is an exception. The majority of observers, including the Congress leadership, did not take Muslim separatism as a real and dangerous possibility until far too late: indeed, in the 1920s and 1930s the League itself did not publicly proclaim Pakistan as its goal. That waited until Jinnah's proclamation at Lahore in 1940. But it was not originally his or the League's idea. The League in general and Jinnah in particular had been sold the idea by its most fervent proponent, at a time when Jinnah's dismay at Congress' high-handedness had persuaded him that there was no other realistic option.

C. Rahmat Ali, Founder President of the Pakistan National Movement, had been promoting his ideas by publishing pamphlets from his base at 16 Humberstone

Road, Cambridge, England. The first edition of his short pamphlet entitled *What does the Pakistan National Movement stand for?* was published in 1933. The following quotes are taken from the 3rd edition, 1942. They include some analysis which is persuasive, and which is far from nullified by the weight of polemical assault.

> When in the current period of the history of South Asia, "Indianism" under the auspices of British Imperialism ... established in 1881 its first political institution, it cleverly called this institution the All-India National Congress. I say cleverly, because by the subtle, but nonetheless unmistakable, implications of this nomenclature, it first designated as India all the lands of South Asia incorporated into the British Empire; secondly it denied the right of the non-Indian nations therein to distinct nationhoods of their own; and, thirdly, asserted its pretentious claim to stamping Indian nationality on the peoples living in those lands which, through such dubious devices, it has made known to the world as the Sub-continent of India.

> Indubitably, therefore, this nomenclature was a trap cunningly set by "Indianism" for non-Indians – a trap which ... they should have avoided. But blindly enough they all fell into it. For in the course of time, when even the great, historic peoples like the Muslims, the Sikhs, and the Rajpoots started their own organisations, they apishly called them the All-India Muslim League, All-India Muslim Conference, All-India Sikh Conference, and All-India Rajpoot Conference, etc. (Rahmat Ali, 1942: 3)

Further below he writes:

> "Indianism" ... has debased the Saracenic civilisation of the Muslims, the chivalrous code of the Rajpoots, the knightly creed of the Sikhs, the martial tradition of the Mahrattas, and in the end attempted to "Indianise" them all, body, mind, and soul ... only if and when an impregnable defence is created against it, can they revert to their original conception of life and regenerate their respective cultures in their national strongholds. (Rahmat Ali, 1942: 5)

The talk of regeneration is of course backward-looking: he wanted to disentangle history. It was reactionary conservatism: but it was also forward looking – preferring 'South Asia' to describe what he refused to call 'the Indian subcontinent'.

Ali understood "Indianism" as the forces of the caste Hinduism of north India – known in India in contemporary parlance as The Cow Belt, or Bimaru (Bihar, Madhya Pradesh, Rajasthan and Uttar Pradesh). His appeal was that all the nations of South Asia should claim their independence from this yoke. The map (Figure 7.4) on the front of his pamphlet made the same point: Pakistan was only one of the nations that should be freed. In the south was Dravidia, in the northeast Bang-i-Islam, in the west Rajistan, Maharashtar, and Guruistan – the latter being his name for a Sikh state.

**Figure 7.4   C. Rahmat Ali's Proposals for Pakistan and Other New
          Indian Nation-States**
Source: Rahmat Ali (1942)

Examining these nationalities in more detail revealed that few other than the
Muslims of Pakistan were in a majority in their own area, so that few had realistic
hopes of a homeland of their own.

'Pakistan National Movement has ... made it a principle to admit the birthright of
each and every nation to [its own territory] ... to support by all legitimate means the
realisation of this right by all such nations; furthermore to acknowledge this right even
in the case of the Sikhs, of the Christians, of the Dravidians, and of the Depressed
Classes [Untouchables: Harijans], who, though morally and numerically qualified
to form distinct nations of their own, cannot at present do that because they are so
scattered that they can neither possess a majority in any province nor claim a part of
a province as exclusively their own ... The Movement ... concedes their right to as

much of the area of the land of their birth as may correspond to the numerical ratio of their people to the total population of the province concerned ... even if the birthright may have to be satisfied, as in the case of the Sikhs, at the expense of Pakistan itself.' (Rahmat Ali, 1942: 7)

The principle of the homelands for the scattered nations clearly seems fanciful: it would have entailed differentiation and mass cross-migration, as the communities adjusted to the lands they were given. Yet, precisely that did happen for many people in North India in 1947, continuing in Bengal until the early 1970s when a modified Bang-i-Islam became a reality. And it is still happening even now for some Sikhs who want Guruistan by another name, Khalistan. There remain a few Tamils, too, who also have their dream of a separate South.

The name Pakistan itself was an artifice, using P for Punjab, A for the Afghan Province (meaning the Northwest Frontier Province), K for Kashmir and S for Sind and taking the -stan from Baluchistan. This was the land it claimed.

The interest in all of this is that the Pakistan Movement clearly saw that the issue was one of nationalities and homelands, not necessarily communities. There was no suggestion here that Bengal be shackled with Pakistan, though Bengal could claim its own inheritance.

The problem in explaining the Partition of 1947 is, therefore, the problem of explaining how Pakistan became the cause of the whole community of Muslims both West and East. The leaders of the two communal groups, the Congress and the League, held power in proportion to the mass support they could generate, and the threat of strikes, riots and disturbances which they could instigate. To sell themselves to the masses, they had promised and were trying to deliver swaraj – self-rule, self-determination. The leaders were nationalistic. To Gandhi and the Congress, swaraj meant All-Indians ruling all Indians. To the masses, self-determination meant 'we people' rule ourselves, where the definition of 'we people' would vary from province to province. A Tamil peasant knew himself as a Tamil peasant, not as an Indian. A peasant in Bengal knew himself to be Bengali, not Indian. Even more than that, a Bengali Muslim knew himself to be a Muslim rather than a Hindu like his neighbour might be, without questioning the basis that both were indeed Bengali. He distinguished himself within Bengal, not within India – whatever that might have been. In short, while the leaders may have been nationalistic, the masses were communalistic. Therefore, to mobilise mass support the leadership had to harness this communalism. Then, the Muslim League might find an enthusiastic ally in East Bengal, fighting for its own determination, not remotely guessing what federation within Pakistan would bring in the future.

In the triangular struggle between Congress, the Muslim League and the British, the different sides of the triangle needed different approaches, which were not simultaneously reconcilable. Between the Congress and the League there needed to be accommodation and mutual tolerance. But there equally needed to be mass resistance to the British – if there wasn't, there was no need for the British to go. It was necessary for Congress to stress the identity of the masses, which was different from the British. Now, the élites may have been nationalistic, and have developed some indigenous sense of what it meant to be 'Indian', to feel the All-Indianness which C. Rahmat Ali had disparaged. But since to the masses swa-raj – self rule – required first and foremost an understanding of what the 'self' meant, so to stress the difference from the British was to stress the difference between the communities.

The Indians may blame the British who divided to rule, keeping princely states apart, and devising separate electorates for the Muslims. But the Muslims blamed Congress – for not in truth being secular at local levels. Casting blame to one side, it is clear now that to expect this sub-continent of creeds and castes, still largely illiterate, and a veritable linguistic Tower of Babel, to have formed a national identity at that time in history was to expect the impossible. Congress might have wanted the Empire the British had created, but it was committed to democratic representative government, not autocratic military power and, therefore, could not achieve its objective.

PART III
The Successor States

Chapter 8
Divide and Quit

Pride and Prejudice: The Search for Unity in Western Europe

A contemporary European reader may perhaps be best able to put into perspective the events in South Asia since 1945 if she first reflects on the events of her own subcontinent. At the end of the First World War older empires in Europe disintegrated, and were replaced by many small nation states. This could have been a retrograde step economically, but it was inevitable in the search for self-determination. After Hitler's failed attempt at the coercive re-integration of Europe within the German Reich, some of the previously sovereign states of Europe regained their sovereignty, whereas others found the Soviet yoke replacing the German. The confrontation between the two super-powers during the Cold War years was nowhere more evident than in the Iron Curtain of barbed-wire and mine-fields that rendered Europe asunder.

In 1945 Britain was the dominant West European military power, but not the dominant Western military power in Europe – a role which the USA had assumed. During the war Churchill had worried about the future of Europe and had assumed that Britain would play a leading role in developing new institutions and new democratic forces; but, by the democratic will of the British, he lost power and office when peace was declared. The new Labour administration was more concerned with building the welfare state at home and disengaging from some of the imperial baggage abroad. During 50 years of often weak leadership, this post-war Europe made slow and faltering steps towards closer integration, a project which has been tested by the collapse of the Warsaw Pact and the rush by the poorer ex-communist eastern countries to find prosperity within the European Union.

The search for unity has been long. In 1948 Britain, France, Belgium, Holland and Luxembourg founded a mutual defence group, the Brussels Treaty Organisation. In 1949, 15 states formed the Council of Europe, which included most of West Europe plus Iceland, Greece and Turkey. Most of the states of West Europe also joined the new NATO in 1949, a defence association with the USA and Canada to protect West Europe against the perceived Soviet threat. France had and still has an equivocal relationship with the military command structure; and, out of the difficulty of accommodating French anxiety, the Brussels Treaty Organisation was expanded and reformed in 1954 as the Western European Union. In 1952, six states (The Netherlands, Belgium, Luxembourg, France, West Germany and Italy) formed the European Coal, Iron and Steel Community – an example of economic (utilitarian) integration. In 1957 the same states formed Euratom, and signed the Treaty of Rome which formed the six into a Common Market in 1958. This market meant that the six states eliminated by 1968 all customs dues between them, and they agreed a common external tariff.

Seven other states in Europe, initially including Britain, formed the European Free Trade Association (EFTA) in 1961. This reduced internal barriers between members but allowed them the sovereign flexibility of different external tariffs. The movement of goods internally was more difficult than in the EEC, monitoring being necessary to prevent foreign goods entering through the lowest tariff member state to be transmitted to other members. In the 1970s most of the members of EFTA either joined or formed close associations with the EEC, and one group within the EEC formed a monetary union, the EMS (European Monetary System), popularly known as the snake. This kept their currencies within tight reins of each other, although floating as a group against other currencies, including those of EEC states which were not members of EMS (the British pound sterling for example.) The EMS was followed by the ERM – the Exchange Rate Mechanism – which again tied currencies within narrow bands as a prelude to attempts at full monetary integration. For a while the British joined the ERM, only to be forced out by speculators and devaluation two years later. As a result of the Maastricht Treaty, on 1 November 1993 the EC became the European Union, with provision for an ever closer political and economic union, a promise or a threat, depending on one's point of view, that has traumatised both major political parties in Britain. In 1999 eleven out of the then fifteen members of the union established a new currency – the Euro – which finally replaced the coins and notes in circulation in 2002. At its launch, the Euro area had a population of 290 million and an economy approaching the size of that of the USA. The Union now has 27 member states, and a population approaching 500 million, the majority of both – but not the totality – within the Eurozone.

New principles have been debated in Europe – a new word has entered the vocabulary of the lay person, even if not their understanding. This is the word 'subsidiarity' – a principle which broadly means that power should not be surrendered to an ever-stronger Centre in Brussels, but that decisions should be taken at the closest level to local concerns and local people as is practicable and appropriate. The application of this principle is much harder than defining it.

Culturally, the chauvinistic states of West Europe are not significantly closer than they were 50 years ago. Certainly the chances of establishing a single official language for Europe are significantly less than the chances of persuading the Dravidian south to accept Hindi as the only lingua franca of India. 'Local' differences in Northern Ireland sustained a sophisticated terrorist movement into the 21st century, and ETA in the Basque country of Spain still remains a threat. In former Eastern Europe nationalism and xenophobia has spawned the vicious Balkan wars of 1990–1999, and the secession of Kosova from Serbia in 2008 is unilateral and only possible with military support from members of NATO.

Pride and Prejudice: Recrimination and Divorce in South Asia

The end of the Second World War therefore marked a moment in history when Europe began yet again to seek a history of integration, the construction of a stable regional system of governance that simultaneously allowed representation without confrontation. By the end of the Second World War in South Asia a very similar set of questions was posed. It was clear that there would have to be some drastic redrawing of the map, with some complex federal or quasi-federal structure, if all interests were to be satisfied. In a sense, what the British had to do was find the kind of solution overnight which Europe has only slowly moved towards in the last fifty years. This is, of course, to exaggerate: the British had indeed seen a future for the Federation in the 1935 Act, but clearly then, before the Second World War, they thought they would have more time on their side. And certainly they had done little to resolve the problem of the princely states.

Lord Wavell, formerly Commander-in-Chief in India, was appointed Viceroy in 1943. He was an able man experienced in the ways of the country, but perhaps not enough of a politician for the dexterities and complexities of an agitated India. In 1945, at the conclusion of the war, he wished to form an interim government (i.e. an Indian Government to which most power would be transferred while negotiations for new constitutions and final independence proceeded) with the co-operation of all parties, and called a conference at Simla in which he acted as mediator between the rival leaders. Jinnah wanted not only parity for Muslim representation in the interim government, but also a policy commitment to the right of the Muslims for self-determination. Wavell rejected the demands, and accepted himself the responsibility for the failure of the conference, although the Congress was sure where the blame lay. But Wavell had also lost, not the respect for his integrity, but the confidence of the communal leaders and of the British Prime Minister Attlee that he was actually achieving any progress. And this time it was progress that the British wanted. The riots and civil disturbances during and after the war had persuaded them that inaction now was a dangerous policy and, besides, too many hopes had been raised about 'after the war' for them to be postponed as easily as they had been after the First World War. The possibility of uncontrollable public disorder was a spectre that began to haunt all the major actors, and which some were from time to time prepared to provoke.

The strategy, but not the tactics, were worked out: there would be fresh elections at the provincial and central level in India, and an attempt would be made to form an interim government. The elections held late in 1945 revealed that the Muslim League did now have strong support at the central level, but at the Provincial level the support was not so clear. In Bengal the League won sufficient seats to form a Government, though with some help of the smaller communities. In Sind the League formed a government again with coalition support. In Punjab the League won nearly all the reserved Muslim seats, but it was frustrated and kept out of power by a Unionist-Congress-Sikh coalition. The Unionists were lead by a Muslim, Sir Khizar Hyat Khan, who had been the previous premier of the Province, and who became premier again. In retrospect this may have been unfortunate: the coalition with the Congress and the Sikhs weakened the appeal of the Unionist Party to those Muslims who might otherwise have been opposed to partition.

To Jinnah, now called the Qaid-i-Azam (Great Leader) the results were not in the least discouraging. The bandwagon had begun to shift his way, and with time and the right confrontations, no doubt more could be persuaded to climb on board. What Congress needed to do was to conciliate, and not threaten – which was the one sure way of committing moderate Muslims to Jinnah's cause. Jinnah's personality and tactics have been examined many times by many authors, fascinated that in the end he could achieve so much from such an unpromising start. After all, he appeared not to have very strong cards: the Muslims were not completely behind him, the (British) Government of India wanted to leave a unitary state of some sort, and Congress, by far the largest political organisation of India, was implacably opposed to him. There was always the possibility that the latter two might in the end impose a solution which denied Pakistan its nationhood. They could tire of offering Jinnah concessions, and, instead of offering him half a loaf, offer him none. But he was a master of brinkmanship, and at the last moment would accept a half loaf, simultaneously demanding more of the rest. The only real card he held was the threat of civil war and chaos: if he could not have what he wanted, then both parties would inherit a smashed heirloom.

From the beginning it seemed an impossible dream that he could get what he wanted – a sovereign state which included not just the predominantly Muslim areas of Sind, Northwest Frontier and Baluchistan, but also the whole of Punjab and the whole of Bengal, even if in the latter two the Muslim majority was not overwhelming. And Bengal included Calcutta, ex-Imperial Capital, still the industrial centre of India and its most important port. But Jinnah wanted it all, and with good regional economic sense. He observed that East Bengal without Calcutta would be a rural slum – not really a workable proposition.

Congress was not immune to some suggestions for enhanced Provincial autonomy, and as the negotiations dragged on they were prepared to see greater or lesser powers given to a lesser or greater Muslim group – that is to say, the more it seemed that power might be devolved under a particular plan, the more they were sure that territorially the Muslim group should be as small as possible, thus extricating Hindus from Muslim domination (which was of course exactly the mirror image of what Jinnah was trying to do for Muslims). If fewer powers were to be devolved, then the Muslim Majority Provinces could be bigger, since the centre could protect the rights of the minorities. But although there were these concessionary ideas in the air, Congress quite clearly felt that it should decide the nature and kind of concessions to be made. What it consistently strove for was for the British to go, and leave them to negotiate after Independence the terms of the deal. Jinnah, of course, knew that he had to achieve what he wanted before the British left – the guarantees of later negotiation were not synonymous with guarantees of later concessions.

A wry comment was made about 'the transfer of power in Burma' in 1948, that there was in fact none: because by the time they left, the British had no power left to transfer, no matter whether the map on the wall was coloured pink or not. This prospect was rapidly drawing in on them in India in 1945–47. The civil and other services at the end of the thirties and after the war had failed to attract many men of calibre, and an overburdened, underfunded and demoralised administration was tottering to a halt. The squabbles of communalism were beginning even to affect army units as well. As civil riots and disturbances increased, the capacity of the state to deal with them diminished. Since the British knew they were going to go, they were increasingly prepared to abdicate the responsibility of finding an agreement, if one were to be found, to the Indians themselves. The more they withdrew, the greater the need of Jinnah to say no to Congress half-promises, and the greater the power of his negation.

In the months of April, May and June 1946, while the weather on the plains got hotter and hotter, a Cabinet Mission from London toiling with the Viceroy and Congress and the League came very close to an agreed Plan, which we will consider in more detail in the next section. A complex three-tiered quasi-federal structure seemed close to satisfying all concerned: following agreement, the next steps would be the formation of the Interim Government, giving most central powers to Indian Ministers, and the simultaneous calling of a Constituent Assembly, which would frame the new constitution in detail within the terms of the broad accord. The Muslim League accepted the Plan, believing it gave them almost-sovereignty in the Muslim Group of Provinces, and Congress appeared to accept it. But Nehru also publicly declared that India would look after minority interests, and that the Constituent Assembly would modify even the broad outlines of the Plan if necessary. This latter observation was tactless – almost a provocation playing on the worst of the Muslim's fears. Jinnah then rejected the plans for the Interim Government, and decided to increase extra-constitutional agitation.

He proclaimed 16 August to be a day of Direct Action. It is not clear exactly what kind of demonstration he expected: meetings in halls, petitions, marches, protests – surely all of these, but surely not what happened. The Chief Minister of Bengal saw a chance of enhancing the size of Action in Bengal by pronouncing the day a public holiday. Large crowds gathered at the Maidan in Calcutta, and were harangued about Hindu overlordship. Since the Commission of Enquiry into the Great Killings that followed was disbanded before it reported, no-one is sure about the extent to which trouble-makers deliberately started the killings. But Muslims turned on Hindus, and in turn the Hindus and the Sikhs, in greater numbers and better organised, turned on Muslims. For three days Calcutta was gripped in a frenzy: half-burnt bodies lying in burnt-out buildings, by the gutter where they had been decapitated, or drifting in the Hooghly river into which they had been thrown. The Governor waited for the Chief Minister to ask for assistance, and when at noon on the second day the request came, General Bucher GOC Eastern Command, refused to allow his troops in small parties into the little alley-ways to stop individual acts of violence, because they would have been overwhelmed.

Instead, by manning the main thoroughfares in large groups and quelling the movement of gangs, the trouble was quelled by the fourth day. The estimates of the dead range from 5,000 to 20,000 – and the number of surviving casualties was a corresponding multiple. The Hindu assault in Calcutta was revenged by Muslims in East Bengal murdering Hindu males, sometimes also raping the surviving wives and forcibly converting them to Islam, and forcing them to eat beef. The riots spread to Bihar, where the Hindus turned on Muslim minorities. And so the killing went on, following the route that at other seasons the monsoon storms take, up the Ganges valley and into Punjab.

The politicians at the centre were clearly dealing with dangerous communal differences that not even Congress could any longer deny or minimise: but if they had thought they were in command of the tiger cubs they now found themselves riding full-grown beasts, and they must by now have been wondering whether the beasts would not throw and consume the riders.

With political stagnation and stalemate, and civil collapse imminent, the Cabinet Mission withdrew. Little had in the end been achieved, although one important ground rule had been changed. Britain had declared that when Independence came, British paramountcy over the princely states would lapse. It would not be handed to the successor government of India (or Pakistan). It would therefore be for the princely states themselves to resolve their future in negotiation with interim or successor government(s). This short-sighted decision muddied further waters which were already extremely murky.

Territorial Options

If it should be the case that India was to be redefined territorially to give a homeland to the Muslims, how was this to be done? The simple answer to this problem might have been to say, divide India to minimise the minorities remaining in the two new nations. But the simple answer did not work. Firstly, even if the religious persuasion of the populace were indeed the only criterion, there was no line that could be drawn, producing even three states or provinces, which did not leave large minorities on the wrong sides of the lines, so rather negating the object of the exercise. The success achieved would depend on the fineness of spatial scale at which the lines were drawn. Partition could occur at the Provincial level, or presumably at any of the lower levels of administration, which were: at the level of the Division, at the level of the District, at the level of the sub-division, at the level of the Taluk, Tahsil, Thana, or perhaps even at the Village.

Secondly, the kind of division acceptable would depend on the kinds of powers allocated at Central or Provincial level. If the Centre remained strong and could intervene in the Provinces, then the Provinces could be larger, and retain large minority groups; for example, Hindus and Sikhs in Punjab, whose rights could be protected by the Centre. Conversely, the more fundamental the rights of the

Figure 8.1   The Seven Regions Scheme for the Federation of India
Source: Coupland (1943, Pt III)

Provinces, the more there should be territorial readjustment to minimise the remaining minorities.

Thirdly, and often running counter to the last point, the more independent the Provinces would be, the larger should be their size to make them economically viable – to avoid the Bengal-minus-Calcutta syndrome which would give political freedom to Muslim Bengalis in return for economic serfdom.

Fourthly, quite simply there were some places where boundary lines could do great damage. Everyone suspected that the division of the irrigation systems of the Punjab could prove economically damaging, even if on other counts Pakistan would be big enough to be viable. The same was true for communications and water resources in Bengal.

Figure 8.2   The River Basins Scheme for the Federation of India
Source: Coupland (1943, Pt III)

It is therefore no surprise to realise that the many schemes that were discussed over the last decades to Independence should represent a span of solutions, of varying degrees of plausibility, and varying degrees of complexity, each trying to find a cocktail which simultaneously satisfied these varying and contrary demands, each one laying emphasis more on one point than another.

Coupland in 1943 wondered whether The Problem might not in fact be as much a Regional Problem as a Religious/Communal one. A Pakistan based solely on religion would lead to wholesale mass migrations, he thought, but which ultimately would have little point, since that would transpose people into alien climates, economies and languages in the name of religion alone. It would be like transporting French Protestants to Norway. There must be many Biharis in Bengal who have learnt the hard way that Coupland had reason to make his observations. It may be remembered that John Bright saw a future for six or seven great states in India, though to my knowledge they were not demarcated on a map. Coupland reviewed two schemes for Regions, one based on an outline by Sir Sikander Khan published in 1939, but with a long history behind it, shown in Figure 8.1. The regions were based on the idea of a new tier of Federation, thus preserving the existing Provinces and States. The Regional Assemblies would have such powers vested in them as thought suitable by the Provinces, and there would be a new concurrent list of powers with the centre. Through this regionalism the different identity and culture of the Regions would remove the minorities from threat of domination by the Hindu Centre. The Centre itself would not have a directly elected assembly, thereby avoiding the monopolisation of power at that level too. The map is not wholly dissimilar to C. Rahmat Ali's map, nor is the scheme so different from the Cabinet Mission Plan that very nearly succeeded in 1946, except for the fact that Indian India is not unified but also broken into the regions numbered 2, 3, 4, 5, and 6 – and bearing in mind the increased power of the South in modern India. Even then, he was not so wide of an underlying truth.

Table 8.1   Percentage Distribution of Communities in India, 1941

British Provinces	Caste Hindu	Untouchable	Muslim	Sikhs	Other	Total
Madras	70.4	16.4	7.9	nil	5.3	100.0
Bombay	70.5	8.9	9.2	nil	11.4	100.0
Bengal	29.3	12.2	54.7	nil	3.8	100.0
United Provinces	62.0	21.3	15.3	0.4	1.0	100.0
Punjab	22.2	4.4	57.1	13.2	3.1	100.0
Bihar	61.0	11.9	13.0	nil	14.1	100.0
Central Provinces	58.8	18.1	4.7	0.1	18.3	100.0
Assam	34.7	6.6	33.7	nil	25.0	100.0
N.W. Frontier P.	5.9	nil	91.8	1.9	0.4	100.0
Orissa	64.1	14.2	1.7	nil	20.0	100.0
Sind	22.9	4.2	70.7	0.7	1.5	100.0
Princely States						
Hyderabad	63.5	17.9	12.8	Nil	5.8	100.0
Mysore	72.1	19.2	6.6	nil	2.1	100.0
Travancore	51.8	6.5	7.1	Nil	34.6	100.0
Kashmir	17.3	2.8	76.4	1.6	1.9	100.0
Gwalior	86.4	nil	6.0	Nil	7.6	100.0
Baroda	68.8	8.1	7.8	Nil	15.3	100.0
Total these and other States	59.3	9.5	13.6	1.6	16.0	100.0
Empire Total	53.0	12.5	23.7	1.5	9.3	100.0

Source: Reworked from Coupland (1943, Pt II, p. 339)

An alternative scheme was based on river basins – bearing in mind that these were the basis of the economy, and that economics must play a part in moulding the success of any nation. Yeatts' scheme, discussed by Coupland, even foresaw a future in which hydro-electric power would be a prime concern of all the citizens of the great river systems. The River Basins scheme (Figure 8.2) produces a different Indian India, but again gives the same sort of result for the Northwest and the Northeast. Coupland's point was that regionalism could chime with Muslim demands for a homeland. But it would only do so if the logic of minimising the minorities in the 'wrong' regions was not pursued, if Hindu minorities were happy to live in Punjab and Bengal – which presumably would depend on the strength of the Centre. The 'chiming' of region with homeland would also have had happy consequences for the Sikhs. They had once been concentrated on the hill-plains rimland of Punjab, and were still centred mostly around Amritsar, but after the development of the canal colonies in the southern arid parts of Punjab, they had spread far and wide, nearly always a dominant economic and social force, but not with the numerical majority that the new game of mass 'democracy' required. 'Their' Punjab would have remained intact under such a scheme.

These schemes were and are logical – but they had no mass basis. There had been no mass contacts campaign for Hindoostan or Bang-i-Islam or Dravidia. The Muslim league might have hijacked some local sentiment in Bengal, but on the basis of religious fraternity, not economic self-interest. Nor was the regional scheme appealing to the Congress, who had aspired to a unitary India as the successor to the Raj.

The Cabinet Mission Plan of June 1946 was in effect a revised regional scheme, in which the central Hindu part were coalesced as one, thereby heightening its domineering appearance. The country was to be divided into three zones, A, B and C, based on the existing Provincial and State boundaries. 'A' would be the majority areas of Hindu India, 'B' Punjab, Sind, NWFP, and Baluchistan, and 'C' Bengal and Assam. The Zones would have almost complete autonomy, and indeed 'B' became known for a while as the Federation of Pakistan. The Centre would retain Defence, Foreign Affairs, and Communications. The Centre, thought the League, would have no power of revenue-raising from the Provinces, being dependent on voted subventions. Congress thought otherwise. The League thought that the grouping of Provinces would be compulsory, though a group as a whole could at some future date secede. Congress thought that the grouping was voluntary, and at a future date a Province could change its group, but that secession and full sovereignty was impossible. Despite these varying expectations, despite contradictions in the clauses of the Plan itself, which sought simultaneously to calm fears on both sides, it did look for a while as though the Plan would be accepted by both sides. This was so because on the League's side it suspected that this was the best deal it might get, short of complete sovereignty, and because it feared the British might otherwise do a deal with Congress alone. For its part, Congress thought this to be a compromise which would, in the short term, allay the fears of the Muslims raised by imminent Independence, while preserving the long-term hope of a unified India. It is hardly surprising that in such an atmosphere the fragile and complex arrangement should have been shattered by Nehru's public affirmation that it was a temporary expedient, and by the wrangles over representation in the Interim

Government, to which Congress wanted to appoint a Muslim member as well as Hindu members, despite the League's vehement objections.

The Great Calcutta Killings were the final death blows to any such plan or later variant succeeding. After them there was no hope of keeping Bengal united in any multi-tiered scheme, in which the Hindus would be a minority in a Region, Province, or Zone which included a tier between them and the presumed Hindu majority of the Centre. For anyone concerned with the maintenance of the rule of law, the potential of mass disruption and disintegration was all too evident, and the imperative had become more and more to keep the communities apart as much as possible – even though that could never be enough to prevent human catastrophe and migration, let alone the damage to the regional economies.

Looking at the figures for British India, just over a half of the population was Caste Hindu. If the Untouchables are added within the fold of Hinduism, the population was 64 per cent Hindu, 24 per cent Muslim, leaving 11 per cent to Sikhs, Tribals, Christians, Buddhists and others. If the 24 per cent Muslim population had been neatly concentrated, the partition would be no problem. In general terms it was of course more concentrated in the north-west and the north-east; but a brief consultation of Table 8.1 will show that Bengal was 45 per cent non-Muslim and Punjab 43 per cent non-Muslim. More detailed scrutiny will suggest no matter how it was done, there would still be significant minorities on either side of any conceivable line. Even a line which was devised solely to minimise the numbers of the remaining minority communities would leave 20 million of the 80 million Muslims in India.

The 'Pakistan Declaration' of Lahore had not mentioned the boundaries of the new state, but had acknowledged that some kind of territorial readjustments might be necessary. Jinnah had been careful never publicly to concede less than the demand for full provinces, but it was widely thought that the League already knew that some parts of the provinces might be lost. The League knew that Punjab would be strengthened if the Ambala Division (very roughly coincident with modern Haryana) were to be conceded to India, which would raise the Muslims to 63 per cent in the remaining Punjab. In such a case the Sikhs would be split down the middle, their concerns trampled beneath the arguments of their two larger quarrelling cousins. The League also obviously knew that if Bengal were divided, a Muslim state or province with far more than 55 per cent Muslims could be created. If, for example the Burdwan Division, lying west of the Hooghly, were to be ceded to India, the proportion would rise to 65 per cent – even though Calcutta would be within the Muslim part.

The calculations began. Dr S.P. Chatterjee published a Calcutta Geographical Society Monograph in 1947 showing that if Bengal were partitioned on a District Basis the rump of Muslim Bengal would be 70 per cent Muslim, and Hindu Bengal 71 per cent Hindu. By partitioning at the sub-division level he could not improve the percentage of Hindus in Hindu Bengal, but he could increase the area of Hindu Bengal and the total number of Bengali Hindus within it. The problem with either of these schemes was that it would have split Hindu Bengal into three blocks

– one in the far east of Bengal (the princely state of Tripura to which he added the tribal area of the Chittagong Hill Tracts), and one in the north-west and one in the west, with Muslim Malda District splitting the latter two. To counter this, he proposed division at the Thana level, by which he managed to find a continuous strip of West Bengal that was Hindu dominated. It is significant that at the regional level it was the Hindus who were calculating how to avoid living under a Muslim majority. It is also important to remember if one looks at Figure 7.2 that, while all the calculating and planning proceeded, the fate of the princely states, 40 per cent of the area of the Indian Empire, was still completely unsettled.

The Decree Nisi

After the collapse of the Cabinet Mission Plan, Wavell was left a tired man with no clear sense of direction. On 20 November 1946 Wavell announced that the Constituent Assembly would convene on 9 December; but it was fairly certain that the League would boycott it, and that it could inflame the disturbances simmering throughout the country. On 2 December, Congress and League leaders went to London with the Viceroy, and after four days of discussions the communiqué simply said that the UK would not impose a constitution on unwilling parts of India. This was an open invitation to the League to boycott the imminent Constituent Assembly. Back in India, Wavell devised a plan of despair: that the British would withdraw province by province, leaving the provincial assemblies to negotiate with each other. One wonders whether, if that had happened, there would not have been, Angola-like, subsequent decades of civil war. In Amritsar and Multan the civil war had started, and no-one was sure how long the army would last before it, too, split on communal lines (although when asked, the question was usually answered confidently at the time). Prime Minister Attlee knew that a settlement was a matter of urgency. He wanted a modern Alexander to cut the Gordian knot, and to all intents and purposes, that was exactly what he got. His choice fell on Lord Louis Mountbatten, a great-grandson of Empress Victoria and acceptable to the princes as the King Emperor's representative, a man as curiously out of his time as was Britain's Raj in India. He was a dashing and impulsive cavalier, but also a man of intelligence. He had made the Royal Navy his career, and had had some heroic failures pressing his destroyers against the odds. He loved tradition and uniforms – the dress of Viceroy would sit easily on him. He had also become increasingly proficient at running large-scale organisations, and had ended the war as Commander in Chief of Allied Forces in South East Asia. He had retaken Burma and was planning the push to retake Malaya when Japan fell to America's new atomic bombs.

He claimed not to have wanted the job, and to have put all sorts of demands on Attlee to make him withdraw the invitation, but Attlee agreed to them all, down to finding and re-commissioning the Avro York which had been Mountbatten's transport as C-in-C. South East Asia. But he also claimed later that his whole life had been a training for this moment. In February 1947 it was announced in London that Mountbatten would be appointed Viceroy, and that he was being so appointed with the specific task of transferring power – that is, it was announced that he would be the last Viceroy. The transfer was to be achieved by 1 June 1948. Mountbatten had essentially gained for himself plenipotentiary powers – i.e. he could make some agreements that would be binding on the British Government. He arrived in India in March 1947, and in fact the transfer was achieved on 15 August 1947, with a rapidity that has been the subject of debate ever since, particularly since some loose ends have remained to this day to sour relations between the successor states. But he and many authors have defended the speed as essential given the collapsing state of law and order.

Lord Louis had one great quality which he himself valued as much, if not more, than his admirers did. He could invariably understand another man's point of view, even if he did not agree with it. He could also communicate this understanding. In this he had a weapon with which he could persuade others to go along with his schemes. And to a large extent his powers of persuasion worked well with Gandhi and Nehru and others of the Congress High Command. But in Jinnah he found an austere aloofness which even he could not crack. He realised early on that Jinnah was implacably set on sovereignty for Pakistan.

He sent his first plan to London on 2 May. It looked like a revision of Wavell's counsel of despair. It provided for the simultaneous transfer of power to the provinces, leaving them to establish the federal part of the constitution after independence. A revised version arrived back in India on 10 May, which Mountbatten promptly showed to Nehru. The latter was dismayed. Here written large was the Balkanization of India. Nehru knew that in the end the Muslims would probably go their own way, but he was adamant that Congress had to have a strong Centre in an unfragmented India.

Next, Mountbatten tried to outflank Jinnah by pursuing the logic of establishing 'this mad Pakistan' to the point where what was on offer for real would prove too little to be attractive to Jinnah. To do so he embraced the plan proposed by V.P. Menon, the central government's Reforms Commissioner, which explicitly provided for the Partition into two. This essentially followed the logic that said that, the greater the devolution of power, the smaller must Pakistan be. Punjab and Bengal would be partitioned too if either community in the provinces wished for it. Jinnah never agreed to anything: he just played for time, knowing that the tide was flowing in his favour. In such circumstances it was inevitable that Mountbatten should have his own priorities – which were that Congress should accept the Plan, that India and hopefully Pakistan should remain within the Commonwealth, that in defence and foreign affairs there should be a pact between Pakistan and India.

Gandhi set his face against any partitioning – vivisection – of Mother India, and on this he finally broke with Nehru, who saw no other option. Or, to put it another way, Nehru knew that any chance for a strong central government in what would become India would only be achieved by letting go of the Muslim areas, and achieving for India the kind of sovereignty that the Muslims were demanding for themselves. Any accommodation would mean a weak Centre, much as in many degrees Brussels is weak compared with sovereign governments in the European Union. As at the regional levels the Hindus now wanted to escape from Muslim domination, so now at the central level too the Congress could not have what it wanted without breaking from the Muslims. This was the extent to which Jinnah had finally turned the tables on the majority community.

On 18 May the new plan was taken to London, and on 31 May Mountbatten returned with the Cabinet's approval. He presented it as the only and final plan to Congress and the Sikhs, who accepted it, and to Jinnah, who said nothing. A formal meeting was called for 3 June at which the principal leaders were present, but hardly allowed to say a word. Mountbatten spoke for them, fearing a row would sabotage this last chance. The night before he had asked Jinnah not to say a word, but to nod his agreement when the time came. After an agonising silence, Jinnah nodded, and Pakistan, of an as yet uncertain size and shape, was agreed. Promptly a detailed document "The Administrative Consequences of Partition" was put before each delegate. The work to be done in so short a space of time left all present somewhat aghast.

That night the agreement was announced in London and Delhi, and the leaders each spoke by radio to their constituents. Independence would be achieved at midnight on 15 August, and power would be handed to the two sovereign states of Pakistan and India, who would determine their own constitutions. The provinces individually would vote for Union with either India or Pakistan, and, having decided, the representatives of the two communities in the Assemblies would then vote for partition or not of the provinces. Predictably, both Punjab and Bengal voted for Pakistan. Predictably, the non-Muslim groups of both Provinces also then voted for Partition. In the meantime, the League effectively became the interim government for Pakistan, and Congress for India.

To draw the new lines, Boundary Commissions were set up for Punjab and Bengal, each comprising four members, two nominated by the League, and two by Congress. Both Commissions were chaired by an English lawyer, Sir Cyril Radcliffe, whose impartiality was guaranteed by his lack of experience in, and knowledge of, Indian affairs. The Commissions were established on 30 June, and on 8 July Radcliffe arrived in Delhi. He had under six weeks to hear the arguments and draw the lines, according to ' contiguous majority areas of Muslims and non-Muslims and, in doing so, to take into account other factors (unspecified). Since the Members were hardly non-partisan, the weight fell on the Chairman. Importantly, it was to be an award, not an adjudication. To that extent he could find latitude for his own invention. And so parts of the new geography began to take shape, while in the wings the Princes argued to and fro over their futures in the new South Asia of majority rule.

Attlee had given Mountbatten instructions to hand over power to a unitary state if possible – but not against the will of any major group. There has been considerable debate over the extent to which Mountbatten did have room for manoeuvre: Hodson (1969) compares his situation with a canoeist descending rapids – really unable to dictate the major course and direction, but able to negotiate the rapids and avoid disaster on the rocks. Although Hodson concludes that this is too deterministic a summary, and that a fiasco could have resulted, it remains a persuasive image. Mountbatten's success lay in the fact that he persuaded the leaders of Congress to accept that Pakistan would be born. His truly great achievement here was to persuade a dissenting Gandhi to remain silent, and not to stir up a storm of protest. He also persuaded Jinnah to accept, however ungraciously, a moth-eaten Pakistan and the partition of Bengal and Punjab. He persuaded a majority of the Princes to accede voluntarily and uncontroversially to either India or Pakistan on the basis of proximity and majority community – but unfortunately he did not quite persuade all. In the end Mountbatten had no choice but to tear along the dotted lines, before the rupture spread even further. Predictably the tear was not to be a neat one. The de-common-marketisation of South Asia had begun in earnest, and would continue for many decades.

Concluding Remarks

At the end of the Second World War and the collapse of Hitler's Reich, individual nations in West Europe rediscovered their tradition of representative government, and from there slowly and falteringly began to find some new system of integration which would provide political stability and economic prosperity. In India one of Europe's Empires was fading fast but, for the first time, with a chance that it would not be followed by a period of instability and petty wars – perhaps waiting yet again another coercive imperial integration. This time the chances were different because some of the seeds of representative and federal government had been sown. Whereas in earlier times factions among the élites could be suborned or co-opted with little reference to the common man, and territory had represented either revenue or strategic advantage, now territory was intertwined with both the consent and the power of the masses. The élites were not biddable simply in their own terms, they now were constrained by the people they also commanded.

Chapter 9
New Lines on the Map

Introduction

The Partition of India into two independent sovereign states had been agreed. Mostly the boundaries of the successor states could be defined by using the boundaries of the provinces or Princely States which acceded to them – in the lower Indus valley Sind, Khairpur and Bahawalpur went to Pakistan and the boundary was fairly easily drawn. But two provinces were to be partitioned – Bengal and Punjab. To those who know the continuity of culture and language in the two parts of Bengal and Punjab, the idea of a partition to separate religious communities seems destructive and almost pointless, except that the violence the communities could inflict on each other in their moments of passion had been displayed too often for it to be ignored. So in Punjab and Bengal new lines had to be drawn – lines which would appear on the map of boundary frequencies, in a sense the basic map around which this book is written (Figure F.1), for the very first time. For the first time in South Asia's history the principle of identitive integration at the mass level was being applied, and it produced new lines which utilitarian and coercive integration had never done before. Someone had to be responsible for these new lines.

This was not the only problem. Although it has just been said that the princely states acceded to the most appropriate state, in fact not all accessions ran so smoothly. So here was another area where the territories of the successor states might be in dispute. Finally, people are potentially mobile. If the new states did not fit the communities, the communities could fit the new states – and migrate en masse to attempt to achieve that end.

These issues and the appalling human suffering that resulted are the concerns of this chapter.

Radcliffe's New Map

Sir Cyril Radcliffe's Boundary Commission's terms of reference were to divide Punjab and Bengal (and Assam) according to contiguous majority areas and other factors, which were unspecified. It was a task that many felt should never have been undertaken. The Governor of the Punjab, Sir Evan Jenkins, who had had to contain the March riots, knew that because towns might have different communities from their hinterlands, no line could be drawn to solve the minorities problem. And if the problem could not be solved, by implication partition could make it worse, provoking riots on a scale not yet seen.

There was another possibility – to move people. The Sikhs were aware that they stood to lose most of all. Fearing that a new boundary would divide them into two, Gianni Khartar Singh on 10 July 1947 asked for a transfer of three quarters of the Sikhs from West Punjab to East, accompanying a commensurate exchange of property. Large migrations had occurred in Europe at the end of the Second World War, with for example the geographical shift westwards of Poland. But there was far too little time to organise such a transfer in Punjab before 14 August. The ghastly irony was that such migrations were about to occur in any event- but not of an orderly type. Fearing the worst, the Sikhs were making preparations to fight, and the extremists to strike at the perceived root of the problem: at one stage a Sikh plot to assassinate Jinnah was discovered.

Aware of the difficulty of the task and the compressed timetable, Radcliffe asked the respective leaders how important it was that the award be announced by 15 August. Both sides said it was imperative, and that speed effectively should over-ride careful attention to detail. In the event, Mountbatten engineered that the award was not published until two days after Independence, which meant that literally no-one was sure exactly where the border was for those 48 hours, as a result of which there were some strange goings on, which are touched on below. Further, Radcliffe had to draw the lines while not knowing initially which of the princely states bordering these two provinces would accede to India and which to Pakistan, although he might have had a fairly strong idea and during the course of his six weeks most of these issues were settled.

In the west he could assume that Bahawalpur and Khairpur, both with Muslim majority populations, would go to Pakistan, and probably he would have assumed that in the fullness of time Kashmir would do so too – because it too had an overwhelmingly Muslim population. Bikaner would go to India. So, somewhere on Bahawalpur's northern edge would be the starting point for his line, on the Sutlej River.

The Boundary Commission met in Punjab in July and for ten intensive days heard submissions from interested parties. Radcliffe did not attend – he was responsible simultaneously for the Bengal Commission too. Instead, he had daily summaries of the evidence taken to him. The Sikh proposal at one level sounded logical enough. If one started from East Punjab and added progressively districts in the west, how far could one go while still retaining a non-Muslim majority in the ever larger East Punjab? The answer was that they could go as far as the Chenab – which would then mean that the most important doabs between the Chenab and Sutlej, the Ravi and Sutlej, and the Beas and Sutlej would all go to India, and so too would Lahore and Multan. There was an historical basis in the claim too, reviving the kingdom of Ranjit Singh, the Sikh King of Punjab before the British annexation. But such a scheme would put many Muslim majority districts on the western edge of East Punjab, and deprive Pakistan of Lahore, a predominantly Muslim city, which was also the hub of communications to Rawalpindi and the North West Frontier. It would also split the Triple Canals Project down the middle. Radcliffe rejected this, and worked in the first instance on the principle of contiguous majority districts (although he was prepared to, and did, divide at the tahsil level, and even below the tahsil.) On this basis his attention was drawn to look in detail at the land between Lahore and Amritsar.

Of this area Coupland had written in 1943:

> The two principal cities of the Punjab – Lahore the administrative capital, and Amritsar, the commercial capital and sacred city of the Sikhs – are both situated in the middle of the Province between the Rivers Ravi and Beas and only thirty-five miles distant from each other. To fix the boundary at either river is plainly impossible: it would mean the inclusion of both cities in either the Muslim or the Hindu State. Between the two cities there is no natural dividing line of any kind. Any boundary set between them would be wholly artificial, geographically, ethnographically and economically. *Inter alia* it would cut in two the system of canals on which the productive capacity of the whole area largely depends. It would also leave the capital city of each Province exposed and defenceless, right up against the frontier. Such an artificial dividing line, despite its obvious disadvantages, might serve, if it were to be merely the boundary between two Provinces in a single federal State. Administrative difficulties, such as that of the canals, might in that case be overcome. But it is no mere inter-Provincial boundary that is contemplated. It is to be a regular international boundary between two separate independent National States. (Coupland, 1943, Pt III: 86)

This spells out Radcliffe's dilemma in a nutshell. His task was, however, to produce an award, and therefore he did not have to, nor would he indeed, give a reasoned judgement between conflicting claims – since that would leave the argument open for rebuttal and appeal. He did explain quite a bit of his thinking in the announcement of the award, but much was also left unexplained. He himself thought that there could be no 'natural' solution, and that he *had* to assume that both sides would cooperate in the running of the canals after Independence. Here, demonstrated in the harsh light of farming and food requirements, was shown the reason why some sort of federation had for so long seemed to be the essential and necessary end of the independence movement.

At the southern edge of the territory Firozpur (Ferozepore) District had a non-Muslim majority, and the district's northern boundary at the Sutlej provided a good starting section for Radcliffe's line. Some of the northern tahsils of the district had Muslim majorities, but within areas that had towns such as Firozpur itself with non-Muslim populations. Firozpor, like Lahore, was also a communications centre for the territory south of the Sutlej, and also a major garrison town, being at that point where several battles had been fought between the British and the Sikhs from across the river itself. As Pakistan must have Lahore, so India must have Firozpur – and hence the Muslim majority tahsils would also go to India.

The river had been harnessed by a barrage near Firozpur with canals on both banks, and thus if the exact line of the river were used for the boundary, the barrage would straddle the new frontier. But here was a curious twist of fate. The river may once indeed have marked the district boundary, but the river had shifted course a

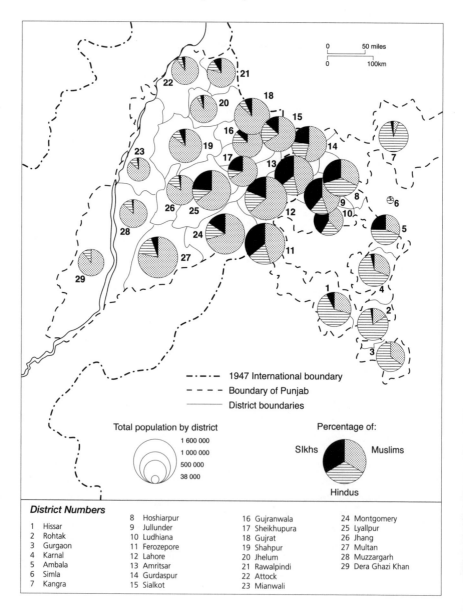

District Numbers

1	Hissar	8 Hoshiarpur	16 Gujranwala	24 Montgomery
2	Rohtak	9 Jullunder	17 Sheikhupura	25 Lyallpur
3	Gurgaon	10 Ludhiana	18 Gujrat	26 Jhang
4	Karnal	11 Ferozepore	19 Shahpur	27 Multan
5	Ambala	12 Lahore	20 Jhelum	28 Muzzargarh
6	Simla	13 Amritsar	21 Rawalpindi	29 Dera Ghazi Khan
7	Kangra	14 Gurdaspur	22 Attock	
		15 Sialkot	23 Mianwali	

Figure 9.1   The Communities in Punjab and Radcliffe's Boundary Award 1947

Source: Based on Michel (1967)

little, while the district boundary had not changed. The result was that a curious salient of Firozpur district crossed north of the river and enclosed the barrage on the far bank as well. When the district was awarded to India, she then had total control of the barrage, which was the headworks for the important Dipalpur canal in Pakistan. Upstream of Firozpur, Lahore district had a curious salient going in the other direction – so that Pakistan came to own both banks of the river for a short stretch, and could presumably start building another barrage there given the necessary will. For the Indians this was a real fear in the 1950s, and was one of the reasons for their rapid construction of a new barrage at Harike, yet further upstream.

If Firozpur, Jullundur and Amritsar went to India this would leave significant Hindu majorities in Lahore District. Partly because of this, and to preserve the integrity of some of the eastern branches of the Upper Bari Doab Canals, the District was partitioned, and in one case a tahsil as well. This provided the southern part of the line between Lahore and Amritsar; but a similar approach does not appear to have been used for the northern part of the line on towards the Ravi.

Amritsar District included a Muslim majority tahsil, Ajnala, contiguous with Muslim Sheikhupura. The princely state of Kapurthala also had a Muslim plurality.[1] Both of these shared borders with Gurdaspur District, the next north of Amritsar, recorded in the 1941 census as being 50.23 per cent Muslim, hardly an overwhelming majority over non-Muslims, but on the other hand since Sikhs and Hindus were the other 49.77 per cent, the Muslims were clearly the major community. On the face of it, Gurdaspur could have gone to Pakistan, and Ajnala Tahsil and Kapurthala State too – leaving a rump of Amritsar surrounded on three sides by Pakistan.

So sure of gaining Gurdaspur was the Muslim League, that it sent its administrators to take over the local government. On 14/15 August they ran the new Pakistan flag up the flag pole. On 17 August, when the awards were announced, most of the District had been given to India, whose newly-despatched officers bundled the Pakistanis out with little pause for ceremony. With Gurdaspur in India there was then no reason to suppose that Kapurthala could join Pakistan – and its ruler was not that way inclined anyway. The Ajnala Tahsil could stay with India too, and push the border to the Ravi, thus maintaining a fair equidistance between Lahore and Amritsar.

The award had other effects too. Although one tahsil of Gurdaspur north of the Ravi went to Pakistan, the head works of the Upper Bari Doab System at Madhopur went to India. Thus India had both the headworks on the Sutlej at Firozpur threatening the Pakistani West Bank Canals, and the control of the system on which Lahore, further downstream than Amritsar, depended, not only for agriculture, but also for municipal water. Had Gurdaspur gone to Pakistan, then Pakistan could not have cut water from Amritsar without simultaneously cutting

---

1 Plurality means 'largest community', but without an outright majority. In this case it was 46 per cent.

off Lahore. And it would too have left both new nation states with one headworks apiece on which parts of the territory of the other depended. Alternatively, using the river as the boundary on the Sutlej at Firozpur would have given Pakistan half of one of these barrages.

No further explanation for the published award was given by Radcliffe – and given subsequent events there have understandably been rumours, which have remained just that – unsubstantiated rumours, about a conspiracy – over Kashmir. Gurdaspur provides the only possible route for a land link between India and the Vale of Kashmir. In August 1947 this should not have been important or a factor considered by Radcliffe, and we do not know that it was, and indeed all the statements are to the contrary. The road from Gurdaspur, through Jammu and Kashmir to join the Sialkot (to be in Pakistan) to Srinagar road via the very high Banihal Pass, had not, at that time, been developed, and in any event given the overwhelmingly Muslim nature of the Kashmir, it would have seemed most likely that it would accede to Pakistan, if to either. But rapidly the road was developed after Independence to give India a land route through to a land which remains the beautiful but contested and divided mountain redoubt of the western Himalayas. This is an issue which is examined further below. Some years later Radcliffe took the secrets of his reasoning to the grave, rightly having maintained a complete silence over how and why the award had been made the way it was.

The Second Partition of Bengal

In Bengal the issue which dominated all others was the assignment of Calcutta. Having decided to which new Dominion it should be given, the contingent questions related to how much hinterland space should also go with it. Given the size of the city and its Hindu majority and its significance for communications in West Bengal, it was perhaps inevitable that it should go to India. Having assigned it that way, Radcliffe's next provisions were not so inevitable, and presumably owed something to the 'other factors' provision. He gave Calcutta its space and avoided the fragmentation of West Bengal that S.P. Chatterjee had feared, by maintaining a corridor running east of Calcutta to the north. In doing so he incorporated significant Muslim parts of Murshidabad into India, and partitioned Malda and Dinajpur Districts as well. In the south, perhaps as compensation to East Bengal for the territory thus lost, he maintained the rather north-south boundary, and he included Khulna and its Hindu majority within East Bengal.

In the East, Sylhet District of Assam, a Muslim majority district with little geographical connection to Assam proper, was allowed to vote to join East Bengal and did so. The terms of reference of the commission were ambiguous about other adjoining Muslim majority sub-divisions – indicating that they could be assigned to East Bengal. But were these contiguous sub-divisions of Sylhet District, or subdivisions of Assam contiguous to any part of East Bengal? Radcliffe took the former view, thus not considering so many other Muslim pockets of Assam.

New Lines on the Map

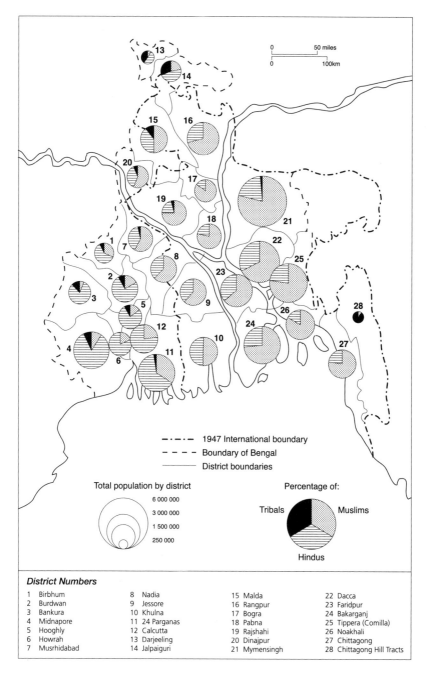

Figure 9.2   The Communities in Bengal and Radcliffe's Boundary Award 1947

Source: Based on Chatterjee (1947)

Lastly, there was the outlier of non-Muslim population in the Chittagong Hill Tracts, inhabited sparsely by tribal peoples, principally the Chakmas. Their leaders had expressed a desire which Congress encouraged to be awarded to India, but Radcliffe took the view that their economic life was so closely tied to East Bengal, that an assignment to India would be unrealistic. Hodson observes on this point that though they were certainly not Muslim, neither were they Hindu, indicating that he thought there was reason to put them in East Bengal. But this mistakes the major point that Congress was making: it claimed not to represent the Hindus, but Indian peoples in general, even Muslims, though it had come to accept that the majority of the latter would be divorced. Certainly being non-Hindu was never a principle for the exclusion of people from India, whereas being non-Muslim has been a problem for the people of the Tracts locked into a Muslim dominated state ever since.

The awards were ready just about in time for Independence Day, 14 August. But Mountbatten knew that there would be dissent from all sides, and he wished not to turn the day of celebration itself into a day of recrimination. Both sides had agreed to honour the award when it came, but he was not sure of their will to do so. By managing to be out of Delhi when the last report was received, he put off the day of reckoning till the 16th. Then the communal leaders were invited to receive the reports and to discuss them. After hours of bitter debate, all agreed to accept them as they stood, and they were published on 17 August. What persuaded the leaders to accept them was not so much the fact that any of them was satisfied, as the growing realisation that their opponents were as bitterly unhappy as themselves: and therefore such an award could not have been prejudiced in their opponents' favour either. In the atmosphere of the time, a diplomatic triumph of sour grapes for all could arguably be counted as simply a diplomatic triumph.

In Punjab, though some irrigation schemes had been dissected, the greatest, the Triple Canals Project, stood mostly intact in Pakistan, though this did not stop a major dispute erupting very quickly. Bearing in mind that no line could have divided the communities fairly, the actual line was adhered closely to the principle of contiguous majority areas. In Bengal the resource element did not figure so strongly, though the problems resulting from the division of the river basins here has now grown into a problem which outweighs that of Punjab. But the line seemed to be neither one thing nor another: East Bengal shorn of Calcutta was born as the rural slum prophesied; and yet it was not even defined to maximise its Muslim majority. With significant Hindu-majority pockets in Khulna and elsewhere, East Bengal started as a 30 per cent Hindu state. None of this reflects on Radcliffe: it reflects on the folly of the communal leaders opting for partition at all costs. It reflects on their inability to learn from 1905. That Partition at least had had the merit of being within the same 'federal' structure.

The Princely States

The formal position that Mountbatten found was startlingly clear. At Independence, British paramountcy over the Princely States would lapse – thereby meaning that in theory they could go their own way, as independent states or forming their own groups, within or without an understanding with the new India or Pakistan. The reason for this state of affairs was simply that there were many conservatives in the British parliament who considered that the Princes deserved to be rewarded for their past loyalty to King and Empire. To these parliamentarians, who might fairly be described as backward looking romantics, paramountcy was not something that could be transferred by the Crown: it was between the Crown and the Princes, and if the Crown were no longer to be represented directly in India by a Viceroy, and if the Crown would no longer be able therefore to fulfil the obligations of paramountcy, then the only possible action was to announce that paramountcy would lapse. It could not be handed over the heads of the Princes to the successor governments, and Attlee's Labour government could not afford to have a divided House on a bill as important as the Indian Independence Act.

But the geographical realities were different. Simply because there had been a paramount power, all of them had economies which were in some degree integrated with the rest of India, and communication lines to them and across them from British India. The majority of the 596 were too small to be viable on their own anyway, and certainly they would not withstand the pressing demands for democratic rule that were encouraged by Congress from without. Mountbatten, a relative of the Royal Family in Britain, was charged directly by the King with seeking a solution compatible with the honour and dignity of the Princes. His personal links with the crown were of great value in this matter. His first ideas were that the Princes should join the Constituent Assembly, much as they would have done under the Cabinet Mission Plan, to work out their own future. But events had moved swiftly beyond that solution. Now it had been agreed that India would be partitioned, and two Dominions would emerge. For neither Dominion was there a constituent assembly that would produce a constitution by 14 August; and it was clearly impossible for the Princes individually to negotiate with the interim governments on a one-by-one basis for particular constitutional arrangements.

A new States Department was instituted specifically to handle the problem. It worked on the presumption that the Princes should be persuaded voluntarily to accede either to India or to Pakistan, on the basis of the composition of their population and of contiguity. The accession would be for the purposes outlined by the Cabinet Mission Plan for federation, namely defence, communications and external affairs. The carrot was that princely titles and rights would be honoured, and that no charge would be put on the states for the purposes listed.

Of threatening sticks, there were many. Nehru had proclaimed in April that any state not acceding to the new India (this statement was made before Partition was agreed) would be treated as 'hostile' – a remark which earned him a fulsome rebuke from Mountbatten. But it was the truth of what would follow, despite the fact that, as Mountbatten emphasised, many of the Princes had large well-equipped and loyal armed forces. The biggest threat was uncertainty over the best timing of negotiation. That some negotiations would have to take place was certain, but would they best be held under British auspices before paramountcy lapsed, or after, between independent states and independent India (and Pakistan)? The latter course Mountbatten saw as leading to chaos and anarchy, with the scales tipped in favour of unrestrained intervention as Nehru had threatened. Nehru indeed continued to maintain that India would exercise paramountcy as a matter of course. Jinnah in turn delighted in the British position that paramountcy would lapse, since that enabled him to negotiate with free States, and to offer them inducements to join Pakistan – thereby enabling his moth-eaten state to poach a little here and there. To that end he offered all kinds of inducements to the Maharaja of Jodhpur – and very nearly succeeded. There would be other muddied waters to fish in too.

Two specimen documents were drafted – instruments of Accession for the three Central matters, and Standstill Agreements covering non-acceded matters, which simply stated that until the final accession details were worked out, all existing economic and political accords between the States and paramount British India would be observed unaltered after paramountcy lapsed, until such time as renegotiation should occur. With this base secured, on 25 July Mountbatten then persuaded the massed Highnesses of the Indian Empire at a meeting of the Chamber of Princes to accept that there was in reality no course open to them other than to sign treaties of accession. Three variations of the treaties appropriate to three classes of states of different size and importance were ready. And all but three Princes had by 14 August acceded either to Pakistan or India. These three were the rulers of Junagadh, Hyderabad, and Kashmir.

Junagadh

With their internal status and income assured, and their external affairs looked after, there had been plenty of opportunities for the Princes to degenerate into pampered eccentrics, of whom there were many. Free to indulge in their chosen proclivities, many did so, though equally there were able and progressive Princes who were the very model of stabilising patronage that the conservatives of Britain always imagined them to be. In the Kathiawar peninsula lived the moderately eccentric Nawab, Muslim ruler of Junagadh, which had a Hindu majority populace, and two vassal minor princely states, Babariawad and Mangrol. The Nawab, though a Muslim who should have found dogs unclean, in fact was devoted to them, and kept eight hundred, each with its own attendant. On one famous occasion a state wedding was arranged for two of them, and a state holiday was proclaimed in their honour. Here was a man unprepared for the political storm that was breaking. His administration initially seemed favoured to India, but a Pakistani agent persuaded him that his dogs would be better looked after in Pakistan; in fact, he had heard that Congress agents were out to poison them. In May, while he was in Europe, there was a 'revolution' in the Palace administration, and a Muslim League Sindhi politician took over. On 15 August, Junagadh formally announced its accession to Pakistan. India was dismayed, but could play a trick in such a confused situation. Babariawad and Mangrol independently declared their accession to India: whereupon troops of overlord Junagadh were despatched to whip the vassal states back into line. India isolated the states with units of its armed forces, which was an effective step, since the states could only otherwise be supplied by sea, although not while the monsoon season lasted.

In the aftermath of Partition there was indeed a joint Defence Council for India and Pakistan, but one which was not destined to endure. At the time of this incident the Council was chaired by Mountbatten, who had stayed in India as its first Governor-General. (He had hoped also to be Governor-General of Pakistan, but Jinnah took that post himself.) He used these two positions above all to prevent India from resorting prematurely to an armed invasion, and although the apple did finally drop into India's lap, it was not before Nehru had made public concessions of potential value to Pakistan. In a press communiqué of 6 October Nehru repudiated Pakistani claims on Junagadh, and stated that the issue should be resolved by a plebiscite of the State's population. On the 1 November Indian troops moved into Babariawad and Mangrol without bloodshed, and the tourniquet on Junagadh was tightened. Internal conditions were deteriorating as State revenues dried up, and economic activity withered. The Nawab fled to Pakistan by air with his dogs (leaving his wife behind, so it is said) and a new administration declared accession to India. Pakistan refused to accept the new position, arguing logically that once the original deed of accession had been signed, its was no longer in Junagadh's power to alter it. To this day it claims the territory as part of Pakistan. It is said that the accession to Pakistan would have been territorially absurd: but this is not manifestly apparent. Being a coastal state and not so far from Karachi, it would have been no more absurd than much of Borneo being part of Malaysia, and no more so in purely locational terms than Ulster within the UK. Investment could have ensured that one port stayed open the year round. Given its population structure it is true it would have been unwise to incorporate it within Pakistan. But that is not quite the point. The rulers did have the right to accede as they wished, even if it might prove difficult to enforce. And Radcliffe had not succeeded in parting Muslim from Hindu, let alone Sikh. Initially, quite what Pakistan was playing for other than embarrassment for India, or simply to hold a card in reserve, is unclear. To say that as early as May Pakistan was preparing to contest for Kashmir would have been to foresee events in Kashmir with greater clarity than at that time was possible. But as the summer of 1947 passed on, the link between what happened in Junagadh and what was happening in Kashmir became much clearer. Nehru's commitment to holding a plebiscite in contested cases was adequate reward for Pakistan's opportunism.

Hyderabad

The vast territory of the Nizam of Hyderabad, one of the world's richest men, who locked himself and his wealth in old newspapers in a small room of his Palace, lay in the middle of peninsular India. This was a state of strategic importance over which the British and French had struggled almost two centuries before. Its size was certainly big enough to make it a viable independent nation; but its position in the heart of the Deccan straddling major lines of communication from the great port cities meant that India could not be indifferent to its claim to independence. In addition, the majority of the population were Hindu.

Nevertheless, the Nizam refused to sign an instrument of Accession, and hoped for total independence. Negotiations dragged on for months after 15 August, with no result, although from time to time it seemed that the Nizam would in the end sign a deed of accession, on terms much more favourable to him than to any of the other princes. But there were dedicated Muslim extremists in the State, who effectively blocked attempts by the Indian Government and the Nizam to reach even such a watered-down accord. As a means of outflanking them, and in any event following policy adopted elsewhere, attempts were then made to induce the Nizam to introduce representative government, which would have isolated the extremists.

In the summer of 1948 clandestine gun-running by air from Pakistan was reported, and trains running across the State were ransacked,[2] in plain contravention of the Standstill Agreement. Thus India was given a pretext to undertake a 'police' action, as opposed to a 'military' action, against the state. Contingents of the Indian army invaded the state, in the face of resistance by the Nizam's army, which held out for a week. Had there not been a stroke of luck in capturing an officer carrying orders for blowing up all bridges, the fighting could have been much longer. Even so, the death toll was nearly 1,000 combatants. In the end world opinion did not seem as hostile to India as Mountbatten had feared it would have been; and the disorder, murder and looting in the State may just have been adequate to justify the use of force. The loser was undoubtedly the Nizam, who could have remained a major leader in his own state had he been able to control the extremists.

Jammu and Kashmir

As Kathmandu in Nepal hides in its own broad valley behind the lofty Mahabharat Mountains, so Srinagar hides in the Vale of Kashmir, behind the Pirpanjal Range, which reach to 4,500 metres high. The southern flank of these mountains is drained by the Chenab, and commanded by the city and district of Jammu. The north wall of the Vale is formed by the Himalayas themselves. Beyond them lies Ladakh and Baltistan (Little Tibet), then the deep trough of the Indus. Beyond the Indus are the mighty Karakoram Mountains, behind which is the vast plateau of Tibet.

---

2 Not by Muslim Razakars, however, but by their Hindu Marxist foes.

The great Vale of Kashmir, legendary for its beauty and its climate, is drained by the Jhelum, the most westerly of the five rivers of the Punjab. It is the heart of that territory which in its much wider extent embraces Jammu and Ladakh and is called Kashmir. It is said that the Kushan king Kanishka (reigned c.AD 78 to AD 102), whose capital was at Purushapura (Peshawar), founded Srinagar after he had forced his way up the Jhelum and annexed the Vale to his kingdom. Certainly under his reign the Fourth Buddhist Council was held in Srinagar, and the area may be presumed to have been mostly Buddhist. Hinduism later staged a revival under the Pandits, and then Sufis and other Moslems converted the majority in the Vale itself to Islam. During Mughal rule the Vale was celebrated for its beauty, and the famed Shalimar Gardens were created. The Moslems retained many customs not associated with Islam: they sing in the mosque. Later still came the Sikhs, who imposed their rule into the hills from Punjab. To many observers there was a fusion which gave to all Kashmiris of any creed their own identity – the culture of Kashmiriyat. Kashmiris of the Vale have their own language, Kashmiri. The British noted with some distain that they were not a martial people, that they could not recruit good soldiers from them, nor provoke them to fight.

Out of the demise of the Sikh Kingdom, described in Chapter 5, Maharaja Gulab Singh of Jammu added the Vale of Kashmir to his kingdom. The new princely state and its Hindu monarch were then bound by the treaty of 1846 to a tributary relationship with British India, though there was vagueness about its northern and north-western border.

The nationalist struggles in British India did not pass Kashmir by. In the 1930s a local leader, Sheikh Abdullah, was vocal in demands for the Maharaja to devolve power to the peoples' representatives. Since there was no electoral system the extent of his popular support was never measured. It also has to be said that there were Islamic leaders as well, who became pro-Pakistani. Abdullah's programme was at least semi-socialist and secular. He and Nehru shared much of their philosophy in common, and it can be said that for much of his political life Sheikh Abdullah believed that Kashmir could find its own statehood within India, although at other times he was more pro-independence. Though Ladakh and Baltistan are predominantly Buddhist, and Jammu predominantly Hindu, overall the population of the legal entity of Jammu and Kashmir in 1947 was 77 per cent Muslim. Inevitably, as in India, the struggle for power began to emphasise communal symbols, and the Maharaja was seen increasingly as an alien Hindu from Jammu. Nevertheless, by 1947 the then Maharaja, Hari Singh, had conceded none of his constitutional power.

In the crucial months of the Spring of 1947 the Maharaja, a vacillating man, avoided confrontation with Mountbatten and his envoys, preferring to have a 'stomach indisposition'. His Highness could not contemplate subjecting his state to Muslim rule, and he could argue with some justification that Pakistan would be theocratic, not secular, and that his Highness' Sikh, Hindu and Buddhist citizens should not be so subjected. Accession to India on the other hand was also ruled out, as it would provoke a revolt by the Muslim majority. So, doing nothing seemed to suit both his character and his purpose. By 15 August no accession of the state had been announced, and de jure he was now ruler of an independent country. De facto he was too, but destined to be so for a short time only. As Mountbatten had warned the assembled rulers, they would not be able in all circumstances to isolate themselves from the turmoil of communal disturbance and demands for majority rule. In the summer of 1947 communal riots threatened the State with collapse. Embittered Sikh refugees from Punjab moved into Jammu, and meted out revenge on Muslims. Muslim peasants from Jammu and Kashmir fled to Sialkot in Pakistan. The Maharaja's army, predominantly Hindu, was charged with re-imposing discipline, but it failed to curb the communal violence in which the majority of victims were Muslim. In Jammu elements of the Maharaja's army massacred Muslim villagers. As the cycle of violence intensified, in late October 1947 a force of 5,000 Pathans crossed from the North-West Frontier to invade the Vale of Kashmir following the only road, up the Jhelum. These Pathans are from the same stock as those who fought the British on the Afghan border, and more recently the Russians in Afghanistan. India, of course, suspected that the invasion had been engineered by the Pakistanis, but Pakistan maintained that it had been spontaneous and that they had not had the forces in position to prevent it. The tribesmen succeeded in blocking access to the valley and were on the threshold of taking Srinagar itself. Muslims in the Maharaja's army began to defect to the rebels and complete chaos threatened. The Maharaja pleaded with India to send help. It has been asserted that the only thing that stopped the invasion from achieving its aims without more ado was that the tribesmen knew they had the Maharaja bottled up in Kashmir, that no help could reach him, and that they could therefore take their time – which they did by looting in Muzaffarabad, and having what was by all accounts quite a binge.

Even at this stage there seems little evidence that India was planning to grab Kashmir for itself. But there were undoubtedly emotional forces at work. Nehru was deeply attached to the State – it was the ancestral home of his family, and of particular interest within Hinduism, because of the Brahmins of India, the Kashmiri Brahmins are ranked highest.[3] And the high mountains contained the headwaters of the major rivers: here possibly was where the Lord Brahma's foot touched earth, where the mythical Mount Meru was the centre of the world. The Indian Government seemed disposed to help the Maharaja even without accession, and it also took the view that the State's eventual fate must depend upon a plebiscite once peaceful conditions had been re-established. Mountbatten persuaded Nehru that India had no right to intervene until the Maharaja had signed a treaty of accession; although mindful of the rights of the majority it was agreed that the accession should be temporary pending the plebiscite. Hari Singh had fled

---

3 In 1945 after his release from 1041 days of imprisonment and after the Viceroy's Conference in Shimla, Nehru's mind turned to 'mountains and snow covered peaks... I hurried to Kashmir. I did not stay in the valley, but almost immediately started on a trek to the higher regions and passes. For a month I was in Kashmir.' (Nehru, 1962: 602)

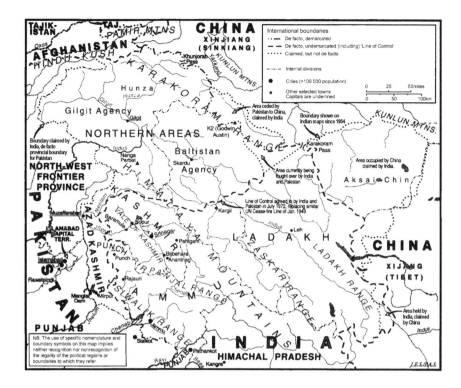

Figure 9.3   The Line of Control in Kashmir

Source: Schwartzberg (1997)

to Jammu, and no-one in New Delhi knew whether or not Srinagar had fallen. In despair he signed the deed, although it is still disputed whether Indian troops were despatched before or after he signed.

Mountbatten had had considerable experience of the use of air power during the Second World War, and there is no doubt that some of this experience was put to use in helping to arrange the one thing the invaders had not considered, an air lift of regular Indian troops into the Vale. Crucially the airfield was secured in the nick of time, and Srinagar itself could now be 'protected'. India had acted alone, and had failed even to keep Pakistan informed until the orders for the airlift had been issued. Clearly the chances of a bi-partisan approach were wrecked. Jinnah ordered the move of Pakistani troops into Kashmir, though he was persuaded to retract, on the basis that the accession of Kashmir to India had been as legal as the initial accession of Junagadh to Pakistan. The two Dominions again seemed on the brink of war, just weeks after pulling back from confrontation over Junagadh. Attempts at reconciliation were ruined by rumours of Muslims perpetrating atrocities on Hindus and Sikhs; and, of course, vice-versa too. Winter set in, making re-supply of Srinagar by land from India almost impossible, and the tribesmen began again to push forward, almost certainly by now abetted by units of the regular Pakistani army in irregular dress. As an alternative to settling the issue by outright war, possibly including an Indian invasion of Pakistani Punjab, the issue was taken by India to the United Nations in New York. The UN resolution of 3 April 1948 called for the establishment of a neutral administration, and the holding of a plebiscite to determine whether the state should accede to Pakistan or India. This UN resolution did not mention the third option – the independence of Kashmir. Both sides rejected the resolution, but did agree to a cease-fire on 1 January 1949. At that time the cease-fire line petered out in mountains, snow and ice above 18,000 feet in the area of the Siachen glacier south of K2, the world's second highest mountain (see Figure 9.3). The original line has since been 'policed' by a UN observer corps, who count the shells fired each way (on a more or less continual basis). The Line has been crossed in hostilities by both sides on many occasions since. But it is still the Line, since 1972 known as the Line of Control (LOC), but with an extra definition. Maps were published in Pakistan showing the LOC extending north-eastwards to the Karakoram pass. In the 1980s the Indians (first to deploy troops here it is thought) and the Pakistanis have fought each other on the Siachen glacier. When hostilities started at high altitude, mountain climbing shops in Europe suddenly sold out of high-altitude climbing gear. Hostilities continue here as elsewhere, although both sides lose more troops to cold and accidents than they do to enemy action. Men cannot serve at these altitudes for more than 3 weeks before bodily deterioration becomes severe and possibly irreversible. Surely this has to be folly of the highest order.

Although there are others, Kashmir remains the principal bone of territorial contention between India and Pakistan to this day, the issue that sours all attempts at normal relations, and which has the capacity to draw both into war with each other. The full plebiscite has never happened, because neither side has withdrawn its forces, so neither side has acknowledged that the peaceful conditions of a reunited State have been re-established suitable for the holding of the plebiscite. Since 1949 the chances of a peaceful solution have been fleeting and few. Such a solution would probably require an act of compromise and concession that no politician on either side has the stature to survive.

Pakistan's case remains that by geography, and by religion, the State should be theirs. Pakistan maintains officially that Kashmir is still independent, and that it is waiting for the formal accession of the whole state. The part of the state now administered by Pakistan is known as Azad (Free) Kashmir in Pakistan, but as Pakistani-occupied Kashmir (POK) in India. It has a population probably above 2.5 million. India's part includes the majority of the population, now more than 8.0 million, and the original capital in Srinagar. Azad Kashmir has equivocal status within Pakistan. The nominally independent government, based at Muzaffarabad, is, in fact, under the control of the Pakistani Ministry of Kashmir Affairs. India's case is equally adamant: that Accession is a fact, and indeed it held a plebiscite (referendum) in its part of the state in 1954 which India claims confirmed the accession.

Sheikh Abdullah became Prime Minister of (Indian) Kashmir, and dominated much of the politics of the next 30 years. But he had a falling out with India, and became banned for while from his own state. By Article 370 of the Indian Constitution, Kashmir has special status, having ceded to India only those powers mentioned in the deed of accession – defence, foreign affairs, and communications. There is no concurrent list shared with the centre, and to emphasise the difference, in all other states there are chief ministers. The title Prime Minister applies only at the Union level in Delhi, and in Kashmir. It became clear that increasingly India saw the Article as temporary, and there has been a creeping annexation. It extended the jurisdiction of the Indian Supreme Court in 1962, extended the functions of the Reserve Bank and the Census, and integrated the electoral systems. The Congress Party of Kashmir has also integrated itself with the Congress Party of India. However, Indian nationals from other states are still not allowed to buy property in Kashmir. A point of real contention for many Kashmiris was the Shimla Accord of 1972 (at the time it was Simla), when the Ceasefire Line was supposed to be transformed into a 'soft' border, which Kashmiris would be able to cross. A weakened Pakistan that had been trounced in the Bangladesh war agreed that Kashmir was a bilateral issue between India and itself, and that the two would work towards a solution. The Accord does not refer to a plebiscite, or the possibility of an independent Kashmir.

For a while in the 1970s and 1980s it seemed as if the status quo might be accepted, and the issue die down. But political drift reaped its own reward. Guerrilla activity in Indian Kashmir started again in 1989, and the crackdown by Indian security forces was insensitive. The Indians claim that militants, many of them international veterans from the Afghan War of the 1980s, have targeted the Kashmiri Brahmin Pandits – and certainly those who were still there (less than 200,000) have fled in fear to Jammu and other cities in India. Since 1989 Kashmir has been living through internal siege, and normal life has been paralysed. The tourist industry has collapsed, with dire economic impacts. There are many different groups involved with different aims: Kashmiris seeking unification with Pakistan, Kashmiris seeking a reunified and independent Kashmir, irregular Islamic militants from outside Kashmir, and sometimes Pakistani regulars, usually disguised as militants. Pakistan usually denies any involvement, and claims that infiltration across the border is beyond its control. But it is clear that regular forces were committed across the LoC in 1999 at Kargil, when the fighting threatened to intensify to a full blown war. In 2002 the confrontation reached fever pitch, many analysts expected a full-scale war which had the potential of escalation to a nuclear exchange, and western countries advised their nationals to leave both Pakistan and India immediately.

The cost of all of this to the people of Kashmir in trauma, injury and death has been horrific. Perhaps 50,000 have died in the last 10 years. Some commentators believe the total could be as high as 100,000 (Hewitt, 2001). More is said later about the cost to the Indian and Pakistani economies and polities, and to their international relations. The significance of the State in terms of its natural resources

– for example, the rivers which flow from these mountains – and the significance of the struggle in other international spheres such as Indo-Chinese relations, or America's 2002 War Against Terror, is something to which we will also return below. If ever there was an untidy result from Mountbatten's speedy tearing along the dotted line, this is it.

Gilgit and the Northern Territories

Far in the north, where Afghanistan, China and greater India meet, the Indus receives two northern tributaries to its right bank. These two rivers, the Gilgit and the Hunza, occupy deep trenches through the mountains, and give their names to the two districts. Both lead to passes, the Mintaka and the Khunjerab, into Central Asia; and indeed it is through Gilgit and Hunza that the Karakoram Highway connects Pakistan with China.

This frontier area came nominally under the control of the Maharaja of Jammu and Kashmir, but the British began to worry in the late 19th century whether the Maharaja had either the means or the inclination to be bothered with 'protecting' the passes. They tried to garrison a Gilgit Agency with the agreement of the Maharaja, but by the end of the century had withdrawn. In the 20th century these anxieties resurfaced, and in 1935 they signed a lease with the Maharaja to administer the territories directly for the next sixty years. After Pakistan assumed sovereignty over the North-West frontier Province and 'protected' Azad Kashmir from India, the status of these territories, also under Pakistani protection, was called into question. Are they or are they not part of the Kashmir issue to be settled? The Pakistani answer is that they are not. Pakistan has incorporated them within its own administration, and the sixtieth anniversary of their lease passed without formal action by the Pakistani Government. But for some Kashmiris, this is another issue with which to muddy the waters. It is not one that usually causes much trouble in Delhi, since the Indian government would almost certainly accept a united Kashmir without Gilgit, acknowledging Pakistan's highway to China. But that does not guarantee that at some future stage the issue might not be raised again as a card to be played at the negotiating table – thence to become part of some greater entanglement.

The Human Flotsam

The new borders had not completely solved the minorities problem. India was born with 11 per cent of its population Muslim, and Pakistan overall with 13 per cent of its population non-Muslim. In East Pakistan (East Bengal) the population was 30 per cent non-Muslim. These are the dry figures: what it meant in reality was something far more horrific. Many of those people who were left in small minority pockets on the 'wrong' side of the new borders found themselves subjected to every conceivable act of violence and degradation. Murder, pillage, arson, rape –

all of these words can fit accurately what happened. Many, particularly those near the borders, abandoned everything and trekked towards their new 'homeland'. In Punjab trains overladen with refugees carried some, trucks and bullock carts others, while many simply made long lines of despairing walkers trailing as far as the eye could see. One caravan was estimated to comprise 800,000 people. All were subject to attack. Trains arrived with their whole passenger complement silent – unable to speak because all were decapitated or stabbed to death. A Punjab Boundary Force of 50,000 men had been established to handle some expected trouble; but the scale of violence was so colossal that it struggled to achieve even a minimum of control, and then it too collapsed as it broke down into its pro-Pakistan and pro-Indian elements. The Sikhs in particular were prominent in some of the worst attacks on Muslims – this was their revenge for their own inability earlier to make political accord with the Muslims to avoid the Partition. A common estimate is that at least 250,000 people died.[4] In the first five months of Independence 12 million people fled both in the Punjab and in Bengal – one of the biggest population movements in history. (By 1950 the total migrants had risen to 15m.) This was ethnic cleansing on a scale that dwarfs the more recent events in the Balkans in the 1990s.

In Bengal, where 16 August 1946 had left many scores unsettled and where the worst trouble was expected, there was little violence. This 'miracle' was as a result of the moral power of Mahatma Gandhi's leadership – the one-man boundary force as Mountbatten described him. Gandhi had gone to Calcutta and threatened to fast unto death unless the communal and gang leaders agreed to hold the peace. As a demonstration of moral persuasion this must have been one of the most powerful ever.

This mass movement left marks on the geography of both countries. In Punjab so many people had moved that allocating new refugees to abandoned land became in effect a way of redistribution and land consolidation. It was part of the modernisation of agriculture that has continued since. In West Punjab the Muslim peasants found that they had been relieved of debt, as Hindu money lenders had fled. In the big cities – Delhi was the largest urban focus for refugees – urban populations suddenly swelled, and the new governments had major problems in feeding the crowds, and keeping them healthy. In Delhi in 1951 24 per cent of the population were refugees from West Pakistan. In Calcutta, unlike in Punjab, the flows each way did not balance, and 2.1 million people converged on the city from East Bengal, against 0.7 million going the other way. It was the turning point that pushed the great former imperial capital on the downward slope to its current symbolic role as the epitome urban squalor. The net effect had been to increase the population of an already over-settled State (Province) by 12 per cent.

The streams were also highly selective in other ways. For the Muslim poor near the Pakistan border, it was perhaps a realistic option to cross to the new homeland. For those living far from the borders in eastern United Province or Bihar, there

---

4 J. Schwartzberg estimates c. 400,000 (personal communication).

was no comparable option. But for the richer urban Muslims of these areas, there was the chance of organising transport and fleeing. Many of the professional urban Muslims from all over India converged on Karachi. It became the city of the new migrants – the Mohajirs as they are known, the people of the great pilgrimage. Jinnah, the greatest of the Mohajirs, seated his new government in Karachi, and the immigrants became a major force in government and the armed services. To the native Sindhis they have remained mohajirs and have been integrated little within provincial Sindhi society. These differences have been entrenched by hostile political relations which have erupted into armed conflict and terrorism, particularly in Karachi. Since 1995/96 it has been periodically completely paralysed by violence and strikes. In the east, one movement of Muslims from Bihar settled in East Pakistan, to find themselves strangers alongside Bengali 'brothers', and after another political break in the future, when Bangladesh seceded, they would find themselves completely rejected, reduced again to refugees in their adopted Muslim homeland.

Selective migration affected the running of every kind of industry and service. The railways lost Muslims to Pakistan, and gained non-Muslims the other way. But the Muslims were mostly the drivers and the metal-working artisans, while the non-Muslims were mostly clerical. Train movements suffered, and coal production and distribution stagnated. The members of the civil service could opt for Pakistan or India, and as they did so and as the British officers left, so the skills remaining to Pakistan and India were reduced and unbalanced. Regiments split, and brother officers parted company, many to take up armed conflict against each other in future wars.

The Divided Inheritance

Physical and financial assets had to be split. From June 1947 onwards a Muslim and a Hindu representative had headed the teams who divided the spoils. Pakistan would get 17.5 per cent of government cash reserves and sterling balances, India the rest, and in return Pakistan would cover 17.5 per cent of the outstanding government debt. The physical assets would be divided 20 per cent to 80 per cent. This meant, for example, the Food and Agriculture Department's 425 clerks' tables would be divided 340 to India and 85 to Pakistan, the 85 large tables divided 68 to India and 17 to Pakistan – and so on for the 850 chairs, 85 officers' chairs, 50 hat pegs, 6 hat pegs with mirrors, 600 inkstands, etc. that were also listed. Libraries were divided; in some cases alternate volumes of encyclopaedias went to the two inheritors. It is even reputed that English dictionaries were torn down the middle at the letter K.

The division of other responsibilities and liabilities also had to occur over the next few years with companies, such as life assurance companies, who had clients in both countries, and assets and head-offices in one. And in tax revenue which was collected at the point of manufacture – such as excise on matches levied at source where they were produced in India. They were sold both in India and Pakistan and thus the tax was raised indirectly in both countries.

Finally, the division automatically placed India and Pakistan in trading relationships with each other. Both became each other's largest trading partner overnight – and yet within a few years they had completely embargoed their trade with each other. They were natural trading partners. Pakistan produced most of the long staple cotton: India had the mills. East Bengal produced all the best jute: the mills were at Calcutta. India had been born into food deficit: West Pakistan was in surplus. East Bengal was born into food deficit too.

The complete breakdown of this natural trading pattern caused enormous hardship to both countries and is something which we will touch on again below. Obviously it seems unnecessary, but perhaps it was a price which Pakistan had to pay, for without economic freedom and independence it would never achieve political freedom and independence. But in becoming economically independent of India, Pakistan became dependent on other outside nations.

Concluding Remarks

The pursuit of self-rule – swa-raj – had revealed more 'selves' than Congress had wanted to admit. However, in admitting to the reality of the Muslim League's mass following, Congress inherited a strong centre in the new India, still the largest self-governing unit the sub-continent has ever seen. The real 'surprise' at the end of the long saga was the other 'reality', of applying the logic of Partition at every scale. Pakistan got its much sought-after sovereignty, for a rather moth-eaten territory, at the price of the vivisection of Punjab and Bengal, and the creation of a bifurcated state whose two wings were separated by a thousand miles of Indian territory, and also separated by language and script, economically unrelated to each other, and linked only by Islam.

Chapter 10
From Two to Three: The Birth of Bangladesh

Introduction

Both India and Pakistan have had to struggle with the problems of regionalism and separatism since independence. In India this has frequently meant the realignment and redefinition of constituent states. In the case of Pakistan it has meant a variety of experiments with the redefinition of, and even the abolition of, provinces. It has even meant the loss of East Pakistan, which has become the sovereign state of Bangladesh. In trying to write about all of these I have experimented with keeping all of these events in one chapter, treating the case of the independence of Bangladesh simply as a special case of the problems of accommodating regional forces. In the end, for clarity and convenience I treat it in this chapter as a special case on its own. In making the story a separate chapter I have also had to decide whether it should come before or after the other stories about regionalism: I have decided it should come first, as the most traumatic event in Pakistan since 1947.

The story of the birth of a nation, in this case Bangladesh, is usually treated as a singular story – a linear narrative of that territory which achieves its sovereignty – and that would indeed be a simpler way of explaining the emergence of the new state. But, Bangladesh is just one of the many regions of South Asia, and the main reason to single out its story as opposed to that of Maharashtra or Sindh or Tamil Nadu is that it has become something which they have not, an independent sovereignty, although there is no a priori argument why it should have a stronger claim to such a status than the other provinces/states just named. Why it happened is a result of the sequential order of specific events. In 1947 East Bengal sought self-determination and independence from Hindu over-lordship. It joined the Pakistan movement and 'got' Pakistan, but only by failing to get the former – independence – as, within Pakistan, East Bengal was always the junior and dominated partner. So the next step logically would be to arrogate from Pakistan its share of the sovereignty that had been granted in 1947, something it achieved by embroiling the whole of Bengal, and therefore India too, in strife in 1971.

Unequal Development in Pakistan

In 1947 India acquired 81 per cent of the Indian Empire's population, but only 72 per cent of its area. This indicates that Pakistan in aggregate was less densely settled, but the aggregate masked the huge inequality between East and West. The East had in fact the majority of the population – 57 per cent, but only 14 per cent

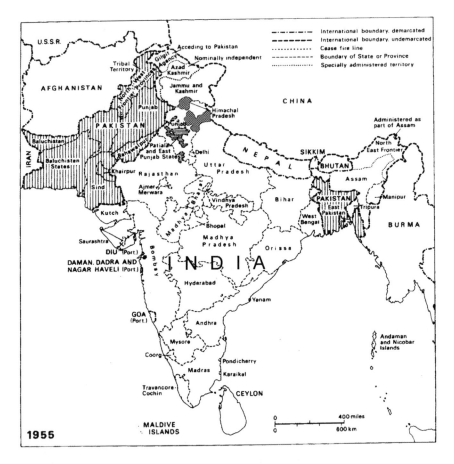

Figure 10.1   Territorial Division and the Successor States, c. 1955

Source: Schwartzberg (1992)

of the area. Its population density was 777 persons per square mile, compared with the West's 222 (and India's 276). Pakistan was less urban than India, but West Pakistan was three times more urban than East, which had less than 6 per cent of its population in urban areas. In 1951 83 per cent of the workforce of East Pakistan was in agriculture, 65 per cent of the West's workforce was in the same sector. In the East few Bengalis had been recruited by the British into the Indian Army; but in the West there were many who had been recruited and who had become the nucleus of the new Pakistani Army.

These figures also reflect the infrastructure: in 1960, 66 per cent of the villages of the West were within five miles of a pukka (metalled) road – whereas in the East only 25 per cent were. In the East water transport was more important – but only 17 per cent of the villages were within five miles of a 'steamer' (regular power boat) landing. The literacy rate in the West was much higher too. This of course was reflected in government service figures; in 1960 of the 2779 1st class gazetted officers of the Government, 87 per cent were of Western (or Western Mohajir) origin. In short, the East, shorn of Calcutta, was indeed a rural backwater, as Jinnah had feared.

Overwhelmingly, the skilled immigrants from India had gone to Karachi, which was the new seat of Government for Pakistan (until 1956 when the plans were announced for the new capital city of Islamabad, also in West Pakistan near Rawalpindi in Punjab). In 1960, 70 per cent of the professions, 72 per cent of the managers, and 64 per cent of the skilled workers of Karachi had their origins in India.

Pakistani economic policy could have been to maintain the parity of the Pakistani rupee with the Indian, and to continue to trade as before. But this is an unrealistic view of the workings of political sovereignty, which would soon see policies that were divergent from India's, in order to foster industrial development and economic independence. In Pakistan's case it was evident that it exported one major crop to the rest of the world, the proceeds from which could be used to raise the first revenue for an industrialisation programme. This crop was the jute of East Bengal. It was a crop for which demand was presumed to be fairly inelastic- that is to say falling prices would not increase world consumption, and East Bengal was overwhelmingly the world's largest producer in 1947. And, if capital were raised from that sector, where should it be invested? From the point of view of a country struggling to maximise its growth rate from inauspicious beginnings, it would be foolish not to place it where it would have the highest incremental growth impact; that is to say where the infrastructure would best support it, where skills might be highest, and where perhaps the internal market might be strongest.

In pursuing such a logic the government would, of course, also guide and protect its infant industries, and use such fiscal and monetary policies as thought appropriate to induce such changes. These in turn meant the creation of a bureaucracy and the necessary licences and permits, which inevitably if unintentionally favoured those with best access to the information and personnel necessary to 'play' the new system. In other words, Karachi in the West became the new industrial growth pole.

Coupland in 1943 had already sketched out the beginning of such a story by considering the jute industry in a divided future India. Under the Act of 1945 export revenues on jute were appropriated by the Central Government, but 62.5 per cent of the revenues were re-assigned to the Provinces generating them.

Most of the jute is grown in Eastern Bengal, and North-Eastern India, without Calcutta, would be able to levy duty on raw jute exported from Chittagong, but its diversion thither within the existing system of communications would not be easy. Moreover, 3/5 of the duty in 1938/39 were levied on jute manufactures, and this industry is located in Calcutta. For North-East India to levy export-duty on raw jute crossing the frontier to Calcutta would be dangerous for two reasons. First the Hindu State would probably levy an export-duty on its manufactures and, since the price in world markets would have to compete with substitutes, this second charge would inevitably reduce the cultivator's profits almost, if not quite, to nil. Secondly, a duty on raw jute would be a direct invitation to the Hindu State to extend its own area of jute cultivation and so threaten North-East India with the loss of its nearest and largest market. (Coupland Pt III, 1943: 97)

An independent North-East India (Bang-i-Islam) might have heeded the caution, but Karachi did not. Shortly after independence Pakistan placed an export duty on jute. India responded by banning the import of jute, and restricting coal exports to East Bengal as well. Chittagong, with an annual capacity to handle 0.5 million bales of jute only, became overloaded, and most of the crop of 5 million bales was stored in inadequate places. India began to expand her own jute production, which had doubled by 1960.

In 1949 the UK pound Sterling was devalued against the US dollar, by a significant 30 per cent. (This was in the post-war period of fixed exchange rates.) The whole of the Sterling area (the majority of the Commonwealth/Empire at the time), including India, devalued in step, principally, in India's case because of the sterling balances held in London. The whole area followed Britain with one exception – Pakistan, which felt that the market for jute was inelastic and that demand would not increase with lower prices, merely it would be paid less. But the effect was to make Pakistani exports to India 30 per cent more expensive, and simultaneously to reduce the relative value of Indian exports to Pakistan. India was stung by the move; partly her political pride was piqued by the independence which Pakistan had shown, but also this was a significant change for the economy of West Bengal. The result was that the trade war which had already started now became a complete trade embargo. There is no doubt which of the two regions of Pakistan suffered most; East Bengal entered a period of economic difficulty which provoked serious unrest.

There was also a second effect on the relations between East and West Pakistan which starts at this period and which became exacerbated over the next two decades. This is an effect which derives from the over-valuation of the currency generally with regard to overseas trade, and not one which relates simply to India (with which trade had virtually ceased anyway). Let us suppose that a true international value for the rupee might have been Pk Rs 7 = US$ 1. The official exchange rate for exporters in the early 1950s was Pk Rs 3 = US$ 1. Let us further suppose that the dollar price represents the true international value of goods exported. Then a Pakistani exporter selling $1 of jute would receive only Rs 3 instead of the Rs 7 which he would have expected if the exchange rate was free. In effect the exporter lost, or was taxed, more than half his true export earnings.

If the same man who exported $1 worth of exports was then able at the same exchange rate to ask for foreign currency to buy imports, he would have been able to buy $1 for Rs 3 and not Rs 7, thereby saving on imports what he had lost on exports, and so would have ended up equal. But the problem was that the exporters and the importers were not the same, either as people, or as regions.

Because an overvalued exchange rate makes imports artificially cheap, there is excess demand. To control this excess demand tariff and quota barriers are used to protect domestic producers supplying goods to the domestic market. The benefit of low import prices for imports is directed at capital goods for new industries, and to their raw and semi-finished materials, by selective tariff barriers. The industrialists and new entrepreneurs were in Karachi; partly because that was where the talent mostly lay, partly because it was the largest port of Pakistan, and partly because with any complex bureaucratic system quick access to the central government was essential.

In these circumstances of controlled imports of consumer goods (even basic ones), and of a major export crop of jute in the East wing, the trading pattern for the two wings (see Figure 10.2) that emerged was one in which the East was in surplus with the Rest of the World, whereas the West wing was in deficit. The surplus of the East and the deficit of the West were then matched by the surplus internal movement of consumer goods from the West wing to the East. The total effect of this system was that, on average, the East was systematically being deprived of much of the true value of its external exports, while the West was systematically underpaying for the true value of its imports. This represented an undisclosed and unaccounted drain from the East, the poorer part, to the West, the richer part.

Such transfers in developing countries are common, from agriculture, seen as the only source of capital, to new industries, seen as a hope for economic and job growth. Sometimes, as in the case of North East Brazil in relation to the South East, there are strong regional components to the inter-sectoral shift of resources. But in Pakistan the regional component was more clearly demonstrated than ever before. And to add to the grievances of the East, the prices they now paid for the goods imported from the West were higher than they would have paid internationally, and the quality was lower.

Much of this account would be denied by West Pakistanis; they would point to projects undertaken in the East, and to the fact that the East too had a captive market in the West – for example in tea from Sylhet which dominated Pakistan's internal market. It is also said that in the 1950s the terms of trade world-wide turned against agriculture and that, therefore, there would have been non-policy causes for such resource transfers. It is true that the gross figure is much more complex, but the net effect was as I have described it. It is also said that by 1970 the West was the major exporter, in cotton manufactures and new engineering goods; but given the investment that the West received it would have been a major indictment if the relative positions of the two wings had not changed.

In nearly every other way the inequality was exacerbated. The first two five-year plans (1955–65) allocated more money to the West than to the East – and this was even without including the mammoth Indus Basin Project (dealt with below). The argument was in a sense sound, in a Thatcherite kind of way. Invest money where it would show the highest rate of return: wealth generation must come before wealth distribution. The income levels in the West were 30 per cent higher than in the East, and rising relatively. Here in the West was therefore

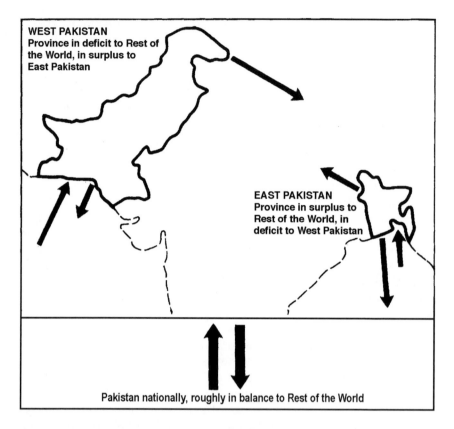

Figure 10.2   Trade Flows between West and East Pakistan and the Rest of the World

the greater internal market too. The West was urbanising faster round its new factory towns: the market was more accessible. The Planners acknowledged dissent in the East, and in the 3rd Plan of 1965 a public statement was made of the intent to reduce regional disparity and to put more resources into the East. Even by the wildly optimistic statements of that Plan, however, the timescale set for the elimination of the disparities was 20 years; and any normal cynic would have known that if it were to happen at all it would take much longer than that. The final twist in respect to the regional imbalance was the inability of the Easterners to migrate to the growth centres of the West. There were no formal internal barriers to equilibrating migration; but separated by a minimum 10 day sea journey and by language, a poor Easterner would have to save at least a year's income just to voyage West in the hope that he might then get a job. By Independence in 1972 there were just 0.5 million Bengalis in the West, out of a population in Bengal of about 70 million.

Table 10.1   West and East Pakistan: Inequality in Per Caput Gross Regional Product 1949–1970 ($1959–60 prices)

	West	*East*	*Ratio*
1949–50	345	287	1.20
1959–60	355	269	1.32
1969–70	504	314	1.61

Source: Ahmad, 1972, p. 34

Language and Representation

In 1948 the Constituent Assembly ordained that Urdu, together with English, would be the lingua franca of Pakistan and would be used in the national assembly. Bengali was excluded as a language from the daily tokens of national identity – the coins, currency notes, postage stamps, of the new Pakistan. In 1952 Prime Minister Khwaja Nazimuddin (a Bengali) proclaimed while in Dacca that Urdu would be the only national language. In the riots that followed the announcement several Bengali students were shot dead by the police. Now in independent Bangladesh one frequently sees throughout the country replicas of an abstract sculpture, looking something like three gallows, which is the national monument at the site in Dacca where the students were killed, the first martyrs of the independence struggle. Language was the first issue to seriously threaten the integrity of the new state. During the 1954 Provincial elections in Bengal several opposition parties formed the coalition United Front, who campaigned on regional issues. Their demands for regional autonomy seriously threatened the Central Government. It was a repeat of the earlier demands which the Muslim League had once made against Congress: for a federation that gave full provincial autonomy preserving only defence, foreign affairs, and currency for the Centre. It also demanded recognition of Bengali as a national language. Given that more Pakistanis spoke Bengali than any other language (such as Punjabi or Sindhi), this was not unreasonable at face value. To the Central government these demands must have seemed as something like a stab in the back, while it was pre-occupied with issues deemed of national importance, but in reality concerning the West wing, over Kashmir and the Indus Rivers. The Muslim League still had the organisation to campaign in Bengal, pleading for the consolidation of Pakistan. But the verdict of the people was clear. The United Front won 97 per cent of the vote.

After forming his ministry, the new leader of Bengal Fazlal Huq then went to India, to West Bengal, and joined in talks aimed at some sort of re-emergent understanding with India, a move which unnerved the Central Government. Within weeks the Central Government dismissed the Provincial government for taking an anti-national attitude. It could not completely ignore Bengali sentiment, however, and under the 1956 Islamic Constitution, when East Bengal was officially named East Pakistan for the first time, the regional nature of Pakistan was recognised in a slightly backhanded manner. The constitution recognised that the central government should reflect the regional division, but did so by awarding an equal 155 seats to both wings, despite the fact that the East was the more populous part. It also recognised Bengali as a national language.

The Military Cost of Pakistan

The confrontation with India over Kashmir immediately after independence persuaded the Pakistani government that defence must have the highest priority. Despite the poverty and urgent development needs of the new country, by 1950 as much as 80 per cent of public expenditure went to defence – and even in the 1960s and 70s it was rarely less than 40 per cent. This meant that the military were being built into Pakistan's political power structure. The armed forces became big enough to be not just the recipient of funds allocated, but to be a determinant of how much they should have in the first place. Throughout the period of union, 80 per cent of Pakistan's armed forces were in the West, with the benefits too for the Western economy of their expenditure. The defence of the East, the Westerners argued, lay in the deterrence provided by the West's ability to march on Delhi; it was claimed that it would only take a few days for Pakistan's tanks to be parked outside the Red Fort.

The 1965 war with India which Ayub Khan had engineered was viewed askance from East Pakistan. In the Thar Desert in March tanks were engaged, with Pakistan possibly making small gains. Emboldened by this apparent success, in September Ayub launched a war against Indian forces in Kashmir, which then spilt over to fighting in the Punjab too. East Pakistan was technically at war with India too, but was essentially left defenceless. But the war in the Punjab, if anything, went India's way, and the deterrent threat against Delhi was exposed as a myth. What saved the East was India's astute perception that it was not fighting the whole of Pakistan, but merely the Western military clique, although there are also claims that China threatened intervention if Bengal were touched.

The next year the Awami (People's) League under the leadership of Sheikh Mujib-ur Rahman, became openly more vociferous in its demands for regional autonomy. His six-point programme was an open demand for secession in all but name. He wanted a completely independent economy and currency, with co-operation between the wings based on their self-interest and not on central government dictats and, even although defence was still to be a central matter, the separate wings would have been able to raise independently their own militia for defence purposes; again therefore placing the central government's role in defence more on the lines of a defence pact. To cap it all one of the demands insisted that representation in the central government would be on the basis of population- which would ensure the East's domination. Given that the power behind the existing military government was Western, the response was inevitable. In 1968 Mujib was imprisoned on trumped-up charges, becoming something of a Bengali national martyr/hero.

After Ayub's fall and the installation of a caretaker government under Yahya Khan, Mujib was released from prison to participate in the subsequent elections of late 1970, which were free and democratic. His Awami League won 158 out of 160 seats in East Bengal, endorsing his claims for regional autonomy, the Six Points which almost amounted to secession. His Awami League became the biggest single group in Pakistani politics. In the West wing, Zulfikar Ali Bhutto led his populist Pakistan People's Party to a dominant victory. It thus became the task of these two men to negotiate the new constitution.

Bhutto and Mujib set out to negotiate the terms of a new constitution; but given the obduracy of the former and the proven political strength of the latter, a breakdown was inevitable. On 1 March, in response to the crisis, Yahya announced the postponement of the meeting of the National Assembly. In Bengal, Mujib called for non-violent non-co-operation and a general strike. What happened next is a matter of dispute; but there is no doubt that some extremely nasty incidents occurred as Awami League supporters set upon factories owned by West Pakistanis and their unfortunate workers. Non-Bengalis were murdered in their thousands, and so too were Bengalis suspected of pro-Pakistan sympathies. Mujib proclaimed the Province under his control, and the Awami League started collecting taxes instead of the Government. The West Pakistani government started the difficult task of building up its forces in the East (difficult since India had suspended over-flights after a highly publicised hijacking incident) and Yahya nominated a new hard-man as commander in the East, General Tikka Khan. On 25 March the army came out of its barracks to re-impose order, by force. On 26 March a clandestine radio staion in Bengal announced the secession of the independent state of Bangladesh. That date has since been taken by Bangladesh to be the date of its independence, although the worst of the killing was yet to come.

In all the years of Pakistani union, the demands made by East Pakistanis were for some kind of recognition as equals in the new homeland. Having lost many of their own Bengali administrators and businessmen to India in 1947, the Eastern wing had become almost like a colony to the West Pakistanis. It was above all Punjabi administrators and officers who ran the province, and West Pakistani businessmen who set up the new jute mills. Their attitude towards their Bengali compatriots was patronising, often arrogant, being insufficiently aware of why the province was the way that it was. It was not simply because Bengalis were inherently ignorant country bumpkins. The Bengalis for their part had something of which they could be deeply proud; their cultural heritage, their poets and orators. Not much of this was immediately accessible to Urdu-speaking civil servants, serving time in this backwater. It was easy to remain detached and above such provincialism – to become in fact a replica of the expatriate administration that once the British had run. The recognition which the Bengalis finally received was the systematic genocide of Bengalis, their intellectuals in particular, by the Pakistani army. Mass executions and mass graves became something of a commonplace. Not a few outside observers mourned the passing of the ideal of Pakistan, but what kind of ideal did this action represent?

The South Asian Roots of Bangladesh

Since 1947 there has been a continual efflux of non-Muslims from East Bengal, the subject of intermittent harassment during times of tension. In 1947 East Bengal was 30 per cent non-Muslim; in 1961 20 per cent; by 1968 10 per cent. Attempts were made by India to control the scale and speed of migration, which did of course cause problems in West Bengal and Assam, the principal receiving areas. (Although many of the Bengalis in Assam are actually Muslims in search of land to till.)

The disorder in East Pakistan in 1971 was of a different order of magnitude, and inevitably it could not be contained within East Pakistan alone. Great streams of refugees trekked into India, and with the refugees went those young men determined to train and fight as guerrillas. India assisted in the setting up of the training camps for the Mukti Bahini (freedom fighters), while it also struggled to attend to the refugee problem. By India's account the final total was 10 million, by the account of others perhaps as few as 1 million – but even that doubtfully small figure still represents a huge influx of refugees for any country to handle, and particularly when they arrive in one of the most densely settled and poorest parts of the country. In the beginning of December 1971, full-scale war erupted between India and Pakistan, and for the first time it was fought both in the East and the West. In the West, Pakistan again faced defeat. In the East the Indian forces and the Mukti Bahini (Bangladesh guerrillas) routed the Pakistanis in Bengal in two weeks. On 10 January 1972 Sheikh Mujib returned from imprisonment and exile to claim leadership of the sovereign state of Bangladesh, and, remarkably quickly, by 12 March, Indian forces had retired back within their own borders.

Inevitably the hostilities were part of a wider international picture – the Americans, Russians, Chinese, and even the British being involved in the international diplomatic manoeuvring around it. (These are matters touched on in Chapter 13.) For their part, the Pakistanis believed the international community had let them down, and that India had shown its true colours – and that it had dismembered Pakistan as but one part of its long-term policy of undoing the partition of 1947. Almost certainly Bangladesh would have been born, even without Indian direct intervention, although it might have been a much longer and bloodier business. At the least, indirect intervention by India was inevitable, given the refugee problem thrust upon it – in the same way that, later, Pakistan could never have been uninvolved at all in the Afghan war in the 1980s, or the fate of the Taleban in 2001/2002. The result has been to create a State in Bengal which is to some extent overshadowed by India. The overwhelming size and wealth of India compared with Bangladesh, and its position upstream on the delta rivers, are enough to ensure that – but India has shown no inclination to pursue a policy which would directly lead to re-integration. For the time being, there are indeed three independent sovereign States that have emerged from the legacy of the Raj.

Concluding Remarks

The forces of regionalism that caused the secession of Bangladesh are not greatly different from those with which India has itself struggled since 1947 – with the delineation of new internal states and boundaries to satisfy Tamil or Gujarati demands etc. These are demands for self-determination, even if within some federating system. The irony of history is that the forces of self-determination which were heightened in the Independence struggle of the early 20th century should have been so easily hijacked by the Pakistan movement in the 1940s. Muslim Bengal's dislike of Hindu Bengali domination was always and simply just that: it was not that Muslim Bengal's love for Muslim Punjab and Sindh led to the rejection of Hindu Bengal. From this perspective Curzon's partition of 1905 can be seen in a different light. It had been popular with the Muslim East Bengal, but deeply unpopular with Hindus in West Bengal who felt their dominion threatened and who were instrumental in its reversal. 1947 represented a chance to repeat 1905, but this time the East Bengalis knew they could only succeed with allies: their allies needed only to be temporary, partners in a marriage of convenience.

The price paid was horrifically high. After the year of civil war and genocide from March 1971, when as many as ten million people fled their homes, when villages were laid waste and fields left unfarmed, a major famine was predicted. It did not happen. Subsequent calculations indicate that the reason it did not occur was because the shortfall in food had already been met by the premature violent deaths of at least 1 million people. The Bangladesh flag, originally a red map of Bangladesh on a field of green, is now a red circle on a field of green: red for the blood of 1971 on the green of the paddy fields and the green of Islam.

Chapter 11

Raj and Swaraj: Regionalism and Integration in the Successor States

Introduction

The two (later three) sovereign states of India and Pakistan had been born – the struggles of the last century (from the Mutiny/First War of Independence to Independence itself) had been resolved at the grand political level, but the new states had yet to show themselves capable of their own cohesive integration and survival while simultaneously functioning through representative democratic constitutions. If they could achieve this, then this would be something entirely new in the history of South Asia, and the Partition of 1947, accepted at the last moment as the only means of achieving this goal, would possibly have been a price worth paying. In India, Nehru and the Central Government took over a well-functioning system of bureaucracy, and Nehru's national following and pre-eminence over any other Congress politician – whose bases were more often regional than national – was some sort of guarantee of national integrity in the short term at least – much as the personal standing of Tito (an ally of Nehru) ensured Yugoslavia's survival for some decades after World War II. India also had development strategies and plans, a result of debates about economic management that had been conducted in Congress for some time before independence. Pakistan had fragmented bits of administration and services – the severed ends of many tentacles – which had to be linked together. It also had to find out what its purposes and strategies were. After all, the aim of the Pakistan movement within the Muslim League had now been achieved. The idea of Pakistan had attracted atheists, communists, poets and soldiers as much as the Muslim clergy. Now, it was necessary to decide if it was to be theocratic or secular, socialist or capitalist.

Both India and Pakistan had to contend with the problem of the integration of the Princely States – most of which had had no internal experience of representative government. India was in a stronger and more favourable position with respect to the rulers than was Pakistan – which in 1946 and the first half of 1947 had been a 'dream state' from which acceding Princes could extract much more favourable treaty terms than they would have done from India. Both India and Pakistan had also to contend with the problem of regional languages – so important to the identity of regional communities; but in India this was a problem with so many regional components that it was difficult to see a single major national split on the issue (though perhaps the Indo-European/Dravidian divide between North and South could conceivably cause a split if very poorly handled). In Pakistan the regional divide between West and East (the Indus Valley versus the Bengal delta) was not only the most significant divide within the new polity, it coincided exactly with a linguistic divide, exacerbated by the use of the Persian/Arabic script in the West and Bengali's own script (related to Hindi's Devanagari in part) in the East. This new Pakistan in two 'wings' a thousand miles apart was described as 'the Camel and the Ox in one yoke' (Tayyeb, 1966): though perhaps the better image might be the Camel and the Water Buffalo. The last chapter detailed how the yoke was finally broken.

The situation for Pakistan thus started as much more difficult, and soon became almost impossible. Mohammed Ali Jinnah, popularly known as the Qaid-i-Azam ('mighty-ruler'), had the same sort of standing amongst his new compatriots as Nehru did with his. His vision, now forgotten or twisted by modern commentators, was for a secular state. What he wanted to defend was not Islam, but Indian Muslim cultures. But he died of tuberculosis in 1948 less than a year after Independence, and his successor, the most significant and able of his close followers, Liaquat Ali Khan, was assassinated in 1950. It was left to a succession of weaker politicians to define what Pakistan was for.

The Integration of the Princely States

India

In the Partition, India had lost 365,000 square miles of British Indian territory, and 81.5 million people. By absorbing the Princely States, it gained 500,000 square miles, and 86.5 million people. The new Dominion (soon to be a Republic) of India had an area as great as British India, despite the loss of most of Punjab, most of Bengal, of Sind, and much else. Thus it was much more compact and coherent geographically, provided that the system of government was similar throughout the territory so formed. In India's case the process of integrating the Princely States ran fairly smoothly, because Congress had existed for longer as a political force than the Muslim League, had inherited the greater part of old India and its central seats of power, and, importantly, because it had a clearer idea of where it was going. It would have no truck with Princes who wished to retain autocratic internal rule, and the various deeds of accession all provided for the adoption of representative government in some form or other. Concessions were indeed given to the Princes, most of whom retained their titles and were given a privy purse, something which stuck in the throats of a socialist government, but which accorded, as Gandhi had insisted, with honouring the treaties drawn up. Some of the very smallest states were simply absorbed into adjacent Provinces. Some were grouped together in new Unions, such as in Saurashtra (the Kathiawar peninsula in Gujarat). Some retained their identity as separate geographical entities, such as Kashmir, Hyderabad and Mysore. In the Unions the Princes elected a Rajpramukh, one of their number who effectively became the State Governor, holding an office in which he exercised power only on the advice of the cabinet, now composed of elected representatives. In the large single states the ex-ruler also became a Rajpramukh. But the Constitution of 1950, which listed the former British Provinces as Group A States, and the former Princely States and the larger State Unions as Group B States, gave the Indian President the same powers of dismissal of Governors in Group A as of Rajpramukhs in Group B. The distinction had become something of a sop to history, and to local sentiment, which could not be completely ignored. The majority of Princes simply retired from public life, with their purses and their other privileges intact (such as preferential access to import-restricted goods). But there was still a last twist to the tail. In 1967 Congress announced it would abolish the privy purses. The resulting legal battles precipitated a general election in 1971 in which Mrs Gandhi had to seek a sufficient majority to make an amendment to the Constitution. She succeeded, and the Princes passed officially into history. But, as in Republican France the Count of Paris still uses his title and some of the public still respond, so in India some of the Maharajas are known still as that, and still have residual wealth and status in society.

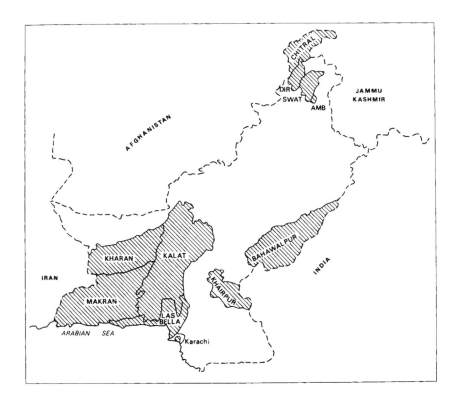

Figure 11.1   Princely States in Pakistan c. 1950

Source: Wilcox (1963)

Pakistan

In Pakistan it was not quite so simple. Jinnah had, in a sense, threatened Congress before Independence by wooing the Princes with better terms than India was prepared to offer: in effect by promising them the same relationship as they had had with British India. The wilier Princes responded by playing him at his own game: Bahawalpur (See Figure 11.2) toyed with the idea of acceding to India, and actually for pertinent reasons. It was dependent on the irrigation water of the Sutlej projects, and feared the worst (as seen in Chapter 12, the ruling Nawab had some reason) if it were separated politically from the headworks. The mountain States of Dir, Chitral and Swat were also wooed by Congress, and indeed had no great affection for the plainsmen of Punjab.

In the event Khairpur acceded on terms which were relatively favourable to Pakistan, mostly because the ruler was declared mentally unfit and a compliant cabinet acted during the ensuing Regency. But Bahawalpur's Amir[1] did manage to force a deed of accession on Pakistan which preserved his full autocratic powers, and indeed did replicate the relationship he had had with British India. One Article even went so far as to specify that no alteration of the accession deeds could be legislated without his signature.

In 1947 Pakistan was of course born in every sense into crisis, with 6 million refugees, Lahore in flames, irrigation water turned off (see Chapter 12), and neither an established seat of power nor an adequately staffed and well run-in administration. Then came war in Kashmir, and an economic war which was particularly unpleasant in Bengal. In such circumstances Jinnah did not have the energy or inclination to steam-roller a Princely State, whose internal administration remained fairly well intact after 1947. But, to some extent fortuitously, the convulsions of Punjab inevitably swept into Bahawalpur, into which Punjabi settlers had followed the waters of the Sutlej irrigation schemes. The riots which ensued proved more than the State could handle, and in appealing to Pakistan for help, the latter forced the first concessions from the Amir, who had to place all his forces except his own bodyguard under Pakistani control, and who had now to pay a levy for defence. Then there followed a major confrontation over smuggling. The Pakistani government was trying to control food prices and availability, while in India prices were rising faster. The 300-mile frontier between Bahawalpur and India became extremely leaky, with the connivance, it was alleged in the Pakistani Press, of the State Administration. Another agreement varying the Accession was signed: this time the Amir assigned more power to his Chief-minister and he himself adopted a more constitutional role as Head of State. In 1949 the Amir sanctioned direct elections to a small Majlis (parliament), though effectively reserving veto powers to himself (via the Chief Administrator), and elections were held in 1951

---

1 This is an unusual title. Until paramountcy lapsed in 1947 he was a Nawab. But at Independence, much to the annoyance of the Pakistan Government, he awarded himself the title of Amir.

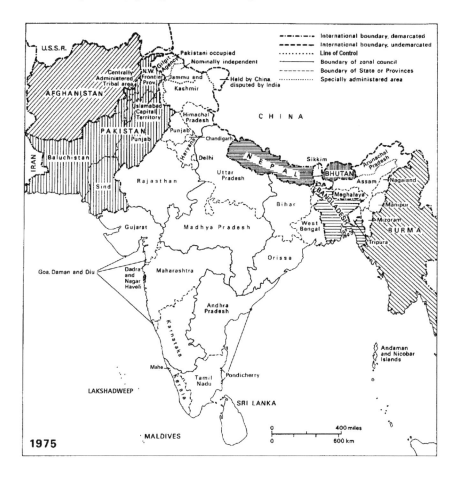

Figure 11.2   Territorial Division and the Successor States, c. 1975

Source: Schwartzberg (1992)

for the first time. Another Constitution in 1952 gave more power to the Majlis, and the Amir became in effect a Provincial Governor, able to disband the Majlis and proclaim Amir's Rule during emergencies only, and this time also only with the agreement of the central government. His role by then was getting close to that of the Rajpramukhs in India, except that his post was still hereditary, and he himself could not be dismissed by the Central Government.

In Baluchistan the Pakistani Government managed to push the princely states together into the Baluchistan States Union (BSU), but the move failed to forge any common identity between the tribes, who in any event remained mostly beyond and untouched by administrative changes between the BSU and the Federal Government. Internally it was still tribal law in a backward and poverty-stricken region. Under the new Constitution of 1956 (see below) the assemblies of the

States, States Unions, and Provinces of Punjab and Sind were replaced by a single assembly for the whole of West Pakistan ('One Unit') and this required amending further the accession of the princely states, effectively ending their individual survival as private fiefs, although not re-defining their borders. In 1958 the Khan of Kalat, most powerful of the former rulers, denounced Pakistan, unilaterally abrogated his accession, and,it is said, raised his own flag in revolt. Minor bloodshed led to accommodation. But during Zulfikar Ali Bhutto's incumbency the area became involved in a much longer and bloodier insurgency against the central government.

In the north-west mountain states of Dir, Swat and Chitral, the rulers' internal autonomy also remained strong for a long period. Partly there was little alternative: even in the North-West Frontier Province itself after One Unit, law and order and administration at lower levels was mostly conducted through traditional tribal councils – 'jirgas'. (In the case of Dir and Chitral the two even managed to wage a short war with each other in 1948 over a bride disputed between two rival royal claimants.) The system of Special Areas which the British had forged essentially remained unchanged, until they were absorbed into Pakistan in the 1960s.

Territorial Redefinition in India and the Emergence of Linguistic States

Writing in prison in 1942 Nehru said:

> Sometimes as I reached a gathering, a great roar of welcome would greet me: *Bharat Mata ki Jai* – 'Victory to Mother India.' I would ask them unexpectedly what they meant by that cry, who was this *Bharat Mata*, Mother India, whose victory they wanted? My question would amuse them and surprise them … At last a vigorous Jat, wedded to the soil from immemorial generations, would say that it was the *dharti*, the good earth of India, that they meant. What earth? Their particular patch, or all the patches in the district or province, or in the whole of India? And so question and answer went on, till they would ask me impatiently to tell them all about it. I would endeavour to do so and explain that India was all this that they had thought, but it was much more. … what counted ultimately were the people of India … *Bharat Mata* was essentially these millions of people, and victory to her meant victory to these people. You are part of this *Bharat Mata*, I told them, you are in a manner yourselves *Bharat Mata*, and as this idea slowly soaked into their brains, their eyes would light up as if they had made a great discovery. (Nehru, 1961: 62)

Perhaps Nehru believed too much in his own idealistic propaganda. What if, for each group of people, *Bharat Mata* could only mean *their* patch of earth?

Effectively by 1950 India had a new constitution, modelled to a large extent on the 1935 Act, a functioning Federal democratic system of government based on the Westminster Parliamentary system, representative governments in all Provincial Assemblies, a coherent national party organisation in the Congress, in power in the

Centre and in the States, and a new foreign policy initiative to pursue in the Non-Aligned-Movement. The Princely States had been absorbed, and a more or less uniform system of government and representation established through the country (except for some centrally administered areas). The expectations of the populace were also high, for 'self-determination' and social and economic advancement.

Although since the 1920s Congress had called for the establishment of linguistic states, the Dar Commission which looked at the issue afresh in the light of Independence, reported to the government in 1948 that linguistic states would probably be the basis for sub-national movements, i.e. those which could ultimately lead to secession. Thus faced with a challenge to the policies till then pursued by Congress, Nehru, Patel and Sitaramayya formed a committee to further investigate the issue, and they came back with the same answer, except that they conceded that there might be a case for separating Andhra from Madras State. Pandora's box had been opened.

If one looks at Figure 11.3 Madras and Bombay appear as central nodal cities in the arc of hinterland that had grown round them. Figure 11.5 shows how they are now eccentric cities, near borders which divide states. What Nehru described as 'tribalism and provincialism' has clearly won the political battle. Language is clearly an emotive issue – as any Belgian or Welsh nationalist knows – but for good reasons. In any State where there is a large linguistic minority, it is likely to find itself dealing with an administration conducted primarily in the language of the majority, of timetables printed in the language of the majority, and of schools giving education in the language of the majority, and of political debate taking place often in the language of the majority. To establish equal opportunity it ultimately suggests that there should be equal access to all these things in the language of the minority. Although Nehru believed that the States were just there to serve the purposes of administration, even that 'just' actually mattered. In 1953 agitation in Andhra escalated into widespread rioting, and a prominent leader began to fast unto death. The administration began to collapse, and that was why the issue was conceded and Andhra was born.

A States Reorganisation Commission was then established to look into the whole question yet again; but by then other communities were also demanding their linguistic region. When the report was accepted in Parliament in 1956, it represented something of a volte-face. By then Nehru had realised that forces that were not accommodated within the constitution, that were not ephemeral, could well break the nation. Fourteen new states of more uniform language were defined, along with six union territories. Demands for a separate state of Jharkhand for the tribal people of south Bihar and north Orissa were not conceded, nor demands for a state in the northeast for the Nagas. Bombay was however not split, neither was the new East Punjab. In the former case the argument was that although Bombay was a majority Marathi-speaking city in a Marathi region, it was dominated by Gujarati wealth. The same problems with which Radcliffe had struggled in the Punjab and Bengal were mirrored here. The result was the instantaneous growth of antagonistic Marathi and Gujarati political fronts, and large-scale rioting. The

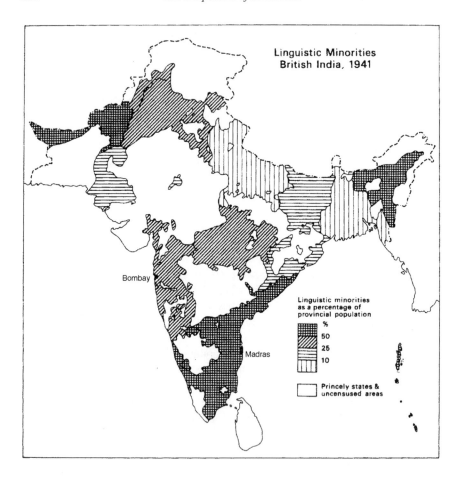

Figure 11.3   Linguistic Minorities in British Provinces of India, 1941

Source: Schwartzberg (1985)

Congress vote in Maharashtra collapsed, much like the Muslim League's vote had collapsed in Pakistan in the 1950s. In 1960, the Government conceded, and Bombay was split into Maharashtra and Gujarat.

The effect (and the cause?) of this process is graphically displayed in the maps and diagrams by Schwartzberg, Figures 11.3–11.7. At the State level the dominance of the major language groups has been continually increased, and in general this has been welcomed as increasing the legitimacy and transparency of government and bureaucracy. At the federal level plurality cannot have increased – after all we are dealing with the same population, and broadly similar constituencies of the Lok Sabha. But there is now a sharper definition of the self-identity of states, and there has been a growth of the representation of regional parties in the Lok Sabha.

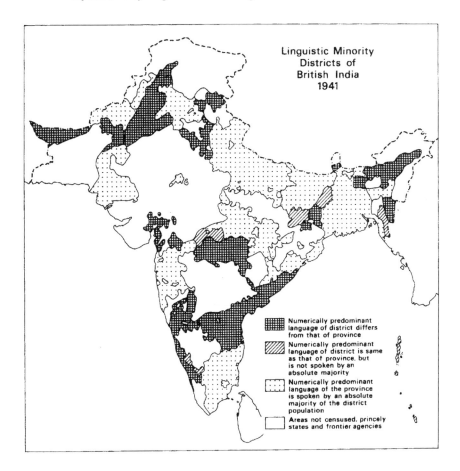

Figure 11.4 Linguistic Minority Districts of India 1941
Source: Schwartzberg (1985)

Increasingly, national parties have been unable to form Central Government without coalitions with some of the 'regional' politicians.

In the case of Punjab the Sikhs had been pressing for their separate State. But the case for resisting here was slightly different. The Hindus and the Sikhs of Punjab both spoke mutually comprehensible dialects of Punjabi, a language which is but little removed from Hindi. The two groups did however use different scripts. The Hindus mostly used Devanagari. The Sikhs, who in practice mostly used the Persian-Arabic script, argued for the adoption of their liturgical script, Gurmurkhi, which is closely modelled on Devanagari. Because the pressure was from a religious group, the central Government resisted. It seemed, after all, a re-run of the arguments of the Muslim League, which had been refuted (though not overcome) by Congress, which had always insisted on the creation of a secular state. But Hindu pressure also grew for separation, and in 1966 the demand was finally conceded with the creation of Haryana in the eastern part of Punjab. Finally, the Sikhs had a state territory in which they were in a majority, although only a small part of the original Punjab of the British Raj. It was not, however, the end of their demands: there is an interesting aftermath to the story. After the partition of 1947, Indian Punjab had lost the capital of Lahore and so built itself a new capital at Chandigarh. This now fell astride the new dividing line between Haryana and latest Punjab. Both sides claimed it, but in an act of even-handed justice, neither got it. It became instead a Union Territory, and capital for both states, and thereby a bone of contention for Sikh demands in the future. We cannot go into detail here, but in other new states the building of new state capitals has had a major impact on the urban map of India, and has also cost the tax payer considerable sums of money.

The divisions made so far are not necessarily the end of the process. In the mid-1990s the hill districts of India's largest state, Uttar Pradesh, demanded their own statehood as Uttarakhand, even if on their own they would be a rather small and poor state a little like Himachal Pradesh. The demand for Jharkhand has waxed and waned over the years. The tribal people here do not form a cohesive group linguistically, speaking a variety of Austro-Asiatic, Dravidian and other languages. Their coherence is in a similarity of life-style and non-Hindu culture. Their movement has also suffered because it has mostly been rural: the new industrial towns of the area are overwhelmingly Bihari, Bengali, and Oriya, and now represent in some districts a majority of the population. The south-eastern areas of Madhya Pradesh had similar problems. It was and is backward, heavily forested, and in rural areas dominated by tribal populations with very low levels of literacy and education. But it also incudes India's largest iron-ore mine, and new urban populations. Nevertheless, in 2001 statehood was granted to all three aspirant areas: Uttaranchal, Jharkhand and Chhattisgarh.

The north-east region is a very different case, and it is the subject of the next chapter.

Linguistic tension has also existed between the states over other issues. Article 343 of the 1950 Constitution specifically named Hindi using the Devanagari script as India's official language, although at the time of the adoption of the Constitution English was also to be used for fifteen years. India of course straddles a major linguistic divide: between the Indo-Aryan languages of the North derived from Sanskrit, and the Dravidian languages of the south. Since the use of English was strong in the educated classes, in inter-state commerce, in higher education, in the courts, and in administration and even parliament, and in many ways it was stronger in the south than in the north, if English ceased to be an official language it would mean that the south had to learn the North's tongue, but not vice-versa. Unease in the South broke out into outright rebellion when on Republic Day 1965 Hindi became the sole national language. The South found incomprehensible Hindi timetables posted, which caused railway officials considerable difficulty. Protest against this 'Hindi imperialism' went far beyond ripping down Hindi posters and burning Hindi books: rioting resulted in police firings in Tamil Nadu and between 80 and 300 deaths, and several suicides by self-immolation. The

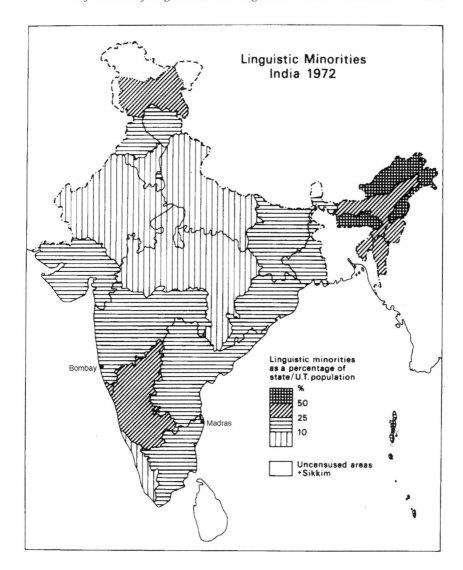

Figure 11.5   Linguistic Minorities in States of India 1972

Source: Schwartzberg (1985)

central government gave way: Hindi remains the official national language of India, but English may also be used for the 'time being' – a time period which is very precisely not defined. In schools, in theory, those in the north learn Hindi, English and another Indian language (but in fact rarely do), those in non-Hindi-speaking states learn their own language, Hindi and English. But many southerners still learn little Hindi. The reorganisation of the states has meant that the use of English as a lingua franca within each state is no longer so important; so its use

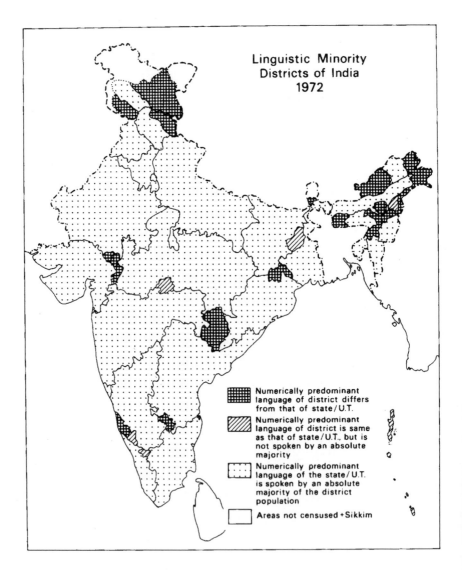

Figure 11.6   Linguistic Minority Districts of India 1972
Source: Schwartzberg (1985)

has declined in, for example, secondary and tertiary education in state institutions. However, English still has a remarkably strong hold particularly in newspapers and book publishing, and with increasingly internationalised media and the advent of satellite TV, for a metropolitan élite it is growing again. Private schools which use English as the medium of instruction are proliferating.

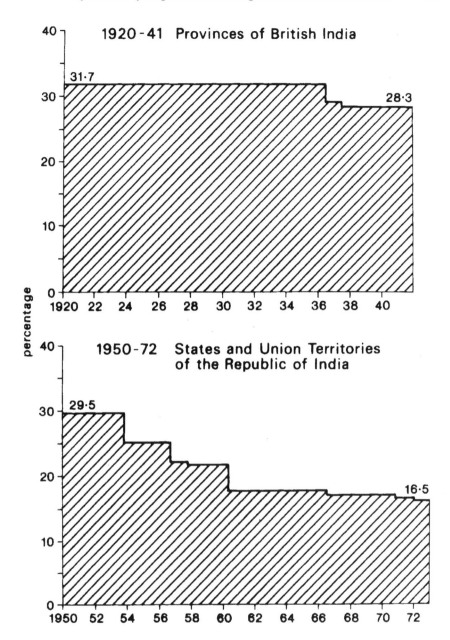

Figure 11.7   Percentage of Linguistic Minorities in Provinces/States over Time

Source: Schwartzberg (1985)

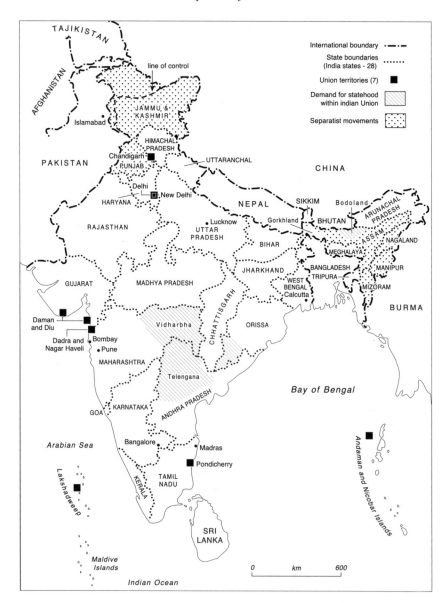

Figure 11.8   Statehood in 2002

Source: Based on Singh (2000)

The Centre-Province Balance and Pakistan's Search for a Constitution

The Independence Act of 1947 provided for the creation of a Constituent Assembly in both of the new Dominions of Pakistan and India, which would draw up new constitutions, in the meantime also acting as the nations' parliaments. Until the new constitutions were provided, the Government of India Act of 1935 remained the basis for the framework of Government. The Independence Act did however vary, obviously, some of the provisions and practices of the 1935 Act, under which the Governor-General (Vice-roy) had held considerable power. The role of the Governor-General after independence was one of constitutional head acting on the advice of his prime minister and cabinet, in the model of the British Monarchy and the Governor-Generals elsewhere in the Commonwealth. Mountbatten had hoped to be Governor-General to both new Dominions, but although India was happy with his appointment (and used his experience and knowledge profitably), Jinnah turned him down, and took the position for himself. In India, therefore, the long process of the devolution of power from the head of the executive to the head of the legislature was complete.

As Governor-General of Pakistan, and while acting under the terms of the 1935 and 1947 Acts, Jinnah could dismiss a Provincial Government with the concurrence of the Provincial Governor. Shortly after independence, perhaps because of the crisis he knew Pakistan would face forging its new national identity, he brought a new section 92A to the Statute book through a compliant Assembly, providing the Governor-General with the power to *direct* a Governor to dismiss a provincial assembly and assume all the powers otherwise invested in the representative institutions. The process of the devolution of power to democratic institutions was thereby in Pakistan's case reversed, back towards something of the model of the Viceroy and the Raj, and if I am mischievous, perhaps dimly back before that too, to the Moguls.

Jinnah, the Qaid-i-Azam, was also leader of the Muslim League. From his vantage point of powerful head of state and political leader it was he and not the Prime Minister who dominated the political agenda, and it was the latter who acted on the advice and consent of the former. Although, after Jinnah's death, Liaquat Ali Khan tried to reassert the power of the Prime Minister, he himself did not survive long, assassinated by a dissident Pathan in October 1951. He was succeeded by Khwaja Nazimuddin, a Bengali, who stepped down from the office of Governor-General. A West Pakistani, Ghulam Muhammad, was appointed Governor-General in his stead.

By 1952 it was already obvious that riots in Bengal against the imposition of Urdu as the sole national language (considered in more detail in Chapter 10) had seriously weakened if not destroyed the status of the Muslim League in the eastern wing and, therefore the prospect was already looming of Bengali demands dominating any assembly elected on a population basis. In April 1953, Ghulam Muhammad used the government's inability to handle law and order in both West and East Pakistan to summarily dismiss Kwhaja Nazimuddin, and to install his own nominee (then the Ambassador to the United States) to office. The Constituent Assembly could, and should, have contested this abuse of its prerogative, but it failed to do so, having spent much of its energy on quarrelling over representation in the new constitution. The pattern of 'strong' government from the centre in the Jinnah mould was now well established, as was also the disregard for constitutional nicety. Given the fact that the divisions between the provinces of West Pakistan weakened the West wing in relation to the East wing, several Punjabis became champions of the idea of One Unit, or the merging of West Pakistan into a single province. The idea chimed with the supporters of strong central executive government. Standing between this emerging 'élitist' force and their goal was the fact that the Constituent Assembly had been charged with devising a new constitution, and that it was still actively if acrimoniously trying to do so.

In October 1953 the politicians reached an agreement under the Mohammad Ali formula, which proposed for the Central Government an upper house, the House of Units, based on representation of the federating units, and a lower people's house based on direct representation by population. Five Units were proposed for the Upper House: Bengal; Punjab; Baluchistan States Union together with Baluchistan Province, Bahawalpur and Karachi; Sind together with Khairpur; and North-West Frontier Province, with the mountain tribal states as well. The Unit which inter-alia linked Baluchistan with Bahawalpur was to say the least of it, strange, and it provoked a lot of hostile criticism, particularly from the Princely States. By 1954 a revised scheme was accepted, which conceded a status for Bahawalpur and Khairpur equal to other Provinces. By September 1954 the Assembly passed a resolution making cabinet advice binding on the Governor-General, and began the task of finalising the Constitution on the lines of the agreed principles. It began to look as if the power of the Governor-General might again be curbed.

In October Gulam Mohammad replied: he suspended the Assembly (while it was in recess) and formed a new Cabinet without parliamentary responsibility. The Amir of Bahawalpur was instructed to dismiss his assembly, and a new Chief Minister, chosen by Ghulam Muhammad, assumed all of the Amir's powers and responsibilities. The steam-rollering of the 'strong central executive' had begun in earnest. Sind objected to integration into One Unit, and its provincial assembly was dismissed. The Federal Court also fell into line, dismissing an appeal against Muhammad's action in dismissing the Assembly, but preventing the framing of a new constitution until a new constituent assembly had met. A new assembly was duly elected, again by indirect means from the existing provinces and states, and in 1955 the outlines of the new Constitution wanted by the Governor-General were confirmed: in particular, West Pakistan (except for some of the frontier tribal states) would be a single unit. However, Muhammad was now an ill man, and forced into retirement, but the centralised power he had accumulated remained intact, bequeathed to one of his strong supporters, General Iskander Mirza.

The new constitution for the Islamic Republic of Pakistan was adopted, and became effective in March 1956. General Mirza became first President, and the scene was now set for a conflict between two equal and obdurate antagonists,

West and East Pakistan (as East Bengal had been officially renamed.) East Bengali sentiment would now have to fight against an unaccommodating Central Government structure. In an attempt to placate and control Bengali opinion, Mirza chose H.S. Suhrawardy, the veteran Bengali politician and leader of the largest group in the assembly (the Awami League) as the Prime Minister. Inevitably his policies aimed at parity in economic as well as political matters, raised the ire of Western politicians. By provoking instability on the floor of the house, Mirza then engineered Suhrawardy's dismissal.

The first elections under the new constitution were scheduled for March 1958, but before they had taken place the old pattern of conflict between President and Prime Minister that had re-emerged enabled Mirza to proclaim an Emergency. He asked the Chief of Staff, General Ayub Khan to administer Martial Law, which he did more amply than perhaps Mirza had hoped for. Ayub engineered a military coup in October 1958, and suspended all political activity. Thus began what was effectively 12 years of military dictatorship, but not without some attempt at yet further constitution making.

In framing the next constitution (adopted 1962) Ayub needed of course to legitimate his power, and to consolidate central control over the regionally divisive body politic. This he did by abolishing party political activity, and abolishing direct representation (which had only ever theoretically occurred anyway) in the National Assembly, instituting what was termed Basic Democracies instead. These were 80,000 units (local boards) into which Pakistan was divided, which were controlled for local government purposes only by elected representatives in conjunction with important local officials such as the District Officer, and which then elected the representatives for the national assembly. An analogy would be for the Monarch in the UK to abolish parliament, and to allow elected Parish Boards both to run local affairs, and to elect representatives (not standing on party tickets) to a new parliament. But there is an even better analogy than that. The process of winding history back, of eliminating all that the British had conceded, had gone yet one step further than the re-invention of the Viceroy. It was in town boards that the British had first invested some local electoral responsibility, and now effectively it was at this scale only that democratic rights were still preserved. Ayub had done merely what Mirza had once hoped someone would do. In 1954 Mirza had said:

> You cannot have the old British system of administration and at the same time allow the politicians to meddle with the civil services. In the British system the District Magistrate was the king pin of administration. His authority was unquestioned. We have to restore that. (Mirza in the newspaper *Dawn,* Oct 31 1954, cited in Inayatullah, 1964: 28)

But history was to show Pakistan, as it had shown the British, that it is not possible to pretend that political aspirations do not exist merely by inventing structures in which they cannot be expressed. Repression can give an air of stability for just so much time, depending on how hot the fire that builds the pressure, depending on how strong the boiler is.

In 1965 Ayub's popularity at home was low and his grip on power weak. He did what military dictators are prone to do at such time – stoke up a foreign campaign that will unite the nation – just as for example Galtieri of Argentina would do 17 years later in the Falklands/Malvinas. Pakistan provoked a war with India over border disputes in the desert in the Rann of Kutch, and shortly afterwards in Kashmir. The war in Kashmir was not a military success for Pakistan, and this diversion reduced rather than enhanced the status of the military government. Subsequent public disorder and rioting led to the downfall of Ayub in 1969. He handed power over to another general, Yahya Khan, who promised elections for a new constituent assembly and a return to civilian rule. The elections were duly held in December 1970. East Bengal, or East Pakistan as it had officially been renamed, wanted no more of the strong Pakistani Central Government. Its war to achieve independence as Bangladesh was outlined in the last chapter.

Regionalism Post-1972 in the Residual Pakistan

With the two wings of Pakistan now irrevocably split, civilian power in the west fell on Zulfikar Ali Bhutto, leader of the Pakistan People's Party. The position he faced was almost as bad as that faced by Pakistan in 1947. It was now militarily weak, and its administration tattered by the years of riot and turmoil. In the next constitution of 1972 the Provinces of Punjab (now incorporating Bahawalpur) and of Sind (now incorporating Khairpur) were reinstated, and Baluchistan and the North-West Frontier also re-established. Bhutto was following in Nehru's footsteps in recognising the reality of regionalism, not simply trying to suppress it. But clearly now was an opportune moment for other latent secessionist demands to strike: and almost immediately Afghanistan started to foment trouble in North-West Frontier by openly renewing support for Pathan demands for a separate Pashtunistan (a demand quietly supported by India and the USSR), and in Baluchistan the smouldering insurrection blazed again, aided by arms flown from Iraq. The suppression of these two movements was violent but successful (for the time being), but involved the assassination of many officials and politicians.

Economic stress, caused partly by Bhutto's interventionist policies and nationalisation programme, but also by such misfortunes as the first OPEC oil shock of 1973 and disastrous floods in much of the Indus Valley in 1975, compounded the difficulties. When Bhutto called an election in March 1977, the ostensible result was an overwhelming victory for his Pakistan Peoples' Party but this was denounced immediately by the opposition who alleged vote-rigging and intimidation on a large scale. The opposition took to the streets, and provoked a return to Martial Law. This provided the opportunity for the Chief of Staff General Zia-ul-Haq to launch a coup, and to assume full powers as Chief Martial Law Administrator in July. So, again Pakistan returned to military dictatorship under a strong central government headed by a new Mughal. Bhutto was executed, after a rather dubious trial in which he was found guilty of ordering the murder of a political foe.

Zia lasted in power from 1977 to 1988, when he died in an air crash presumed to have been sabotage. He followed almost the same tracks as Ayub had done, calling for elections to a National Assembly with very limited powers, with candidates standing on a non-party basis. For a short while he devolved some power to a civilian administration under Prime Minister Junejo, but dismissed it in 1988 at the first signs of independent actions by the Government. Zia also had to contend with increasing regional hostility in Sind, and above all the effects of the Russian invasion of Afghanistan in 1979, and the subsequent wave of Pathan refugees into Pakistan. To some extent the Afghan War proved Zia's salvation, since he became the bulwark through which US aid was conducted to the Mujahadeen in Afghanistan. But the long-term effect has been to introduce weapons to Pakistan on a plentiful scale, to produce what is known as the Kalashnikov culture, so that any dissident group in Sind, Baluchistan, the North-West Frontier and even in Punjab now has an armed wing.

Zia also played one other card very strongly. Given the regional dissent and the political instability from which Pakistan has suffered, Zia sought in Islam, the original reason for founding the State, a new force for national identity and cohesion, and also legitimacy for party-less politics. By further pushing the adoption of Sharia law and Islamic economic concepts, he hoped to unite the people and the Mullahs with him. Although the West sees in public floggings and amputations a barbaric past, these have not been wide-spread practices, nor as prominent in their impact in Pakistan as the attempt to push women back towards an inferior status and into purdah. In 1988 female demonstrators against his programme were beaten by the police in public view. Although the Qaid-i-Azam's (Jinnah's) great dream for Pakistan is often used in public appeals, his conception of Pakistan as a secular state seems for the moment to have been quietly forgotten. In 1991 a Supreme Court ruling made the Koran the official supreme source of law and thereby took Islamicisation a further step along the road. For these reasons Zia is now blamed with cultivating the conservative Islamic forces from within which extremists like Al Quaida have been able to find new recruits to terrorism.

Political power from 1988 to 1999 was resumed by elected governments, though none of the four governments completed its term in office. Both the two governments of Benazir Bhutto (1988–1990 and 1993–1996), Zulfikar Bhutto's daughter, and the two governments of Nawaz Sharif (1990–1993 and 1997–1999) were dismissed on the grounds of corruption. Nawaz Sharif's fall from power in 1999 occurred after he reined in the power of the President (with applause from Benazir Bhutto), and meddled in the Supreme Court. He was ousted in a coup by the army Chief of Staff, General Pervez Musharraf. History does not, of course, repeat itself: but the spirals of near-repetition are sometimes striking. Instead of executing Sharif as Zia had Bhutto, Musharraf forced Sharif into exile. For this unconstitutional behaviour, Pakistan was briefly suspended from the Commonwealth. Then, Musharraf's popularity with the USA and the West in general soared after Al Quaida's attacks on New York in 2001, and the subsequent war by NATO in Afghanistan. Pakistan again became a strategically positioned ally.

Dictatorships in Pakistan have a finite life-span, usually finished when public resentment finally dares to confront police and army suppression and boils over into mass protest in the streets. In 2007 and 2008 a combination of such protests, and the West's concern that Pakistan's military forces were not prosecuting the war against Islamic fundamentalists in Northwest Frontier province with sufficient commitment, led to domestic and international pressure for a transfer of power back to a democratically elected government. With US support, Benazir Bhutto returned to Pakistan to lead the campaign of the Pakistan People's Party, only to be assassinated, probably by the extremists that the US hoped she would be able to confront when in power. Sharif also returned, and in the elections of February 2008, the regional dimensions again became apparent, the PPP having a national showing but being strongest in Sindh, while the Muslim league of Sharif held the majority of seats in Punjab.

This simple regionalism has its subplots. The 9 million Mohajirs, the original and subsequent generations of the migrants from India to Pakistan in 1947 (and since), mostly settled in and around Karachi. They want more autonomy for Karachi, and a greater share in power in Sindh itself. The movement has turned into a guerrilla campaign, which in 1995 left more than 2,500 dead in street fighting in Karachi – one of the most violent episodes in Pakistani politics since the independence of Bangladesh in 1971/72. Jinnah was a Mohajir, and it is the frustration of his dream of the assimilation of all Mohajirs within a greater Islamic culture that has bred this bubo within Pakistan's search for stable representative government.

Instability, lawlessness, economic incompetence and corruption were the reasons given by General Pervez Musharraf when he seized power in 1999. Instability, lawlessness, economic incompetence and corruption are the reasons given for ousting him in 2008. It is another repetition of Pakistan's horrendous struggle since 1947 to build representative political institutions congruent with a national purpose and identity.

Concluding Remarks

Nehru's anxiety that the concession of linguistic states would promote sub-nationalism has to some extent been justified. With Chief Ministers each head of their own parochial pond there is acute awareness of the difference and rivalry between the states, and competition for the allocation of central resources. There are also sometimes very divergent policies on, for example, land-law and agriculture. But one must remember that most of India's states have populations much bigger than the states of the European Union, and no-one need be reminded of the rivalry and jealously guarded sovereignty there. The view from the centre cannot afford to be relaxed, but the long term effects of the concessions made to regional demands may well be to encourage a strong sense of the benefits of voluntary federation. It takes time for a people's identity and loyalty to develop: and in 1947 it was mostly not national, no matter how Gandhi was revered throughout All-India. Such loyalties are also hierarchical: to family, to village, to town perhaps, to region/state, and lastly to India. The local loyalty need not reduce the national loyalty. To many outside observers the sense of All-India is growing in the ordinary population. It was felt in the China War, in the Pakistan wars, and when Mrs Gandhi and her son Rajiv Gandhi were assassinated in 1984 and 1991 respectively.

Outside observers also agree that the continued integration of India is not guaranteed. Separatist movements in the South have had strong following: prominent amongst these was the DK or Dravidian Federalist party of Tamil Nadu (and later other southern states) which openly advocated independence as much as the Scottish nationalists have in Britain. They have taken power at the State Government level, and even formed small blocs in the Lok Sabha in Delhi. Interestingly, the more power they have achieved, the less extreme their demands. It seems as if forces which can, in the end, have some effect within the Constitution are less likely to try to operate outside it. Separatist movements in the Northeast are dealt with in the next chapter.

The problem of Kashmir remains, and it symbolises still the original reasons for the founding of the two homelands. It is the only Muslim majority state in India, and therefore a symbol of secular India and the invalidity of the two-nation theory.[2] Though Congress has remained avowedly secular, it has gradually seen its grip on power lost, and its stand against secularism weakened. In 1992 Hindu fundamentalists finally managed to demolish a mosque at Ayodhya, which had been built on the ruins of a Hindu temple said to be safeguarding the site of Lord Ram's birth. This presaged a new outburst of Hindu-Muslim rioting, at its worst in Mumbai. The Hindu nationalist party, the Bharatiya Janata Party, has grown in both appeal and moderation, and in 1997 replaced Congress as the party of central Government, the third time Congress has lost power in New Delhi since Independence.[3] There is no doubt that this represented a shift in India's understanding of itself, and in the way in which it is integrated, even if the BJP is mostly a party of the north, which relies on a myriad of coalition partners. I will say more about this in the concluding chapter. Here I wish to stress that India should be celebrated as a success. It has survived as the world's largest and most complex democracy, and the military have remained subordinate to the civil power throughout. Its representative system of government has been under strain, but its has also shown the capacity to evolve. It has built a national sense of identity,

---

2 Indian Punjab now has a Sikh majority, and some of the small states of north-east India have Christian majorities.

3 For two-and-a-quarter years after Mrs Gandhi's emergency of 1975–1977 Congress was ousted by a coalition known as the Janata Dal.

of being Indian – even if that identity has shifted (hopefully not dangerously far) from a wholly secular ideal.

Pakistan's political history is a complete contrast. Crudely it has been summed up as the three A's – Allah, the Army and America. The first has demonstrated no desire to accommodate democracy, the second has shown itself generally antithetical to democracy, and the third has frequently subordinated democracy to other foreign policy objectives during the Cold War and the war against al-Qaida. Would things have been different if at the beginning in 1947 three successor states had been formed? This is an exercise in counter-factual history, so perhaps the question has to be re-phrased; what impact did the tensions between East and West Pakistan have on the search for a constitution? The answer is clearly, enormous. The struggles between East and West Pakistan were one of the biggest justifications for the first military coups, which gave the Pakistani armed forces their addiction to imposing their own financial settlements from the public purse. There then remains the supplementary question: would West Pakistan, without the Bengal problem, have followed a smoother path of constitutional evolution? Let us suppose that the problems with the Kashmir would have been the same, with the Princely States would have been the same, Indian pressure the same, Afghani opportunism the same, and Sindhi and Baluchi resentment the same, Mohajir promotion and exclusion the same, then the history of the last decades suggests that the problem of Bangladesh may have made things much worse, but the problems would have been bad anyway. Pakistan has an urbanising and industrialising economy, which should further the utilitarian integration of the state, but the tortuous history of the search for a balance between the Centre and the regions suggests that stability may not be assured. Allah has been developing his roots (though religious parties did poorly in the elections of 2008); the Army will only tolerate subordinate democracy. America is rightly detested by most Pakistanis, because it never intervenes for the interests which Pakistanis define for themselves. Instead, the US and the West intervene for Pakistan's strategic utility on the Afghan, Chinese, and Central Asian borders.

Chapter 12

The Forgotten Sisters: India's Northeast

Introduction

On map F.1 with which this book started, the most permanent of South Asia's borders runs northwest-southeast along the Terai between Nepal and modern India, and then almost due south along the eastern edge of Bengal. Contrast that map with Figure 11.5, which shows a constituent part of modern India beyond that border. Figure 11.5 also makes clear that this is a linguistically complicated area, with a number of small states, and that the whole is hanging onto the rest of India by the narrowest of threads. In modern political slang, the states of Arunachal Pradesh, Assam, Manipur, Meghalaya, Mizoram, Nagaland and Tripura, are referred to as the seven sisters. The strip of Indian land which connects them to West Bengal is only 21 kms wide at its narrowest, and is popularly known in India as the Chicken's Neck. The neck is of course narrow because what was once a constituent part of India – East Bengal – became first part of a hostile state as the East Wing of Pakistan, then the independent and sovereign state of Bangladesh. Squeeze through the neck by road or rail between Nepal to the northeast and Bangladesh to the southwest, and you arrive in the valley of Assam, the broad wet plain of the Brahmaputra. In the days before rail and road, when the British first established regular communication between Bengal and Assam it was by steam boat. Following the river upstream, in Bengal the river craft would be travelling north, then they would go round a very distinct bend to enter Assam, where the river is much more west-east.

The massif that dictates this bend is the Shillong Block (modern Meghalaya). In geological terms, this is a somewhat remarkable feature. Remember that the Deccan is the surface expression of that part of Gondwanaland that collided with the southern flank of Asia, and that this Indian part of the earth's crust is bending and dipping under the Himalayas. But in the northeast two large faults or fractures, one on the north front and one on the south front of the Shillong Block, have formed as the crust is buckled, and the rock between the two has been and is being pushed up, rather than bent down to slide under the Himalayas. One of the greatest earthquakes recorded in the northeast happened in 1897 when the northern front lifted by more than 10 metres along the Oldham (northern) fault. Its magnitude has been calculated at 8.6 on the Richter scale. All masonry buildings within an area the size of England were destroyed, and the quake was felt in the southern tip of India.

Since the plateau has earlier been under the sea, some of the geology that is exposed on the plateau surface is more recent than that of the Deccan, and includes secondary and tertiary sedimentary rocks on the southern flanks. This front of the plateau receives the heaviest rainfall in the world, so it is no surprise that erosion of these more recent rocks has produced rugged and hilly terrain, overlooking Sylhet and Cachar to the south and the vale of Assam to the north. The vale is by contrast flat. The gradient of the Brahmaputra is only 17 cms per kilometre at Dibrugarh, near its head, and as little as 10 cms per kilometre at Guwahati. Given the huge discharge and great sediment load of the Brahmaputra, this means that the river floods, deposits silt, and shifts its course. So the vale is almost flat north-south across its width, even if the river and its many tributaries have left depressions and channels here and there with attendant swamps and bogs. North of the vale the jagged Himalayas rise steeply, heavily forested at lower altitudes, with sparse grazing at higher altitudes between the tree-line and the snowfields. Near the border with Bhutan there are passes that lead to Tibet and Lhasa. In the east the mountains bend round in an arc, and splay in different ranges into China and Burma. There are some passes over the mountains into China. From this nexus, the Patkai Mountains strike south-westwards, along the south side of the vale of Assam. The mountain system continues, along the Burmese Indian border, touching the Cachar Hills at the eastern end of the Shillong block. The Patkai mountains form a complex series of parallel ridges and valleys; they are high, rugged, wet, and naturally clothed with thick jungle, much of it bamboo. They are what stand between India and Burma.

In sum, India's northeast is the world's greatest cul-de-sac, a vale skirt by some of the most impenetrable mountain terrain on earth. The vale has been home to its own Ahom civilisation: the mountains home to such a great variety of different tribes that they have sometimes been referred to as a museum of mankind. The distinction between these two parts is immediately clear to a contemporary traveller. In the vale signs are written in the Assamese script (a variant of Bengali); and where the Roman script is used it is mostly for English. In the Hills, it is the Roman Script that dominates, used for a multitude of languages which bear no relationship to Bengali or English. These are languages which have been given a written form only since the advent of missionaries in the nineteenth and twentieth centuries. And the missionaries have reaped a great harvest: by the 1940s up to 30 per cent of the Nagas were Christian, and now the Northeast is the main region of India in which Christians are in a clear majority. (Christianity is particularly significant also in Goa and Kerala.)

Settlement

It is clear from the distribution of ethic and linguistic groups that the early settlement of northeast of India was from Tibet, China and Burma, and that these settlers reached further east into India than is obvious today. Indeed there is a strong mongoloid element in the physical composition of many Bengalis. The classification of the languages the settlers spoke is controversial: some authorities put the languages in a broad Sino-Tibetan group, others in a narrower Tibeto-Burman group. In their own account, the Garos (who inhabit the western party of the Shillong Block – modern Meghalaya) came from Tibet 'because it became drier and less fertile' (Rongmuti, 1933). The northeast of India offered jungles in which to hunt, and the possibility of slash-and-burn agriculture, known in India as jhum (or jhumming.) Tribes sustained by combinations of these activities were not sedentary, and social organisation rarely developed beyond the village or tribal level. It is also clear that inter-village and inter-tribal warfare was common, often in pursuit of slaves, and that human heads were counted as trophies of war.

The vale was the seat of the kingdom of Kamarupa (the name survives as the district Kamrup, north of Guwahati), which seems to have had a sporadic and uncertain existence from about the 4th century to the 12th. The more successful and expansive monarchs seem to have accepted Hindu priests and absorbed Hinduism to a degree. But the majority of the population remained committed to their own animist beliefs. In the 13th century a leader from the Shan tribes of modern north-east Burma descended into the vale, and began the establishment of the Ahom dynasty, whose history is fairly well-known, because noble families had their deeds recorded in chronicles known as burranjis, many of which survive. Hinduism made inroads, though the Ahoms themselves kept versions of their own faiths and ceremonies until the 18th century. During the reigns of the Ahom kings, Assamese finally became the court language, instead of Ahom (a Tibeto-Burman Shan language). Even in the vale, the population remained mixed, with many large tribal groups like the Bodos, retaining their identity. The economy relied on sedentary agriculture, and particularly the cultivation of paddy fields. This meant higher population densities and the tax base to support urban centres. It also meant potential plunder for the hill tribes. The history of the Ahom kingdoms is indeed a tale of constant war on the borders, and punitive raids into the hills.

Resistance and Submission

It is also a story of defence against incursions up the Brahmaputra valley. During the Mughal period, and particularly under Aurangzeb, the most determined attempts were made. The aim was imperial expansion, and the quest for elephants and rare plants. Flotillas of armed boats were used on the river to combine with land armies. Some achieved initial success, but none were able to impose enduring rule on Assam. Sometimes the invasions met with rapid military defeat at the hands of the Assamese; sometimes they withered as they penetrated further into this alien land.

The history of the rest of Mir-Jumla's [general sent by Aurangzeb] Campaign is not unlike that of Napoleon's ill-fated march to the territories of the Czar. Rain, pestilence and constant night attacks of the Assamese did as much injury to the Mughal commander as snow, frost and Russian attacks on the heel did to the French monarch. (Bora, 1942: 83)

Where military might failed, other forces intrusive forces prevailed. The king Sib Singh (d. 1744) was fully committed to Hinduism and fell under the sway of Brahmans.

> Thanks to his support, Hinduism became the predominant religion, and the Ahoms who persisted in holding to their own beliefs and tribal customs came to be regarded as a separate and degraded class. The Deodahis and Bailongs (tribal priests) resisted the change with all their might, and succeeded, for some time longer, in enforcing the observance of certain ceremonies, such as the worship of the Somdeo. But the people gradually fell away from them, took Hindu priests, and abandoned the free use of meat and strong drinks. The change was a disastrous one. By accepting a subordinate place in the hierarchy of Hinduism, not only did the Ahoms lose their pride of race and martial spirit, but, with a less nourishing diet, their physique also underwent a change for the worse. (Gait, 1905: 189)

Gait, a long serving ICS officer in the Northeast who later became Lieutenant Governor of Bihar and Orissa, might well be displaying an imperialist anti-Hindu sentiment, and the British preference for martial spirit, but that by the 18th century the Ahom kingdom had passed its zenith was not in doubt. Some of the tribes, adherents to a monotheistic sect centred around monasteries, erupted in what is known as the Moamaria rebellion at the century's end. The ensuing warfare devastated and depopulated much of the kingdom. The Ahom monarchy survived, partly because of the support of a small contingent of East India Company sepoys despatched in 1792 from Bengal under Captain Welsh. The force deployed muskets and small field guns against arrows, spear and matchlocks with lethal effect. This could have been the start of a more permanent British involvement, but the Governor-general in Calcutta was apprehensive about the costs and other implications, and the British withdrew.

Inadequate men succeeded to power. 'In July 1794, Assam was deliberately relegated to anarchy and civil war.' (Mackenzie, 1884: 3). About 1815:

> One Badan Chandra was chosen as [the new Bar Phukan – chief minister] ... Before long, reports began to come in of his oppressive behaviour and gross exactions, while the conduct of his sons was even more outrageous. One of their favourites pranks was to make an elephant intoxicated with *bhang* (derived from cannabis), and let it lose on Guwahati, while they followed at a safe distance, and roared with laughter as the brute demolished houses and killed the people who were unlucky enough to come in its way. (Gait, 1905: 230)

South of the Patkai mountains there is another vale, that of the Irrawaddy in Burma. Here the power of the king of Ava (near Mandalay) was growing as that

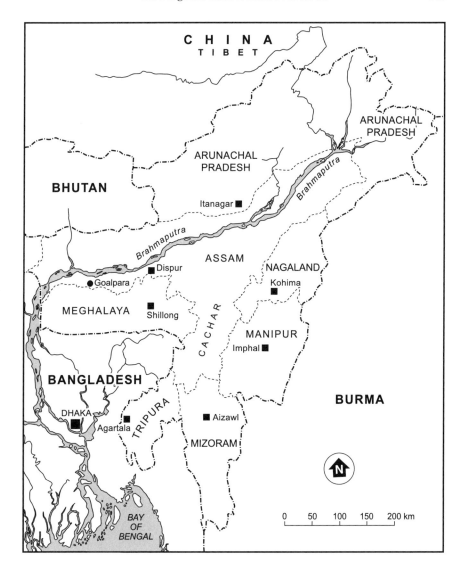

Figure 12.1   States of Northeast India

of the Ahoms was waning. As rebellions again broke out in Assam against the weakened Ahom state, the Burmese took advantage of the situation and invaded upper Assam in 1816. They plundered much of the country, and withdrew with reparations. They also overthrew the monarch of Manipur and pushed towards British Bengal across the Arrakan mountains. In 1819 they returned to Assam in greater force, and occupied the vale downstream to Guwahati. It is reported they acted with extreme savagery, flaying captives alive, beheading others, and burning imprisoned crowds alive. Their ambitions were not limited to Assam: they had designs and claims on Bengal too, which they thought poorly defended. They pushed South from the vale into Cachar (see Figure 12.1), threatening Sylhet, and they pushed west of Guwahati towards Bengal.

In 1824 the British declared war and determined to evict the Burmese from the vale. They started by moving up the Brahmaputra valley, but because of the swamps, jungles and fever; pauses during the long rains, and difficulties of supplies; they took more than a year to do so. A Captain Wilcox noted that he had seen in 1825 a fleet of commissariat boats, with desperately needed supplies, take twenty-five days to make their way only thirty miles upstream from Goalpara (Gait, 1905: 351). The British also tried to reinstall the Raja of Manipur:

> but serious obstacles were encountered in the shape of the mountainous character of the country, the clayey nature of the soil and the unusually heavy rainfall. Large numbers of elephants, bullocks and other transport animals were lost, and in the end, the attempt was abandoned and the force broken up. (Gait, 1905: 339)

However, shortly afterwards, the Raja re-established himself, thereby maintaining his own sovereignty.

By the treaty of Yandabo in 1826 the Burmese recognised British control of Assam and the independence of the Raja of Manipur. So, a quarter of a century after they had declined the invitation to be the power behind the Assamese throne, and the Ahom state had collapsed, the British now found themselves in charge of a ravaged territory. The solution they settled on was to make western Assam directly administered, and then to create a Native State in eastern (Upper Assam) where a protected ruler would manage internal affairs in return for an annual tribute. This did not last long: the native state and other territories had all become directly incorporated by 1843. In the western part, the Ahoms' old system of corvée labour was initially replaced with a poll-tax: but this hardly raised any revenue, so the tax was in turn replaced with land rents, which were sufficient it seems to maintain the military detachments and maintain law and order. Mr David Scott was appointed first Agent then Commissioner, and it was he that supervised many of the administrative and political changes. He died in 1831 at the age of 45, and of his life and labours Alexander Mackenzie (1884) observed:

> He was one of those remarkable men who have from time to time been the ornament of our Indian services. Had the scene of his labours been in the Northwest or Central India, where the great problem of Empire was then being worked out, he would occupy a place in history by the side of Malcolm, Elphinstone and Metcalfe. (Mackenzie cited in Gait, 1905: 348)

So, even while the region was being incorporated into India, it was already, in comparative terms, overlooked, to be forgotten.

The Tea Industry

A prospectus, printed in 1839 for a proposed 'Assam Company', clearly enticed an adequate number of investors, as that is the year that the company, still trading in 2008 under the same name, gives as its date of foundation. The prospectus gives a sketch of the geography of Assam, a brief history, comments on the tribes and British relations with them, comments on the way of life and communications (pointing out the benefits that would be derived from steam navigation), and a list of products. This includes rice, cotton, opium (and how the people use it), lac, silk, rubber, gums of various sorts, timber, iron, coal, limestone, salt, petroleum and gold dust. The principal focus is on the possibilities of tea production, but the prospectus suggests that the company's name ought to be the Assam Company and not the Assam Tea Company, because of the range of products for trade.

The British market for tea grew substantially throughout the 18th century, and tea had become China's principal export. Since tea drinking had for long been part of Chinese culture, cultivation was quite widespread, but one of the important origins was in the mountains of Yunnan, east of Assam. The export trade to Britain involved transport to the Chinese coast, and shipment round Indo-China into the Bay of Bengal.

In 1823 an enterprising trader Robert Bruce reached Rangpur while it was under Burmese domination. He contacted a Singhpo chief who gave him specimens of a local tea-plant, that the tribe also used. There was considerable confusion about whether this was the same tea plant as in China, or a different one. But whatever the case, the prospect of tea plantations in British hands, and a shorter voyage to England, were attractions that sparked the East India Company in Calcutta to action. Very early on there were scientific analyses of the soils involved, and those from Assam were found very similar to those from China.

> The two peculiarities in these soils are, first, that they contain no carbonate of lime, and only traces of phosphate and sulphate; and the next, that their iron is almost wholly in the state of carbonate of iron, a widely different compound from the simple oxides." (Anon, 1839: 17) The indigenous plants were scattered sporadically. "The requisite quality of the soil, which is comparatively of rare occurrence, will account for the manner in which the plant is distributed in spots, or distinct colonies, instead of being uniformly diffused with the common vegetation. (Anon, 1839: 18)

Seeds and Chinese tea producers were brought from China: but, in Assam the imported plants fared less well than what did indeed prove to be indigenous varieties. (Later, near Darjeeling, some of the Chinese stock prospered.) So the early tea gardens were scattered too – widely spaced estates surrounded by jungles, and hill tribes, who would from time to time raid some of them.

These tribesmen were hardly the right stock to labour on the estates; and the number of landless peasants in the vale was too low for them to provide the requisite labour. As a result a contract trade developed, bringing indentured tribal people from Chotanagpur (modern Jharkhand) to the tea 'garden' – a grimly euphemistic expression given the horrific conditions in which the labour force was kept. They were transported in appalling conditions to be given little shelter, medical help, or food. Introduced into a new climate and environment in which all the worst jungle diseases were rife, as many as a third of the workforce might die in six months. Since the labour force never reproduced itself, the trade continued. The echoes of shame and scandal have not died to this day. Tea production grew rapidly, as did exports, by 1900 being 1/3 the value of cotton's exports. But, in a classic case of the declining terms of trade, continually rising production forced down prices. Cultivation did not stop: simply the conditions of the labourers remained appalling, if not quite so mortal.

To get the produce out, steam boats were introduced, and then steam railways. The Assam and Bengal railway, which opened in 1905, was built from Chittagong through Sylhet and the Cachar Hills to the Brahmaputra and thence to Tinsukia. Pathans from that other frontier, the Northwest, were brought in, because they were good at blowing up rocks to make cuttings.

> Mr Wilde, accompanied by his wife went by the present trunk road and by rough tracks with bullock carts and elephants to Lanka, from there by narrow jungle trails to the camping site. One of the party was attacked by cholera, Mrs Wilde tried to get him back to Gauhati. His grave is conspicuous on the left of the line just before reaching Lanka [twenty miles northwest of modern Lumding] from Gauhati. Mr Wilde continued in charge of Hatikhali during construction but was murdered by Pathans in 1897. He and Mr Peddie [death unexplained] lie buried deep in the jungle some miles from the present Hatikhali station … . An old-timer's story said that for every sleeper laid in the hill section and Lanka-Manipur Divisions, a man died. (Prendergast, 1944: 20)

Relations with the Hill Tribes

Figure 12.2, is a convenient simplification of the tribal landscape surrounding the vale. For example, Nagaland is shaded to suggest some sort of Naga homogeneity: but there are thirty or so Naga tribes, many of them speaking mutually incomprehensible languages, even in adjacent (and mobile) villages. The languages themselves are part of the Tibeto-Burman sub-group of about eighty collectively known as Kuki-Chin-Naga – but the relationships between all of these are not clearly worked out.

To use the names of current political territories, between Assam and Tibet/ China lies the state of Arunachal Pradesh – at one stage in independent India's history known as the Northeast Frontier Agency – the counterpart to the Northwest Frontier Province on the Afghan border. This state incorporates the southern flanks of the Himalayas, across which there are thirty-two passes into Tibet, all currently closed. To the south, between Assam and Burma lies Nagaland, Manipur and

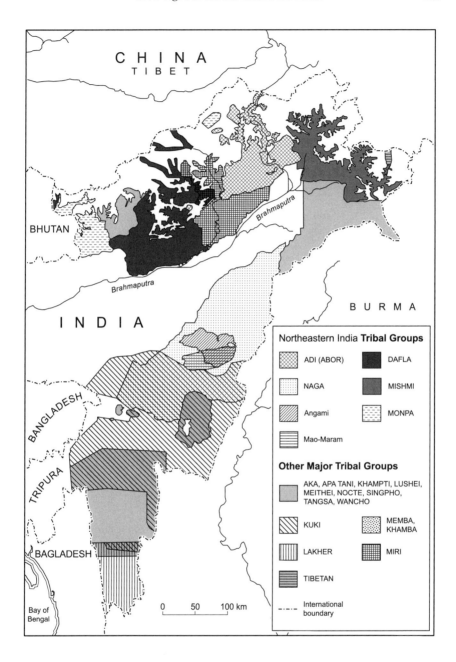

Figure 12.2   Spatial Distribution of Languages in the Northeast

Source: Rose and Fisher (1967)

Mizoram, mountainous and jungle clad countries, with a few very tough routes into Burma. Between Bengal (now Bangladesh) and Burma lie Tripura and again Mizoram and Manipur. South of the western part of the vale of Assam is the hill state of Meghalaya, the Shillong Block that stands between Assam and Bengal.

Like Nagaland, these are all tribal territories, that is to say, they have been (and are) inhabited be peoples who by way of life, culture, and level of social organisation presented the British with unique political and administrative problems within India. The way of life for most was hunting, warfare (including head-hunting, although there were some tribes attributed with more peaceful qualities), trade and shifting agriculture. In rare cases in appropriate conditions, rice was cultivated. In some areas some tribes were 'Royal' while others were 'Slave'. The tribes did not subscribe to mainstream religions. In their levels of social organisation, they might recognise their own language groups, but rarely were more than a few villages under one chief, so it was very difficult to negotiate with them collectively.

The legendary savagery and independence of the tribes does not mean that they were in the Andaman or Amazonian sense 'uncontacted.' Most of them straddled trading routes between Tibet and Bhutan and Assam, and between Burma, Assam and Bengal, and they derived income and necessities like their metal dhaos (machete/dagger) from trade in high value goods, like rock salt from Tibet, cotton, and animals from the plains. Some of them were also engaged in a valuable trade extracting rubber from the jungles.

> The people of all the tribes known to us are great traders, and parties of them are continually to be met with in the cold season, when they bring down cotton, chillies, ginger, &c., and take away salt, iron, dhaos, fish and pariah pups … Unfortunately, too much money is now squandered by Nagas on opium, and many of them particularly from the nearer tribes, are confirmed eaters of the drug. (MacKenzie, 1884: 400)

Where did the boundaries of India stop? In a world of international law (about which, as one British official complained, the tribes knew nothing), different states negotiated borders with each other. Looking north from Assam, the next recognisable state was Tibet – and so the border would have to be negotiated with the Government in Lhasa. This raised questions about the relationships between sovereignty and jurisdiction – because the Government in India went on to claim sovereignty over territory inhabited by these independent tribes, over whom they had no practical jurisdiction and no administration.

This ambiguity applied both to the northern Himalayan border with Tibet and China, and with the southern border with Burma. In the latter case the British had completed their annexation of the vale of Burma by 1885, but again exercised little or no practical control over the tribes in the border mountains. But what they did not have they were prepared to give away. At Independence the British would cede that same sovereignty over the heads of the tribals to independent India and East Pakistan (and to independent Burma in 1948).

One way of dealing with the tribes would be simply to ignore them and their territories. But they interdigitated. Under the Ahom kings of Assam, the corvée system, known as Paik, meant the peasantry donating labour and money to the aristocracy. At the margins of the plains, many were relieved of this obligation, because of the Posa system, by which instead they paid tribute to tribesmen who descended to the plains once a year. The latter collected cotton cloth, animals and money. The tribes even maintained agents on the plains to handle the arrangements and to act as interpreters. As British observers noted, these were not arbitrary and maximal exactions: the rate of levy was acknowledged on both sides.

In addition to the Posa levy, there were indeed sporadic lethal raids on the settlements in the plains. The motivations varied: sometimes mere opportunism, sometimes a tribe may have been displaced by others in the hills, sometimes there was famine. The bamboo forests of the Patkai and Arrakan hills flower every forty or fifty years, and the bamboo dies.[1] This extraordinary and still poorly understood event happens synchronously over the whole region. Rats gorge on the seeds, multiply geometrically, and then eat all other crops and available food. At other times there was resentment against what the tribes thought was incursion on their territory – which happened with increasing frequency from the 1860s as the tea industry boomed. They also feared the illnesses brought by the intruders.

> The hillmen had, it seems, been much troubled by an epidemic, which they believed to have been imported from the plains. (Mackenzie, 1884: 31 – referring to events in 1872 in Lakhimpur)

The Inner Line Regulation, 1873

> It is not open to us on the Abhor frontier [the Himalayan region between the vale and Tibet] to have recourse to the policy of permanent occupation and direct management … To annex the Abhor Hills would only bring us into contact with tribes still wilder and less known, nor should we find a resting place for the foot of annexation till we planted it on the plateaus of High Asia; perhaps not even then. (Mackenzie, 1884: 55)

The administrative response of the Government was to define an inner line, far short of the claimed international border, where the administration effectively ceased. The Regulation gave power to the Lieutenant-Governor to ordain which British subjects or foreign nationals could cross this line, and any conditions on the holding of land or the conduct of trade. The regulation also specifically mentioned the preservation of the elephant population. Beyond the line, there would be no interference in the way of life (and death) of the tribes.

---

1 While writing this chapter, this event happened again. The latest flowering was first reported in November 2007. By March 2008 a million people were said to be starving in Mizoram, and the state was pleading for help.

The Nagas of the unadministered areas were free to commit raids and take heads to their heart's delight within their own habitat, and a blind eye was turned upon their frolics, provided they committed no trespass. (Rustomji, 1983: 27)

On the other hand, if it was British heads that had been collected, perhaps from a survey party, or after a raid over the line, then the British were apt to think this less frolicsome, and the first duty of the inevitable punitive expedition was to collect the self-same heads, to deny the trophies to the tribes. Simultaneously such expeditions would burn down suspected villages.

In 1874 Assam was constituted as a Chief Commissionership separate from Bengal. This, together with the concept of the inner line, was the beginning of a geographical duality which has caused difficulty ever since. The Backward Tracts, as the tribal areas became known, were in theory a part of the geographical definition of Assam. However, any acts passed by the elected legislature only applied within the inner line. Matters beyond that were at the discretion of the Chief Commissioner (later Lieutenant Governor) who could and did make regulations without reference to the Assamese cabinet. There is an echo of this state of affairs in the current constitutional 'otherness' of the northeast. Just as the Indian Constitution's Article 370 prohibits non-Kashmiris from buying property in Kashmir, so Article 371A privileges the property rights of indigenous tribal groups of the Northeast, and legal Indian immigrants (e.g. Bengalis, usually in urban areas) often have very limited rights.

The Japanese Invasion of World War II

The Japanese gambit in the second world war was to establish an Asian "Co-prosperity Sphere", an empire rather like the European ones, with a Japanese industrial core, and colonies which would be both markets and sources for raw materials. After expansion in China (from the 1930s), war was fully joined with the USA and the European colonial powers in 1941. Japanese forces quickly took the French colonies in Indo-china, Dutch Indonesia, and British Malaya. By the end of 1941 and the beginning of 1942 they were probing at Burma, and thinking about India. All along the east coast of India from Madras to Calcutta there was apprehension about a possible maritime invasion. However, British resistance in Burma collapsed, and the Japanese moved in as 600,000 people fled, many of them of Indian descent, mostly through the mud and jungles of the tribal mountains, into Bengal and Assam. At the time it was the largest mass migration in history. 'As many as 80,000 may have died from disease, malnutrition and exhaustion'. (Bayly and Harper, 2004: 167)

India, of course, would represent a huge prize. After the early smarting defeats, British resolve was strengthening, at the same time as Japanese lines of communication were ever more extended. Behind their lines, allied naval power was beginning to cripple Japanese shipping. By 1944 it was literally, for them, now or never. They attacked in Arrakan, and were soundly repulsed. Next they tried attacks on Imphal in Manipur and Kohima in Nagaland.

> In Operation Imphal 12,000 horses and mules, 30,000 oxen and more than 1,000 elephants crossed the Chindwin [river]. The scale of animal fatality was colossal. (Bayly and Harper, 2004: 382)

Next it would be colossal human fatalities. In both cases the assaults failed. Then Indian, British and allied armies pursued starving and abandoned Japanese soldiers back through Burma, killing all and any they could find.

The hill tribes played their part in this: most had not liked the Japanese style of occupation, and:

> To the hillmen, these new conditions required epochal decisions. How far should they support their old masters against the new invaders? The chiefs and many of their followers were relatively satisfied with the old order. However much they resented or even resisted the initial imposition of the Raj, they came to find the British presences not too intrusive and even gave them some advantages. In many cases they had come to dislike the assumptions of the plains politicians that they would easily merge into the new Burmese or Indian nations, forfeiting their political privileges and long-cultivated special identities. (Bayly and Harper, 2004: 198)

McMahon, the Outer Line, and the Chinese War

The Government of India claimed sovereignty beyond the Inner Line, to the borders with Burma (which was actually governed as part of India between its annexation and its constitutional separation in 1936) to the East and South, and the borders with Tibet in the North. The latter meant negotiating with the Tibetan government in Lhasa, and that in itself raised a whole number of other issues.

The Tibetans acknowledged the nominal overlordship of the Chinese and accepted their 'Viceroys' in Lhasa, but actually they administered their territory and people through their network of Buddhist monasteries and priests. In 1900 the Dalai Llama was the real and effective head of the government (by contrast, often it was a regency Council that held power), and he had a close advisor, a Buddhist Mongolian Buriat called Dorjieff, who was nominally a Russian citizen. This meant that the British entertained the same suspicions about Russia vis-à-vis Tibet as they had had about Russia and Afghanistan. All sides played off against all sides:

> The whole situation was complicated and difficult. We have the Indian government pressing a forward policy on the cabinet; the cabinet endeavouring to restrain the eagerness of a masterful Viceroy (Curzon) and anxious not to offend the sensibilities of Russia; the British minister at Peking trying to put pressure on the Chinese Government; the Chinese hating our interference altogether, unable to coerce the Tibetans and

anxious to conceal their inability to do so from the British government; and finally Russia protesting that she had no political designs, but obviously uneasy about British intervention in Tibet. (Roberts, 1952: 532)

When dealing with the British, the Chinese and the Tibetans were happy to shift their responsibilities, pointing the finger at each other for causing difficulties or for the inability to fulfil an obligation.

The British and Russians came to an understanding not to intervene internally in Tibet's affairs, but that still of course left the question of where Tibet began. In 1904 Colonel Younghusband was put in charge of an expedition to Tibet to come to an understanding on issues of trade, borders, and envoys. He exceeded his orders, fighting his way into Lhasa and in the process killing unnecessary numbers of poorly armed Tibetan soldiers. He imposed a severe treaty on the Tibetans, which was subsequently unilaterally watered down by the British Cabinet.

The Chinese determined to reduce the power of the Dalai Lama, and to exercise more effective control over their province. In 1910 the Morning Post in London noted:

A great Empire, the future military strength of which no man can foresee, has suddenly appeared on the North-east Frontier of India. The problem of the North-West Frontier thus bids fair to be duplicated in the long run, and a double pressure placed on the defensive resources of the Indian Empire ... China, in a word, has come to the gates of India, and the fact has to be reckoned with. (Cited in Maxwell, 1970: 42)

Independent India was to find out 50 years later how prescient was this remark.

The British began to mount expeditions through the tribal areas, to map as much of the ground as they could, although they also conveniently had the excuse of punishing those involved in the murder of a British official (Noel Williamson) who had travelled beyond his designated mission. In 1913 they hosted a conference at Simla with Tibetan and Chinese representatives, with the idea of getting the Chinese to concede the autonomy of 'Outer' Tibet as they had conceded Outer Mongolia as a buffer between Chinese Inner Mongolia and Russia. The Foreign Secretary of the Government of India was Henry McMahon, who had previously worked with Durand to complete the line on the Northwestern Frontier. His brief was to negotiate a convention that the Tibetans and Chinese would both sign. By 1914 this had not been achieved, and McMahon, exceeding his brief like everyone else seemed to do in the Northeast, signed with the Tibetans, while the Chinese envoy was in the room next door. The latter knew something was signed, but not what. What was signed was an agreement on an alignment of the border, itself the product of earlier secret meetings that the Chinese had not known about.

Between Eastern Bhutan and the tracts of the Abors was a salient of territory known as the Tawang Tract, under the sway of the Buddhist monastery at Tawang. The cultural and economic ties of the Monpas and Sherdukpens who live there were more with Tibet than the plains, although they do not consider themselves

Tibetan and did not like Tibetan overlordship. Quite how McMahon persuaded the Tibetans to cede this tract to British India is not known, but Maxwell (1970) suggests it may have been in exchange for the British getting the Chinese off the Tibetans' backs. So, the British drew their new borders on their maps with some confidence, as near the watershed line as possible, and with due regard for other minor matters. In practice, they extended authority over the Tawang tract very hesitantly. As late as 1941 if British officials went to the area and found Tibetan officials there, collecting taxes or whatever, the British were instructed not to interfere. Only in 1943 did the British finally complain to Lhasa about 'illegal activity' in the area.

When India achieved independence in 1947, the Congress Government in Delhi accepted the 'reality' and legality of the McMahon line without much questioning. Equally, they accepted the idea of the 'Backward Tracts', and that they be administered by the Governor. In reality this meant nothing more than the occasional 'flag march' through the more accessible territory, avoiding direct contact with the tribes as much as possible. Then, in 1950, the Chinese invaded Tibet. This was not the army of imperial China, but the army of Mao's communist government, that wanted to tidy up its national inheritance. India viewed this with apprehension, and in 1951 strengthened the number of administrative posts in NEFA (the Northeast Frontier Agency), and moved officials into Tawang, while expelling their Tibetan opposite numbers. The government simultaneously realised that it had to enhance the administration of the other tribal areas, which were to the rear of any border to be defended. In 1954 Nehru appointed the legendary anthropologist Verrier Elwin, who had spent years with the tribes of Madhya Pradesh, as Advisor for Tribal Affairs in NEFA, in order to improve the quality of 'contact'. Given that employing tribal labour on road-building was out of the question, the government opted for developing administrative centres by air supply. Flag-showing on foot continued, even after the massacre of one patrol of the Assamese rifles in 1953 by members of the Tagin Tribe. A new service was established, the Indian Frontier Administrative Service, in status equal to the IAS, but with a career structure that would train and keep experience within the northeast frontier. In 1956 Khampa tribesmen in Northeast Tibet started an uprising against the Chinese. The rebellion spread, and by 1959 the Dalai Lama and his government threw in their lot with the rebels, proclaiming Tibetan independence. The Chinese put down the rebellion brutally, and the Dalai Lama and thousands of Tibetans fled into exile in India, via Tawang.

The Chinese government wanted negotiations to settle the border issues both in the northeast, and in the western Himalayas at Aksai Chin. There is good reason to suppose that they would have accepted the McMahon line, even though they repudiated the validity of the original treaty, if India first negotiated a number of detailed clarifications. India ignored this possibility, and adopted Nehru's infamous 'Forward Policy', through which it made its own unilateral demarcation on the ground, posting its own military pickets where it thought appropriate. One of these was six days march north of Tawang, positioned in 1962 by a junior officer, far beyond the McMahon line, as he sought a more comfortable camp-site for his men. Although the Indians had completed a road to Tawang itself in 1962, it took five days to drive from the plains to the town, providing that the road had not been wiped out by a landslide. But that same post was only a few hours from the head of an all-weather road built by the Chinese on the drier and flatter plateau of Tibet. The Chinese reaction to the Indian deployment was to surround the Indians, but not to attack them. The Indians then brought up reinforcements, and the army chiefs were ordered by the government to evict the Chinese. The military commanders were aware that they were in a very weak position if a shooting war started, but did not stand up to the relentless political pressure behind the forward policy. When in 1962 the succession of skirmishes flared up into a real war, India's forward units were very rapidly trounced by the Chinese, and a wholesale ramshackle retreat occurred, to the plains and beyond.

The road from the plains to Tawang started on the north bank of the Brahmaputra at Tezpur, upstream of Gauhati. The administration in Tezpur started a scorched earth policy, destroying the currency reserves, and evacuating the young, so they could not be indoctrinated by the Chinese. Much of NEFA was occupied by the Chinese, and Assam was left defenceless.

> In a broadcast to the nation, a broken and disillusioned Nehru lamented the country's plight. To the people of Assam, the next stage in the Chinese advance, he tendered his profound sympathy and heart-felt condolences ...
>
> The Assamese felt outraged and have never quite forgiven – or forgotten – the hopelessness of Nehru's call. (Rustomji,1983: 137)

In Delhi it was even wondered whether Calcutta was now defenceless. But the Chinese announced a unilateral ceasefire, and withdrew, to what they considered the proper demarcation of the McMahon line.[2] They had concluded their punitive incursion and achieved their aims. The Indian administration moved back into NEFA quite quickly, and was relieved to find that the tribes people had not been indoctrinated against their return. In 1987 NEFA was given statehood as Arunachal Pradesh.

Insurgency in Nagaland

The other North-eastern international border, that with Burma, has produced a different kind of war, and not one at Burma's instigation. Here was territory such as the Backward Tracts of the Naga Hills over which the British claimed sovereignty, but whose inhabitants, beyond the Inner Line, had refused to sign a

---

2 This refers to India's Northeast. In the sector west of Nepal the Chinese did not withdraw from Aksai Chin.

treaty with the British, partly, it is said, because being illiterate they could not trust what they were signing.

The advance of Christianity, particularly after the Second World War, promoted by American Baptists, and the adoption of the Roman alphabet and the bible in only three of the Naga languages, began to give the Nagas something they had never had before – the first glimmerings of a common identity. It was inevitable after the mayhem of the Second World War, when two opposing imperial armies had fought over the territory, that there would be some stock-taking at the end of it. In 1945 the British deputy Commissioner helped the establishment of a Naga Hill Districts tribal Council, which in 1946 renamed itself the Naga National Council – and which started publishing a slim newspaper called the Naga Nation. The title did not represent the past – because there had previously been no cohesive identity to the many tribes and languages – but it was another of the glimmering signs. Where group identity is not yet forged, intrusion from the outside can often stimulate greater solidarity. The leader of the Naga National Council, Zapu Phizo, tried before Independence to come to an understanding with India's national leaders, including Gandhi, but none of them would concede India's claim to inherit British Sovereignty. Shortly before India's Independence, Phizo and the NNC proclaimed Nagaland's independent sovereign status. For quite some time this momentous announcement amounted to very little, since there was no Indian administration to speak of in the backward tracts. Phizo and the NNC did however collect taxes, and organised their own form of administration. In 1955 the Indian response to a 'parallel government' (it wasn't, because the Indian Government did not function locally), was to use the army to put down the rebellion. The strength of opposition should not have surprised them, but action soon resulted in a full-scale war. It was to last fifty years, to tie down massive numbers of Indian troops, and despite a current precarious ceasefire, perhaps it has not yet ended.

The government adopted a forced programme of barbed-wire 'protected' villages – much like the British did in Malaya. But the people were not sedentary paddy farmers, they were hunters and shifting cultivators, who bitterly resented the drastic interventions in their life-styles. Simultaneously, 'development', was to be accelerated:

> Whereas in the rest of the country there is at least a semblance of control on the expenditure of public funds, for the hills there was a blank cheque. The overriding consideration was speed, and if a contractor demanded an exorbitant amount for constructing a road or building, it was conceded; for the tribal must not on any account be allowed to get the impression that the government were stinting money where tribal interests were concerned ... A vested interest was created in the promotion and acceleration of so-called 'development project'... And the contractor, whether tribal or plainsman, the administrator, the politician, all became increasingly involved in the ever-spreading web of graft and corruption. (Rustomji, 1983: 53)

Further, the major earlier source of 'development', the Christian missionaries, were being pushed out.

> It did come as a surprise, however, and as a shock, when foreign missionaries, who were non-officials, were found being eased out of parishes in the hills that they had been serving for the better part of their lives. India, according to her constitution, was a secular state. The first, and virtually the only people who had taken pains to help the tribal people in the field of education and health services had been the missionaries. (Rustomji, 1983: 62)

> It was not understood that, although Christianity was first preached to the tribal people by foreigners, it is neither foreign to India nor anti-national. (Rustomji, 1983: 64)

In 1962 Assamese was made the official language of Assam, and outrage erupted in many of the tribal areas. In 1963 Nagaland was officially separated from Assam and became a State in its own right.

Neighbouring Manipur has its independence movement too. It was a Princely State under the British Raj, and the circumstances under which the Maharaja acceded to India is subject to dispute. Days before the transfer of power, the Maharaja signed an agreement putting external affairs, defence and communications in the hands of India, while he retained internal autonomy. In 1949 he signed a merger agreement with India while visiting Shillong, and probably while detained incommunicado and against his will. The agreement made Manipur a Union Territory under the aegis of the central government.

As in Nagaland, so here, it was not hard for the rebels to get arms. Not only were many left over from the Second World War, until 1971 East Pakistan was a 'natural' source, since it was in Pakistan's interest to keep one quarter of the Indian army busy, so far from the West Pakistan border. China also aided and abetted the rebels, providing training and arms, for a similar reason, until the mid-1980s when it finally ceased its policy of exporting revolution. But even then, arms continued to flow from Yunnan in China, as manufacturers sought to make profits from sales to the insurgents. They came by land though Burma, and also by sea through Bangladesh. Militant groups like ULFA covered their costs by selling weapons further – to Nepal and to central India (Bhaumik, 2005).

India's twin track policy of massive military intervention and hasty development was pursued here too. In 1972 both Manipur and Tripura were granted full statehood, and so was Mizoram the last of the Seven Sisters in 1985 – a tiny state with less than 1 million people.

The Naga independence movement has so far failed, as much because of factionalism within the Naga movement as because of Indian pressure. The most significant of the rebel groups the National Socialist Council of Nagaland (NSCN-IM) signed a ceasefire with Indian forces in 1997. Slowly it leaked out that the ceasefire covered 'Nagalim' – the territory claimed by the Naga movement and which extended to all territory in which Nagas lived and operated – therefore including northern parts of Manipur. This brought outrage in Manipur, and rioting in the streets. In the 'Disturbed' areas where the army operates, it has been given draconian powers. Neither has there been sanctuary in Myanmar (Burma), with whose forces India has cooperated in their mutual suppression of tribal movements.

In the latest monthly statistics for September 2007 from Indian Army Eastern Command (http://www.eastarmy.nic.in/combating-militancy/militants-killed. html), no insurgents were killed in Nagaland, but altogether over the Seven Sisters in that one month the army claims to have killed 218 insurgents and captured 1429 others. Even in these relatively peaceful times, this level of violence is astonishing. It is also of course difficult to know how exactly to interpret the numbers without more context. Is it good if the army claims a large number of militants killed because it shows they are doing their job (but may be also stoking resentment): or is it good if none have been killed because it shows peace has prevailed (or the army is incompetent to fight)?

The Remnant Vale

The Vale of Assam, now the whole of the State of Assam given that its has lost its hills, has seen rising violence too. New electoral rolls were prepared in 1979, but many Assamese were convinced that they included many ineligible illegal immigrants. Protesting Students set up the All Assamese Students Union, which wielded real political power, and staged bandhs (shut-downs) from time to time to paralyse the state. In 1980 an opposing All Assam Minority Students Union was formed, to protect the interests of Hindu Bengalis and Muslims of Bengali descent. Confrontation between different groups was leading Assam into anarchy. In the midst of this, 2,000 illegal immigrants were massacred by Lalung (also known as Tiwas) tribespeople in 1982 at a place called Nellie. The 1982/83 elections were held under a security crack-down. The AASU split and lost some momentum, but another more militant organisation emerged, ULFA, the United Liberation Front of Asom, prepared to take on the Indian security forces. In 1985 an Assam Accord was signed between the Union Government and representatives of some of the dissenting movements. Among other things it stated that Assamese language, culture and heritage would be protected, and also that the issue of illegal immigration would be settled in phases, which would include expulsion of more recent immigrants. But no-one had both the will and the capacity to enforce it, hence the emergence of ULFA. In 1998 the Governor of Assam sent a report to the President of India, requesting amongst many other things, a barbed-wire flood-lit fence around Bangladesh, identity cards, the expulsion of recent illegal immigrants, and the restriction of earlier illegal immigrants, making them stateless and unable to own any property in Assam. He even went as far as suggesting economic aid to Bangladesh 'improving the family's income, women's education curbing population growth and educated women becoming a bulwark against the spread of Islamic fundamentalism.' (Governor of Assam, 1998: 16).

Around Assam's borders, confrontation continues, often following time-honoured patterns, though the language of report is modernised:

> Naga hoodlums are allegedly letting loose herds of wild elephants to scare away Assam villagers from their land and property in the disputed area in Sivasagar district along the Assam-Nagaland border. So much so that the Assam forest department has lodged a formal complaint with the Nagaland government to "stop the hoodlums" from disturbing the wild elephant herds at Geleki reserve forest, which straddles both sides of the border. (*The Telegraph*, 2007)

These years of turbulence have impacted heavily on the economy. Between 1991 and 1997 Assam experienced a decline in the size of its economy, even underperforming Bihar. The proportion of the population living beneath the poverty line in the northeast probably ranks highest in India, but statistics are hard to come by. The area still produces, uses and trades in drugs, and is awash with small arms. Despite the army's presence, India's sovereignty is still compromised. Many people pay 'taxes' to the rebel groups, which run parallel administrations – or more often in most of the territories outside the big towns, the only administration. The Chief Minister of Nagaland even defended this state of affairs: 'The people pay them with trust. When the people want to pay, how can the state government stop them?' (quoted in Hazarika, 2004).

Conclusion

Assam and the northeast was never integrated into any South Asian imperium prior to the British. Their annexation of the vale of Assam came comparatively late, and it remained the Cinderella of British India. If the Assamese were culturally distinct from the Bengalis, the tribespeople of the hills were completely 'other' – ethnically, linguistically and in ways of life and levels of 'development'. Living in impenetrable mountainous jungles, the idea of British sovereignty over them was a convenient and self-acknowledged delusion, never pressed too hard, although the regime permitted the influx of missionaries that promoted a kind of non-Indian modernity.

Mountainous regions like these rarely make good borders. The 'scientific' borders along the watershed, along a mountain ridge, may seem attractive and 'logical' but usually the people of the plains are very different from the people of the mountains. They understand how to live in harsh conditions, and often they inhabit both sides of the trading passes. But the densities of the populations on the plains are high: here is where numbers and taxation sustain large forces. There is a difference between mountains as borders, and mountains as suitable lines for military defence of the plains below. It was defence that committed the British to their heaviest intervention in the northeast, against the Japanese. It was defence that committed India (in this case unwisely even from a military point of view) to the mountains of NEFA to confront the Chinese.

Independent India is trying to go beyond this. It has attempted here what it has attempted in every part of India – the building of nationhood and an Indian identity. The attempt has often been clumsy, and too often brutal. Even now the status of 'local' people is different from 'immigrants' from within what is theoretically the same nation state. But India will not give up, despite the human and economic drain involved. Development of the region is made more difficult by the separation of East Pakistan/Bangladesh, and the new international border which has hindered trade and stimulated arms smuggling. The Northeast is indeed beyond the 'Chickens neck'. In recent years India has adopted a Look East policy, and fostered new trading relations with Southeast Asia. The border with Burma has been opened in a few places, though the trade that results is desultory. There are hopes that transit rights across Bangladesh can be negotiated. However, all of this points towards a sub-regional analysis – that the problems are those of the Northeast of South Asia, not just of India – and that in turn makes Delhi feel nervous.

Few people in Delhi or much of the rest of India give it much thought: few of the national papers carry stories about the region, and few maintain reporters there (Sonwalker, 2004). As it started with the British, so it continues, forgotten by all but a few in Delhi. Those who support the huge expense in hanging on to the tribal areas may have many motives: some, undoubtedly, to do with development and modernization. But there is also the stubbornness of not surrendering 'our' territory, and particularly not while there are outstanding disputes elsewhere, as in Kashmir. There are also the geopolitical considerations, that one day the Look East policy will need a land route from China through this region, that Burma has natural gas, that the strategic defence of India starts at the mountain crests and the water-wealth of the mighty Brahmaputra.

However, it is hard to see how utilitarian integration with India will improve, and even harder to see how identitive integration with greater India may be achieved. On the other hand, even if independence were achieved, what future could such isolated and small states have on their own? Yet, who should have the right to hold such reservations? The British once told Gandhi that if India got independence the place would be a mess, to which he responded, 'Yes, but our mess.'[3]

---

3 I have not been able to track down the original source. This is a riposte which Gandhi gives according to the biographical film made by Richard Attenborough.

Chapter 13

The Power Upstream

Introduction

Chapter 6 described how investment on a large scale in new technologies began to change the geography of India in the latter part of the nineteenth century; in particular, attention was paid to the development of communications and to large-scale canal irrigation. Of all the areas where such irrigation was developed, one stood out for the extent to which schemes and rivers were interconnected, and settlement patterns radically changed. This was the land of the Five Rivers – the Punjab, India's granary and source of high quality cotton. In Chapter 8 the problems that Sir Cyril Radcliffe faced in drawing up new international frontiers in Punjab and Bengal were discussed, along with his definitive (but unexplained) conclusions – the new lines drawn on the map. Radcliffe's line dividing India from Pakistan in the Punjab had been drawn between Lahore and Amritsar across the canal systems precisely where Coupland had said an international boundary would be wholly inappropriate. Radcliffe hoped that both sides would cooperate in the running of the integrated canals. Since not all arrangements could possibly have been ready by 15 August 1947, several commissions continued to sit for a specified period to sort out any remaining business and any tangles that might arise. The Arbitral Tribunal, which was to arbitrate on operational disputes, had a designated lifespan until 31 March 1948.

At that time India and Pakistan were locked in an undeclared war over Kashmir, the source of so much of Punjab's water, and India also suspected Pakistan of gun-running to Hyderabad. On 1 April 1948 India, alleging Pakistani intransigence over re-signing a standstill agreement, shut off the water flowing from the Firozpor (Ferozepore) headworks into Pakistan's Dipalpur Canal. Eight per cent of Pakistan's culturable command area (of the country, not just of this scheme) was deprived of water; so too was the city of Lahore. In addition hydro-electric power supplies were cut. Observers flying over the two Punjabs had no doubt where the border lay: to one side the seedlings of the new kharif crops were ready for the monsoon, to the other was a brown and barren land. India claimed that the waters in the rivers originating in her territory were hers, something which had never been suggested in the Radcliffe award; and in any event her action was contrary to the 1921 Barcelona Convention on international resource use. The Indian Government was a signatory to this treaty, which expressly forbade actions to the detriment of the natural conditions of a neighbouring state. Indeed, in India there were also many precedents against the action taken. The principles adopted during the building of canals had conceded that prior users must be duly

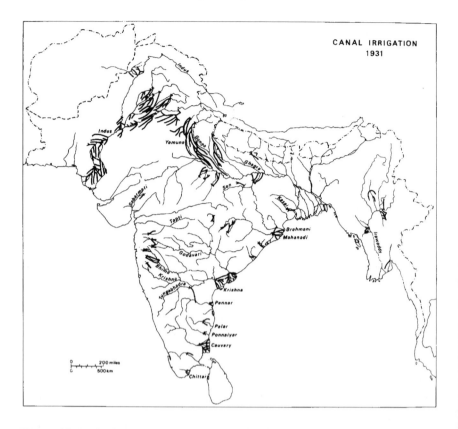

Figure 13.1   Canal Irrigation, British India 1931
Source Schwartzberg (1992)

considered and not disadvantaged by any new developments, and in addition there had been a principle that development should benefit the maximum number of users regardless of territorial boundaries. The first of these two principles had indeed held up new projects in Punjab for many years in the 1930s and 1940s while Sind pressed its case that it would be deprived of too much water.

Be all that as it may, India was using water as a political weapon.

In Bengal no canal systems of a comparable kind had ever been developed. Nevertheless, water is a key to the economy of the region. In particular the rivers of the delta are used as sources of irrigation water during the low season flow; and they stay fresh and uncontaminated by salt-water bores from the Bay of Bengal only if an adequate flow is maintained. Despite that, and again unilaterally, India built a barrage at Farakka, opened in 1975, upstream of the East Pakistan border to divert some of the low season flow down the Hugli (Hooghly) River, past the port of Calcutta, which was silting up. This barrage and competing schemes for the development of the Ganges-Brahmaputra Basin are a principal bone of contention between Bangladesh and India today, although other similar disputes are also emerging in the Northeast.

This chapter looks at the water disputes brought about or accentuated by Partition.

Hydro-politics in the Indus Basin

The water flow of the rivers of the Indus basin is curious, in that the flow of the western rivers in the more arid part is so much greater than the flow in the eastern rivers. This is a reflection of the size and nature of the catchments in the Himalayas, for, once the rivers reach the plains, they are essentially exotic, that is, not receiving contributions to their flow from the plains themselves. (This statement has to be qualified, in that during the wet season there is a contribution from the Punjab plains and from lower Sind.)

In 1850 plans were laid to construct the Upper Bari Doab Canal (3,000 cusecs) – the UBDC – to irrigate the doab between the Ravi and the Beas. Despite some problems with the scheme, such as its over-steep gradient and consequent erosion, the Sirhind Canal was sanctioned on the Sutlej in 1869 and opened in 1882, and with its gradient arbitrarily fixed at 30 per cent of that of the UBDC, it began to clog with silt. However, concerned with the protection of the food supply against drought, the Government accepted the learning process the engineers and administrators were passing through, and sanctioned ever more investment. In 1891 the Chenab Canal for the lower Rechna Doab was sanctioned, at 8,000 cusecs irrigating 1m acres. All these developments were characterised by the fact that they were run-of-the river schemes; that is to say that though they had diversionary barrages at their headworks, they did not involve water storage. Large dams in high mountains were to be a development of some distant future.

Table 13.1   Water Availability and Use in the Indus Basin, 1947

Discharges in Millions of Acre-feet (m.a.f.)

Indus and Kabul	110	
Jhelum	23	
Chenab	26	
Ravi	7	
Beas	13	
Sutlej	14	
Total	193	
	Use m.a.f.	*Irrigated Acres.*
India	9	5m
Pakistan	66	21m

Source: Michel, 1967: 33

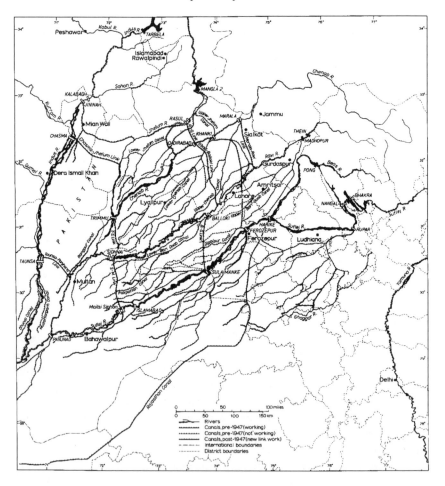

Figure 13.2   Irrigation in East and West Punjab, 1947–75

The most important function of the canals to begin with was provision of water during the low-water season, the 'winter', when, on the plains, wheat is grown. (Nowadays the canals have to provide supplementary irrigation during the wet season too.) Thus, to begin with the total amount of water used from the available year-round flow was in fact small, but could be no bigger given the technology of the period. Therefore, at a comparatively early date (the late nineteenth century), the schemes in the East Punjab were already running short of water. But there remained the rivers in the west, with their much greater flow. The Triple Canals Project of 1905 sought to solve a host of problems, by linking the rivers together and effecting a new transfer of water from West to East. The Upper Jhelum Canal fed Jhelum water into the Chenab, thereby releasing water in the higher Chenab for transfer via the Upper

Chenab Canal to the Ravi. This in turn meant that the Ravi now had sufficient water for a Lower Bari Doab Canal. Now there was truly an interdependent system. No part could operate without affecting other parts. In the 1920s work began on the Sutlej Valley project, including the Firozpor headworks of which we have already heard, which further increased demands on the eastern rivers.

Downstream in Sind there were major new schemes developing too, and, in particular the very large Sukkur Project which coincided with the Sutlej Valley Project. This meant that conflict over low-season water availability in particular was increasing, and in 1941 an Indus Commission was instituted to report on disputed plans by the two Provinces. From the 1920s onwards the plans began to involve the projected construction of large storage dams in the Himalayan foothills on the Ravi, Beas and Sutlej at Thein, Pong and Bhakra respectively. (Some of the water thus stored was earmarked for use in the arid lands of Rajasthan (Rajputana).) The report found that both parties could have much if not most of what they wanted, provided there were sufficient expenditure on the storage dams, and that Punjab should pay a contribution to those works downstream in Sind, which would be necessary to maintain the existing system there. (This came about because of the large number of inundation canals in Sind which relied upon the water level of the Indus maintaining a particular height. Thus, although there would be sufficient flow to satisfy the quantity requirements of these canals, the actual level of the river would have to be raised by new barrages.) Political disputes and the Second World War served between them to delay implementation of any settlement until it had become an international one of a different kind, in which one Punjab was at odds with the other Punjab and Sind, provoked by India's act of 1 April 1948.

India's attitude was that the proposals of the 1942 Commission had never been ratified; nor had a subsequent agreement in 1945 between the engineers (but not the politicians) of Punjab and Sind. She could therefore proceed without delay in implementing Bhakra, and by building a new barrage at Harike upstream of Ferozepur, divert the waters south into Rajasthan. She could even with tunnels divert water within Kashmir from the western rivers to the eastern ones, affecting a transfer before the water even reached Pakistan or the fringe of Kashmir under Pakistani control (Azad Kashmir). It meant therefore that Pakistan was forced to the negotiating table holding virtually no cards. One month later in Delhi an Inter-Dominion Agreement was signed, according to which India (technically East Punjab) agreed to restore the flow to West Punjab until the latter had developed alternative sources of supply; and during the period India would charge Pakistan a proportion of the running costs of the headworks and delivery system and in addition a charge for the water itself (seignorage). The last provision was disputed bitterly, but in practice was paid from a deposit by Pakistan to the Indian central government.

Although this represented an agreement of sorts, clearly it was not one which gave Pakistan confidence in the future. Both sides began new works to develop their resources, and, in the case of Pakistan, to extricate itself from the hold in which it was locked. It rushed through the development of a new canal of 5,000 cusecs from the Upper Chenab Canal to cross the Ravi in a siphon just north of Lahore (Figure 13.2). Later (1954–56) a new link between Marala on the Chenab and the Ravi was built (The M-R Link) with 20,000 cusecs, which meant that the Doabs downstream of India were virtually guaranteed protection.

In Sind, no longer a provincial backwater but containing the seat of Pakistan's government, there was an equally rapid development of the Ghulam Muhammad barrage at the head of the Indus delta, and later at Gudu in north Sind, upstream of Sukkur. (The Ghulan Mohammed barrage has been successful in diverting water into 5,000 miles of canals. But drainage of the flat delta is difficult, and the scheme is plagued with the problem of salinisation.)

Nearly all of these schemes had their origins in proposals considered before independence, but delayed for political or financial reasons. The difference post-1947 was the fact that they were no longer being built as part of an overall integrated system within the undivided Indus Basin, and it was particularly the case that the key part of the proposals prior to 1947, Bhakra, was indeed being built, but to reserve water for India alone.

The political dialogue between India and Pakistan had, of course, not stopped. In 1952 the two sides met in Washington under the auspices of the World Bank. It was still assumed even then that there could be a negotiated outcome in which the river basin would be a shared resource. But finally in 1954 the talks collapsed without a resolution. It had become clear that there was not enough water in the basin for both sides to do all that they had planned in their respective territories, and so an integrated plan in which both co-operated to achieve the maximum use of the basin actually became an impossibility. In the vacuum thus created the World Bank tabled a proposal (it had no power to make an award or an adjudication) to do the only thing left that was possible politically: to divide the waters completely and to allow each to do what it felt best with what it received. The entire flow of the Indus, Jhelum and Chenab would go, after a transition period, to Pakistan, and the entire flow of the eastern Ravi, Beas and Sutlej would go to India. By suggesting that the division was of water flow, not of assets such as headworks etc., the proposal side-stepped the thorny issue of divided Kashmir.

Although at face value it would appear that the proposal would give India all that she had wanted, in fact it did not, because it meant that she would have to forego any projects to divert Chenab water within Kashmir, and as a consequence she could only develop the Bhakra-Rajasthan project fully by building further dams on the Beas and Ravi. Further, India would be asked to contribute to development in Pakistan of new link canals made necessary by the proposals. Pakistan could see the broad advantages of the proposals, but felt it should be given assistance to build storage dams in its territory, not just new link canals. Thus although India accepted the proposal as the basis for detailed negotiation immediately, Pakistan held out until the Bank added an extra clause that allowed for the investigation of such issues in the event that the flows proved deficient.

The new terms showed the extent to which Pakistan's position had improved since 1948. Although there is no suggestion of any direct linkage and causality, the changed circumstances reflected Pakistan's changed role in the world. In 1954 Pakistan started to receive American arms supplies, and also became a member of pro-Western defence pacts, CENTO and SEATO. In 1956 Nehru professed the ideal of non-alignment in a public speech in the USA. India also developed sympathetic ties with the USSR from the early 1960s. We will consider these developments further in Chapter 13 below. In 1954 the World Bank was dominated by American expertise and American money, more so even that at present, and in such a climate it is easy to see how there would be shifts in the understanding of Pakistan's position. In 1955 the meetings between the two parties resumed. Negotiations dragged on over several years, and even the principle of awarding the three eastern rivers to India and the three western to Pakistan went back in the melting pot, as the Indians became aware of the costs of the Pakistani proposals, which continually added development to replacement. The solution this time was for a ceiling to be placed on the costs that India would face, plus loans to help it with the Pong and Thein dams, while also finding a consortium of countries who would help Pakistan financially with her schemes. By August 1959 the consortium, of the USA, Canada, the UK, West Germany, Australia and New Zealand (all connected with one or more of NATO, CENTO or SEATO) had agreed to find $1 bn for the cost of the schemes, half of it to be in grants, and all such grants going to Pakistan.

In September 1960 the Indus Waters Treaty was signed. It provided for the division of the waters by river, as proposed in 1954 (but with provision for existing minor works and their extension in the Indian-held upper reaches of the western rivers); for the funding of new works on both sides, but principally in Pakistan; and also in incredible detail and at great length for a ten-year transition period, during which India would continue to supply water to Pakistan. This specified how much water would be made available not only canal-by-canal, but also by time period within each season, and the extent to which changes would be made contingent on variations in water availability. What it specifically and explicitly avoided was recognition by either side of the other's rights in Kashmir: the formal position of countervailing claims was not resolved in the slightest.

Since then the Indus Basin Project of Pakistan has become one of the largest integrated schemes ever attempted by mankind on this planet. Two huge new dams have been built, at Mangla on the Jhelum, astride the border between Pakistani Punjab and Azad Kashmir, and at Tarbela on the mighty Indus. The latter is the largest earth- and rock-filled structure on earth. The resources employed in the dams, new barrages, new canal colonies, have been sufficient to accelerate inflation in Pakistan, rendering the original cost estimates inadequate. But, despite the wholesale new developments, Pakistan never acknowledged to East Bengal that this project should be counted in the balance of resources spent in the national plans in the West and East – because it was always claimed to be re-development rather than new development.

On the Indian side, the Bhakra-Beas-Rajasthan project and its associated works in Punjab have also been on a monumental scale. Prior to Partition East Punjab had much less of the irrigated land, and used far less water – 9 million acre feet of water on 5million irrigated acres. The completion of this scheme should mean that 15 million acres are irrigated.

The Rajasthan Canal, renamed the Indira Gandhi Nahar after Mrs Gandhi's assassination, more than 500 miles long, is the largest single canal project ever conceived. It is planned finally to irrigate 7.5 million acres, mostly in what is extremely arid desert regions within 40 miles of the Pakistani border. Although the main canal was completed by 1987, settlement has been occurring much more slowly than originally intended, mostly for political reasons, as debates and law suits have disputed who is eligible for settlement. By the mid-1980s perhaps 1 million acres had been settled and were productive. As there was tension between pre-Independence Punjab and Sind, so now there is tension between Indian Punjab, Haryana, and Rajasthan. The water used comes from the Pong dam, and crosses Punjab in a feeder canal, which was the subject of attacks by Sikh secessionists in the 1980s, who claimed that more of the water should be used in Punjab. It is also possible to see the canal in a military light: there are suggestions that it could have been built further east with a larger westward command area. As it is, it has undoubtedly changed the defensibility of the border with Pakistan. What was once open desert suitable for tank warfare similar to the terrain in the Rann of Kutch where there was fighting in 1965, is now terrain crossed by roads, and most importantly the canal itself, with a number of bridges. Nor is that Pakistan's only interest: being so close to the Pakistani border the Indians have been unable to develop drainage away from the area, so that waterlogging and salinisation will probably occur in many places on the Pakistani side of the border too.

Quite clearly there can be no human geography without politics, since politics is an inherent part of human group relationships. Yet one cannot help but wonder about some other utopian South Asia in which this great resource area had not been split. The grouping of the Provinces under the Cabinet Mission Plan would have made it possible, perhaps, for an Independent Federated South Asia (greater India) to continue the integrated approach that was being explored in the 1940s. But the experience of the struggle between Punjab and Sind during the British Raj and between Rajasthan and (East) Punjab now suggests that the prospects for agreement were always poor. Serious inter-state squabbles over water in India, resulting in public riots, police firing and deaths, are fairly common. The scale of the 'best technical solution' that nature proposes in South Asia often seems beyond the capacity of the political systems of management.

Dividing Bengal

Sir Cyril Radcliffe's other task was to draw a boundary line across the Ganges-Brahmaputra Delta, the largest delta on earth, and the most active, yet part of a well-developed regional economy with communications focusing on Calcutta. The new international border, just like the border in Punjab, would cause no great difficulties if the successor states were committed to the shared management of communications, infrastructure and environment – if indeed they continued to act as in a wider common market. But the bitter disputes that erupted between the successor states, over Kashmir and then over water in Punjab, and then devaluation (see Chapter 8) and Pakistan's export taxes, culminated in the fall of the revanchist curtain. Although economic and political links were severed, it was impossible for these states to pretend that they were no longer neighbours, that there were no ties that still bound them together, for the simple reason that they shared an environment that had no cognisance of sovereignty. In Punjab the water flow may have been divided between the two states, but floods still continue to flow from India into Pakistan whenever the discharge is too high. In Bengal, not only does Bangladesh (East Bengal) suffer from major floods originating outside its territory, it too is dependent, for some purposes in the dry season, on water from upstream states. The differences between the two cases, of Punjab and Bengal, are that Bengal can have no headwaters of its own, that there are four upstream states (China, Bhutan, Nepal and India), and the fluvial/estuarine environment of the delta is even more complex and dynamic than the hydrology of the Punjab. In addition, the major technical developments which are at the root of some of the first disputes post-date, rather than pre-date, Independence in 1947. The recent story of Bengal is mostly about one major contentious project – the Farakka barrage built in India on the Ganges – and a myriad of other problems of environmental management connected with flow augmentation, river training, flood prevention, and the minimisation of saline bores.

Farakka Barrage

Because the delta is so active, the geomorphology and hydrology of the area is constantly changing. Very roughly speaking, since the British first established their ocean port city of Calcutta 120 miles up the Hugli, one of the western distributaries, the flow of the rivers has been shifting more and more to the eastern distributaries. The mouths of the western distributaries are now known as the moribund delta – here the mangrove forests are thick. The eastern shore is very different, with bald new islands emerging from beneath the seas, but unstable ones liable to be shifted by the cyclonic storms that dash up the Bay of Bengal. Sometimes from the air it is very difficult to see where sea and land actually meet. The silt laden sea is a muddy brown which simply changes to a slightly different hue where it becomes a wet mudflat, and that in turn becomes lightly vegetated salt marsh. The belt of saltmarsh is usually small, since at the first opportunity some-one will try to claim it for agriculture, no matter how exposed the site may be to the hazards of tidal waves.

Calcutta was for long India's largest and most important port and manufacturing centre. It is still overwhelmingly the most important port for the whole of east and north-east India. Its population, for the urban agglomeration, is given as more than 13 million in the 2001 Census. Since its foundation, whether or not the port can maintain good communications with the open sea has been a matter of prime economic and strategic concern. So, whether the navigability of the Hugli to the sea has been deteriorating over time or not, has been a matter of anxiety and observation for much more than the last century. In all this time that the anxiety has been expressed, and the experts have delivered their reports, there has been no unambiguous nor unanimous conclusion about exactly how the condition of the river is changing. As a 'minor' distributary in the delta it has stretches where shoals form, and the navigable channel can get too shallow. For many decades dredging has been used to deepen the channels where necessary. But this does not constitute unambiguous evidence. Over time, sea-faring ships have got bigger, and their draughts deeper. And, as in any river, shoals can shift from place to place. The problem seems to revolve around a series of very complex questions – both in terms of data and in terms of theory – which include the following: i) has the discharge of the Bhagirathi linking the Ganges to the Hugli (Hooghly) been declining over time, particularly in the low season? ii) have tidal bores increased in the Hugli, bringing greater salinity further north? iii) do tidal bores bring greater sedimentation with them? iv) is augmentation of the fresh water flow in the Hugli 'flushing' the port of Calcutta and keep the Hugli open? v) has it been cheaper than other options (e.g. new ports downstream or more dredging)? vi) what impact does water abstraction have downstream of the barrage?

Table 13.2   Use of Ganges Water

Flow at Farakka (in millions acre-feet)

	Dry Season	Wet Season	Total (November to May)
Requirements	50	322	372
Bangladesh	42	6	48
India	17	1	30
Nepal	–	–	30
Excess/Deficit	-9	304	264

Source: Abbas, 1982

None of these questions have been answered unambiguously. The problem mirrors the current uncertainty over global warming and the 'official' response rather well. Most official reviews since 1850 have concluded that there is not much evidence of any long term decline in the Hugli, but on the other hand most surveys have suggested an increase in salinity at Calcutta and an increase in tidal bores: in other words there are lots of theories giving rise to concern, but very slight evidence. In these circumstance the Indian government effectively decided in the 1950s to go ahead with the Farakka Barrage – something along the lines of a precautionary principle. However, it did so without the experts knowing whether or not this would 'flush' the Hugli, and even despite the fact that some of them thought it would *increase* sedimentation. There were also suggestions that the real cause for the increase sedimentation downstream of Calcutta was the damming of the Damodar river, which previously flushed the lower parts of the Hugli – a problem wholly of India's own making. But the barrage would not just be about water: it would also carry the railway line round East Bengal to the land corridor to Assam and the north-east – reforging a link which had been broken when East Bengal inherited the Hardinge Bridge on the former Assam line from Calcutta. A public announcement was made of the intention to build the barrage in 1961, and the project was completed by 1975. It can take up to 40,000 cusecs from the Ganges, which at some periods of the dry season in low rainfall years is almost a majority of the flow.

From the Bangladesh viewpoint the diversion of water in the dry (low-water) season has a number of deleterious effects. Those most commonly cited are a loss of water for irrigation and urban use. Over the last few decades, as population has grown fast and the green revolution has taken off, the demand for good irrigation water from the rivers has grown substantially. In Bangladesh the reduced low season flow also allows saline bores to penetrate much further upstream in many rivers than hitherto, again reducing irrigation potential, and also possibly harming ground water. Table 13.2 shows the figures given by Abbas (1982), to demonstrate that there is already a deficit in the dry season, without taking into account the future needs as population increases and development occurs in the region, in particular increasing demands for irrigation water upstream in the Ganges plains. The government of, first Pakistan, and then Bangladesh, as the downstream and smaller party, has struggled to sign a treaty for the equitable sharing of the Ganges. The dispute is almost the highest priority issue in foreign relations for Bangladesh, and a subject of emotional fervour for the people. Several times the government has been near to organising a mass march of hundreds of thousands of people the few short miles from the border to the barrage. There is no doubt that there would be that number willing to march if asked.

Crow (1995) divides the history of negotiation into four phases. i) From 1951 to 1971 the project was discussed and planned. Pakistan protested but was impotent. ii) From 1971 to 1975 the barrage was built and implemented, without international agreement: attempts were made to sign a treaty sharing the waters between India and Bangladesh, but failed. iii) In 1977 a short-term agreement was signed between India and Bangladesh over the allocation of low season flows, running for five years. During this time it was imagined that the negotiation would broaden to encompass not just the Ganges but water sharing in the whole delta. The idea was for international co-operation in the management of both the Brahmaputra and the Ganges – which would enable water from the former to augment the latter. This would have been a westward shift of water a bit reminiscent of the eastward shift of water in the Punjab in 1905. But the Indus Waters Treaty of 1960 severed the scheme along the lines of political sovereignty. Bengal is now divided along lines of political sovereignty, and yet they were proposing to run an integrated scheme regardless of these borders. In the event there has been serious difficulty

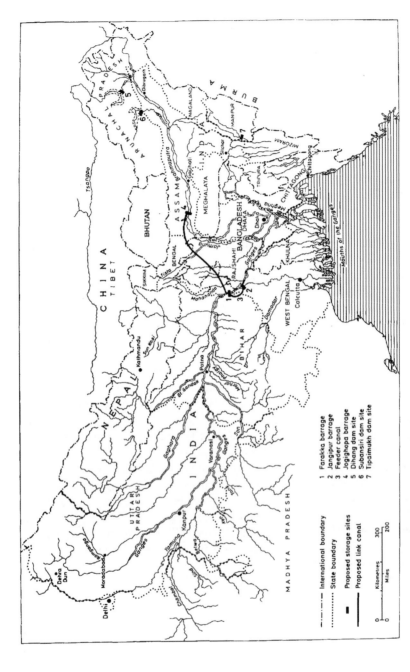

Figure 13.3 The Ganges-Brahmaputra Basin and the Link-Canal Proposal
Source: Abbas (1982)

in reaching an agreement, hence there have been a number of ad-hoc extensions of the 1977 agreement in the form of Memoranda of Understanding (MOU) pending a new 'final' solution. iv) From 1983 the emerging 'reality' is that a treaty dividing permanently all the waters of all the common rivers of Bengal will enable both countries to develop 'their' share on their own territory as they best see fit. This then replicates the 'reality' of what happened in Punjab over the Indus Waters.

Stage iv) has not been completed. The MOU ran out, although its provisions continued to be respected. But the core issue of Farakka has, for the time being, apparently been settled. In 1996 India and Bangladesh signed a Ganges Water Treaty, for the sharing of water at Farakka. The new arrangement is as follows: if the Ganges flow at Farraka is 70,000 cubic feet per second (cusecs) or less, both countries are to receive 50 per cent; with a flow of between 70,000 and 75,000 cusecs Bangladesh receives 35,000 cusecs and India receives the rest; with a flow of more than 75,000 cusecs or more India receives 40,000 cusecs and Bangladesh receives the rest. Further provision is made for the situation where the flow falls below 50,000 cusecs, but this is much less clear.

In the first year following the treaty itself, the dry season of 1997, the inflow hit a low of 46,000 cusecs raising a question mark on sharing. No-one knew who under the treaty would sort out the 'special' situation. Calcutta got its share of 35,000 cusecs for 10 days and stood to get just 11,000 cusecs for the next 10 days. Then common sense prevailed, and each got 50 per cent of the flow. This amount did not satisfy the Calcutta Port Authorities. At this level of flow they predict that Calcutta will silt up and the port die.

There are other ways of increasing the low season flow – through storage dams in the catchment – and of diverting water from the Brahmaputra. Figures 13.3 and 13.4 broadly summarise the conflicting strategies of India and Bangladesh in the early 1980s. Bangladesh wanted many storage dams in the Ganges basin, the biggest on the Nepalese border, to augment the low season flow during phase iii) identified above. (It would also allow a new navigation canal to link the new port of Mongla with Nepal, though crossing a neck of Indian land.) At various stages Bangladesh attempted to initiate tri-lateral talks with all three countries, although India resisted, insisting that it would be the vehicle through which the downstream-upstream issues could be discussed. The Indian proposal was to augment the flow at Farakka by diverting flow from the Brahmaputra. This would involve a canal half a mile wide right across Bangladeshi territory, with both of the significant control points in Indian hands. This was unacceptable to the Bangladeshis.

Farakka and West Bengal's Millennium Flood

In September 2000 there was a peculiarly sudden and destructive flood on the upper Bhagirathi, just down stream of the feeder canal leading from Farakka. It occurred after torrential rain unleashed by a tropical depression on a landscape already sodden by the monsoon. More than 20 million people of West Bengal, overwhelmingly poor rural people, lost their homes and all their possessions.

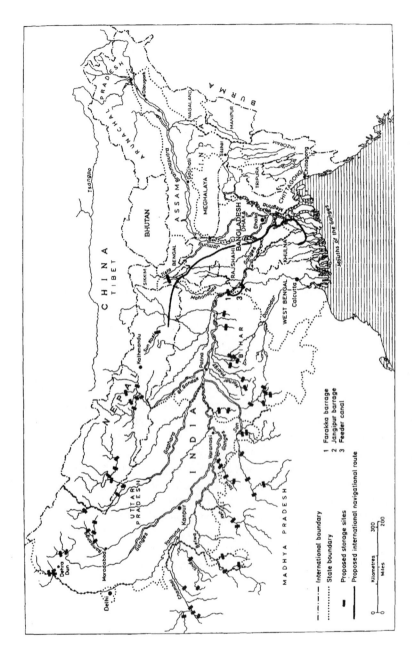

Figure 13.4 **The Ganges-Brahmaputra Basins and the Proposals for a Nepal-Bangladesh Canal and Other Developments**

Source: Abbas 1982

Several thousand died. A description of the flood and its multiple causes can be found in Chapman and Rudra (2007). Part of the reason why the flood was so destructive is that the landscape is crossed by many embankments, for roads and for railways, nearly always without adequate culverts and bridges for drainage, since they are more expensive than simple embankments. Other embankments are for flood protection, but they can fail in a high flood at unpredictable places in a catastrophic manner. The feeder canal from the Ganga to the Bhagirathi runs across several tributaries which used to flow into the Ganga. Since the barrage was built, their flow is captured by the feeder canal, thence into the Bhagirathi. During this great tropical storm the tributaries were all in record spate draining water from the plateau of Jharkhand. For this and other reasons, there is little doubt that Farakka significantly contributed to the misery of the local people in those dark days. The flood moved slowly south and east, into Bangladesh, where a further three million people were severely affected.

Floods in Bangladesh

In 1987, and particularly in 1988, there were very severe floods in Bangladesh. One initial response from the world's press was that it was all the fault of farmers in the Himalayas cutting down too many trees. Actually, there are three types of flooding in Bengal, which can happen independently, or (a true calamity) simultaneously. These are flash-floods from adjacent higher ground, usually around the edges of some of the Pleistocene terraces. They come fast and go fast. The second is river flooding when the major exotic rivers over-top their banks. Third, there is rain-water flooding, common everywhere, when the monsoon rain falls faster than it is drained. This is usual in most of the bhils (beels) and other back-swamps, and the whole of the Meghna/Sylhet depression routinely turns into a lake during the monsoon. None of this is necessarily too much water. The impact of the floods depends on the extent to which there is a 'normal' expectation of them and adjustment to them. Agriculture is finely tuned: flooded swamp land is planted as waters recede. In deep-water areas (predominantly in Bangladesh) there are long-strawed rice types that can grow up to 12" in 24 hours, and ultimately to 12' high, to keep abreast of rising floods waters. But throughout history there have been major floods, often when rainfall floods are backed-up by coincident river floods, which have caused major damage. On these occasions crops may be inundated long enough (total submergence for more than four days is usually fatal) to be killed off; human lives may be lost; draught animals may be lost with consequent impacts on the next land preparation; capital equipment (ploughs, byres, houses) may be lost. In addition, roads and railways may be damaged.

Flood protection levees have therefore been built in various places, some big, some small, some totally enclosing polders, to give some insurance against such losses. The problem is that the rivers shift their courses, and barriers built far back may even then find themselves under attack. If such barriers work for some years, and then are breached, the result can be worse than if they had never been built. Given the very high and ever-increasing population pressure in Bangladesh, people start to live in lower areas which appear safer after a barrier has been built, and hence these people are, almost literally, sitting ducks in the event of a breach of the defences.

The 1987 floods (Figure 13.5) (a map of the actual floods is not possible since there were no cloud-free days for satellite or aerial photography) were predominantly local flash floods and rainfall floods, whose drainage was impeded by exceptionally high river levels, particularly in the Ganges. The floods of 1988 were predominantly river floods caused by exceptionally heavy rainfall in the Himalayas and the Shillong block. The Brahmaputra (Jumna in Bangladesh) effectively became a river 50 kms wide.

In 1998 there were even worse floods, which inundated much of the country for the longest periods of submergence yet. The monsoon that year did not seem to stop: the rains continued well after their normal date of cessation. The floods caused extensive damage to urban and industrial infrastructure as well as to farms; but the death toll was lower than for the 1987 and 1988 floods because the government was better prepared in providing clean water and food for the population. In 1999 there was again extensive and late flooding.

The floods are seen at least partly as a separate issue from the Farakka problem, because the only major works upstream that could mitigate them (and then only some of them) would be the colossal storage dams which would have to be managed tri-laterally by India, Nepal and Bangladesh (and conceivably by China and Bhutan too). And one reason why Bangladesh has recently begun to shift its negotiating stance is that it has conceded that the dams might be too big, too costly, too far off in the future, and too unreliable in terms of operational performance (much of the rainfall that causes flooding would occur downstream of the storage sites anyway.) The reason why Farakka can be partly implicated is because there are claims (in my view unsubstantiated) that siltation in the Ganges has increased downstream of Farakka, so the bed has risen, making high season floods more likely. Then, in theory, if the Ganges rises higher, the Brahmaputra can be caused to back-up higher as well.

Several countries and country groups are involved in planning to obviate as far as possible repeats of Bangladesh's flood disasters, or at least to alleviate their worst effects. The G7 group of developed industrial countries have involved themselves in attempting to devise 'a solution' to the floods problem, and French and Japanese proposals were also forthcoming for different embankment schemes. The World Bank then became involved in producing a co-ordinated Flood Action Plan with the Government of Bangladesh, publishing the first overview and proposals in late 1989 (World Bank, 1989). The plan was abandoned in 1995.

The proposals cover many different aspects of the problem. They include studies of the engineering and siting of embankments, the development of compartmentalised polders, river training, and approaches to improving the management of water control, flood warning and flood preparedness, and enhancing flood refuges. The compartmentalised polders would allow selective flooding of

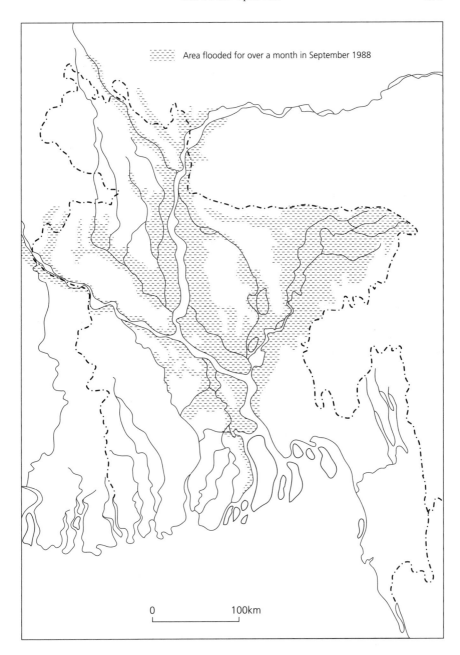

Area flooded for over a month in September 1988

0   100km

Figure 13.5   1987 Floods in Bangladesh
Source: Brammer (1990a)

land in the event of extreme flood events. All of these proposals therefore constitute 'within country' solutions; again acknowledging the difficulty of international co-operation. However, it is even unclear how it would relate to water transfer from the Brahmaputra to the Ganges via a link canal inside Bangladesh.

No-one should trivialise the complexity of the problems faced. Simplistic 'solutions' such as the massive re-afforestation of the Himalayas advocated by the western media after the 1988 floods are not helpful. (And the extent of deforestation in the first place is doubtful.) We do not know the extent of accelerated erosion in the Himalayas, if there is any at all. We should not forget that landslides following earthquakes probably shake most of the sediment into the rivers, that these are after all the youngest and highest mountains on earth in the area of highest rainfall, that natural erosion rates are extremely high, and that the Bengal delta exists precisely because the rivers have been laying down silt and flooding for the last 60 million years. Nor should we forget that the climate is unstable, and that our records of the last 100 years reflect a paltry sample of what can happen.

The Ganges-Brahmaputra-Meghna Basin

The Power of Nepal and Bhutan

Since the first edition of this book, there has been an increasing stress by many NGOs and professional associations, and government think tanks, on considering the basin-wide problems of water use and availability. The considerations embrace demographic, economic, political, and legal aspects, set in the context of international rivalry. The largest part of the flow of the Ganges at Farakka stems from the Himalayan tributaries it receives from Nepal. Nepal is a mountainous and less densely settled state than the adjacent areas of the Ganges Plains, and it has enormous hydro-power potential, supposedly as much as 83,000 MW – or very nearly as much as all of India's installed capacity in 2000, amounting to 98,000 MW. Four major perennial rivers (the Mahakali and Karnali, which form the flood-prone Ghagara in India, the Gandak and the Kosi) offer opportunities for mega-dams that could provide huge exports of power to India, and provide flood protection possibly extending as far as Bangladesh, augment the low season flow at Farakka, and provide irrigation on the plains.

The mega-projects have their detractors, fearful of excessive social and environmental costs, and the risks of structural failure during an earthquake. To Bangladesh and many Nepalese, it becomes rational to talk not of one river basin – the Ganges – but of the GBM – embracing the Brahmaputra and Meghna as well. On the whole India wishes to keep its relationships bi-lateral and resists this encroaching concept. The scale of these projects is so great and the money stake so large, that they could (have begun) to affect all parts of Nepalese politics. Some observers believe that Nepal's young and struggling democracy could be undermined by external pressures. The struggles over pricing policies are critical.

Table 13.3   Major Dam Proposals for Nepal

	Capacity MW	Dam Height (m)	Reservoir storage millions m³	Submerged area (ha.)
Karnali (Chisapani)	10,800	268	16,210	33,900
Mahakali (Pancheswar)	2,000	288	4,800	4,750
Mahakali (Poornagiri)	1,065	156	1,240	6,500
Kosi High Dam	3,300	269	9,370	19,500

Source: Rasheed, 1995 and Subba, 2001

What are the costs to Nepal, in land submerged and people dislocated? How can the flood-control benefits in downstream India be accurately valued and monetised? What will be the price of electricity in India in, say, 2020, in a subsidised and regulated or free market? Can it be almost as high as the next best alternative – derived from coal-burning thermal stations on the plains?

Bhutan has an estimated hydro-power potential of about 20,000 MW (Subba, 2001). It could only ever use a fraction of this for domestic purposes. Small to medium projects in the range of 300 to 1,000 MW have been built or are under construction, with Indian aid and expertise, but the power demands of Assam are as yet much below those of the middle Ganges plains. Bigger project are also entangled in the same pricing arguments as between India and Nepal.

The GBM and India's River Linkage project

To many observers, even more alarming is the prospect being debated in India of the River Inter-linkage Programme. Viewing India as a whole (see Figure 1.3), the northern and eastern areas, and the southwest, are wetter (marked by tropical moist or tropical monsoon forests), while the rest of the Deccan and western India is dry. The rivers that drain the Himalayas are perennial, whereas only the largest of the Deccan rivers have a perennial flow. The logic seems inescapable: build canals from the north taking water to the south, on both sides of the Peninsula. By far the biggest source of 'untapped' water is the Brahmaputra. The idea of transferring this water has been around for quite a long time, and Figure 13.3 shows such a proposed link, across Bangladeshi territory. In this context the link would enable India to release more water down the Ganges to Bangladesh – in effect meaning that Bangladesh would not 'notice' Farakka so much. The river inter-linkage scheme has, however, changed the position of the canal. Even though Figure 13.6 is sketchy, it can be seen that the proposed new canal is wholly within Indian territory. It would be easier to avoid demands from Bangladesh to influence its operation – but it would also be even more expensive and environmentally riskier than the original 'international' line. The canal would be half a mile wide,

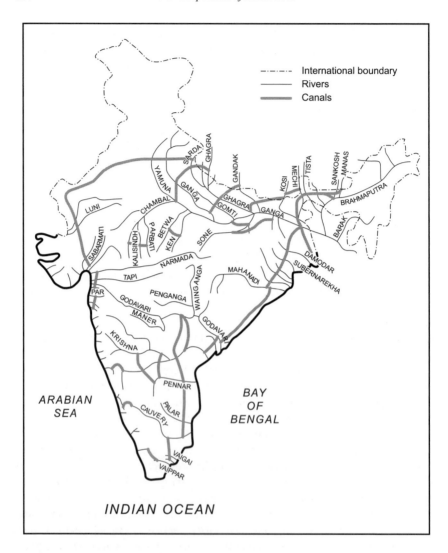

Figure 13.6   India's River Inter-linkage Project
Source: National Water Development Authority (2007)

and would have to cross the mighty and flashy, sediment rich rivers like the Teesta, to the barrage at Farakka, thence across the Ganga, to new canals going south. By off-setting the rivers against each other, effectively 'substitute Brahamputra water' reaches Chennai as 'extra' water.

The scope for environmental disaster seems almost limitless. Although for the management of most environmental concerns, the political division of the Bengal delta is a tragedy, in this case if the division halts this project, then it is a blessing.

India's standing with Bangladesh on these issues is hardly helped by her recent resolve to build the Tipaimukh Dam in Manipur, on the Barak river, upstream of Sylhet.

Concluding Remarks

This chapter has dealt with issues that lie at the interface of environmental management and national sovereignty. The definition of the problems faced and the derivation and implementation of 'solutions' depends on the resolution of political, economic and technical assessments. The economic assessments normally revolve around the idea of viability – that benefits outweigh costs. But this is not always the case. The British accepted, and so have independent India and Pakistan, that investment in irrigation might prima facie appear uneconomic, yet in terms of providing food security, social stability and a longer term and more diffuse return to the economy, such schemes could be acceptable. This is, of course, a political judgement. What can be economically viable also changes, obviously, with changing technology. The feasible technology of environmental engineering has embraced larger and larger scale projects. Considering for example the issues raised in this chapter, the scale increases from inundation canals, to barrages, to small dams and then to large multi-purpose power/flood-control/irrigation dams, or river training and embankments on a massive scale. The largest scale works require financing from the highest political levels. In the cases we have considered of the Indus Waters Project in Pakistan or the Flood Action Plan in Bangladesh these have required international support. But, even at lesser scales, the political structure of society determines the parameters within which decision-making occurs. Thus, no matter in which order these issues are addressed, political factors come out as the dominant ones. Political feasibility underlies the resolution of technical and economic proposals. The great integrated irrigation schemes of pre-Independence Punjab were conceived and executed under British hegemony. They could and would survive only so long as an equivalent hegemony survived – and that did not happen. In Bengal, too, the sequence of events has shown that any grand scheme based on the sublimation of political sovereignty to technical optimisation has remained a dream.

In terms of the identitive, utilitarian and coercive forces of integration, it is quite clear that, although there have been and are environmental utilitarian imperatives for integration, these have never held enough sway with the broad body politic. We do not know, because it was never measured and probably could never have been, how far the publics of each half of Punjab or Bengal saw their welfare interlinked with their other half. Everyone knew that there were linkages; but perhaps because these could not be quantified, they would always rest second to a precisely defined and clearly seen goal – political self-determination (even if it actually turned out to be something of a mirage). In other words, utilitarian bonds invariably come second to identitive bonds.[1] What of coercion? Some observers of the world's hydro-political scene predict that the next wars will be over water, not over oil as in the last war against Iraq. Water scarcity, they predict, could ignite conflict between Turkey, Syria and Iraq, or between Egypt and Sudan. In India conflict between the states over water has resulted in police firings and deaths – between Tamil Nadu and Karnataka, for example, over the Kaveri river, or between Haryana, Uttar Pradesh and Delhi over the Yamuna. The irrigation-hungry and power-hungry plains of North India can cast longing eyes over the water resources and dam sites of Nepal's Himalayan foreland (see Chapman and Thompson, 1995). What role might such issues play in international relations in South Asia? I suspect and hope that the answer is that water will not ignite international armed conflict – although not necessarily for enlightened reasons.

In the west, as a result of the Indus Waters Treaty, each side can mostly act as it wishes with the water it has been allocated. There are conflicts – over shifting river courses on the border and the consequent river- training works, over drainage and local floods – but these are not of major international significance. Having said that a young Pakistani Brigadier attending a year-long course at the Royal College of Defence Studies in London in 1990 wrote that disputes with India over the Indus waters could result in future conflict (Ramesh 2005). That Brigadier Musharraf became a general and President of Pakistan, and his anxieties seemed borne out by Indian plans to build the Baglihar dam on the River Chenab in Indian Kashmir. The World Bank arbitrated the subsequent dispute between India and Pakistan, both of which agreed to the resolution. However, this did not happen until after the dam had been built, and resulted in fairly modest reductions to Indian designs.

In the east the scale of the problems from Bangladesh's point of view is significant and large. But India is the upstream power with the environmental whip-hand and does not need military force to implement many of its unilateral solutions. Moreover, India is such an overwhelming and dominant military power that Bangladesh could never challenge it in armed conflict without a major military ally to back it. That could only happen in the almost unimaginable scenario of Bengal being used as a launching pad for another invasion of South Asia. It would be a totally unimaginable scenario had it not happened once before, two centuries ago. In the case of Nepal and Bhutan, it will probably be in their long-term interest to sell energy, irrigation and urban water, and the services of flood control to the populations of the Ganges plains and Assam. These are their best export resources.

---

1 This ought to be a salutary lesson for European enthusiasts in Britain who assume that the majority of the public understand how closely our welfare depends on unity within Europe. It is important not just that the British learn how to exploit Europe: they also have to learn how to be European, otherwise any 'mad-cow' set-back can cause a knee-jerk popularist reaction and an exit from the Union.

Chapter 14
The Greater Game

Geopolitics

The opening chapter of this book relates the sub-continent of South Asia to other continents through a discussion of geology and plate tectonics. Most of the rest of the book concentrates discussion on South Asia itself, although external factors such as the invasion of Mohammed Ghori, or of the first Mughal Babur, or the advent of the British and their subsequent pre-occupation with the North-West Frontier, are reminders of the significance of the relationships between South Asia and the outside world. This chapter looks at how the international relationships of South Asia have developed in the years since Independence. The word 'international' here subsumes two different kinds of relationships: those between the South Asian states themselves, and those between South Asian states and countries outside the region. Until the last moments of the Raj, the British had mostly managed to contain any difficult regional relationships between the parts of their Indian Empire. They were, however, concerned with the political, military and economic relationships between India (South Asia) and the outside world. They had more or less succeeded in keeping other external powers at bay, although they perpetually saw a bogey in the guise of Imperial Russia and later of post-revolutionary Russia in the northwest, which they feared would extend its power like a latter-day Mogul through Afghanistan and then on to the Indus and Ganges plains. This anxiety accounted for numerous military interventions in Afghanistan. The rivalry with Russia was what Kipling termed 'The Great Game', and the phrase has been used ever since to hint at some stratospheric geo-political game of chess, beyond the vision of mere earth-bound mortals. On the other flank of India in the northeast, during the Second World War the Japanese did take Burma and threatened at one stage to break into India too. The Himalayas, and empty and arid Tibet beyond, provided a strong defence line in the north.

There have been attempts to provide theoretical perspectives on these events and relationships, one of the first and still one of the most influential being Mackinder's theory of the heartland, the Pivot of History. The original paper, published in 1904, still bears reading: my attempts to summarise salient points do not do full justice to the original, which the reader is urged to consult in full. In essence, some of his paper seems empirical, detailing the history of the peripheries of the Eurasian landmasses – Europe and China – as at least in part responses to the pressures of successive waves of Asiatic nomads – Huns, Avars, Bulgarians, Magyars, Cumans, Mongols, Kalmuks etc. But other parts are more deductive, asking why the heartland of Eurasia has remained something of a citadel, radiating

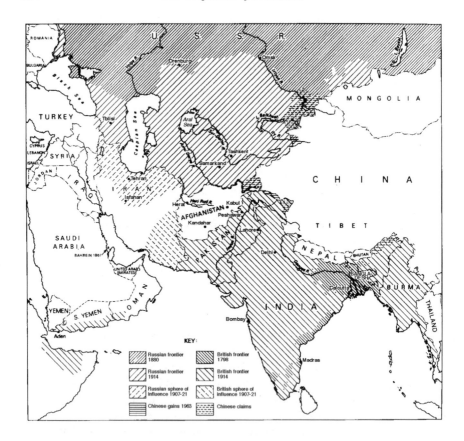

**Figure 14.1   British, Russian and Chinese Spheres of Influence in Asia,
1800–1965**

influences out, but not admitting reverse invasions. The answer lies partly in
what kinds of movement are possible. For the peripheral world, river and ocean
transport was and remained, until the railway age, much easier than land transport.
In the heartland of the steppes, the horse ruled, and physical geography denied
the peripheral world the opportunity to use its own technology to penetrate the
interior. This was because the Eurasian heartland is a vast tract of the earth which
drains to the frozen Arctic ocean or to inland seas, so the European and other
maritime forces have no way to penetrate the river mouths and thence inland.
The distinction between the Pivot area and the Outer Crescent (see Figure 14.2)
was maintained even in the railway age – and indeed the railways reinforced the
separateness of the two arenas of deployment:

The Russian army in Manchuria is as significant evidence of mobile land-power as the British army in South Africa was of sea-power. (Mackinder in de Blij, 1967: 289).

Commercially the two arenas were distinct too:

In the matter of commerce it must not be forgotten that ocean-going traffic, however relatively cheap, usually involves the fourfold handling of goods – at the factory of origin, at the export wharf, at the import wharf, and at the inland warehouse for retail distribution; whereas the continental railway truck may run direct from the exporting factory into the importing warehouse. Thus marginal ocean-fed commerce tends, others things being equal, to form a zone of penetration round the continents, whose inner limit is roughly marked by the line along which the cost of the four handlings, the oceanic freight and the railway freight from the neighbouring coast, is equivalent to the cost of two handlings and the continental rail freight. English and German coals are said to compete on such terms mid-way through Lombardy. (Mackinder in de Blij, 1967: 289)

The Czarist empire was therefore essentially a land empire built around the resources of the pivot area. It may seem odd, or an exaggeration, to consign the rest of the world to a maritime Outer Crescent, to label the vastness of America, Africa and of Australia as maritime, but if the distribution of population is observed, for all these areas the bulk of the population is at or near the coast, and the coastal urban centres are interdependent on each other. This was where the British radically altered the geography of India, adding port cities to the periphery of what had been a land empire.

Mackinder's geo-political vision has survived, and been revived in different forms by later writers. Cohen (1963) derived a very similar deductive political regionalisation of the world, by applying two concepts, the higher level concept of the geo-strategic region, and the lower level concept of the geopolitical region. His geo-strategic regions are multi-featured in cultural and economic terms, but are single-featured in trade orientation and are also distinct arenas within which power can be projected. Mackinder's Outer Crescent is re-named the Trade Dependent Maritime World, and the Pivot area is renamed the Eurasian Continental power. Between these two are the 'Shatterbelts' of Southeast Asia and the Middle East. There is no doubt that such a Mackinder-Cohen view of the world informed US, British and Czarist/Soviet foreign policy for years. It is still an influential thesis.

The Trade Dependent Maritime World has strategic nerve centres. Since the advent of the supertanker, one of these was the Cape of Good Hope, round which the vast majority of Middle Eastern oil reached western consumers. It was therefore in the interests of the Soviet Union to be able to disrupt this trade with submarine threats, and in the interests of the western powers to deny the Soviets access to the great oceans, to keep them bottled up in identifiable Arctic ports such as Murmansk, locked in the Baltic, and under surveillance in Vladivostock. What the West did not want was a move south to a warm-water port – in the

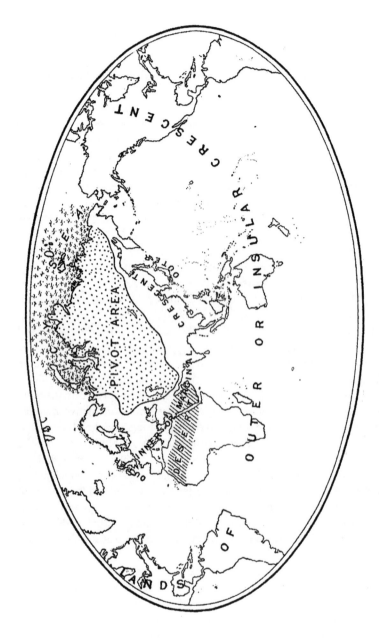

Pivot area- wholly continental.　Outer crescent- wholly oceanic.　Inner crescent- partly continental, partly oceanic.

Figure 14.2　Mackinder's Geopolitical Vision

Source: Mackinder (1904), De Blij (1967)

mouths of the Indus for example. Hence the containment policy of the USA was to ring the pivot area with the NATO, CENTO and SEATO defences, and hence the bitter determination of the USA not to allow the Soviet invasion of Afghanistan in 1979 to succeed. Hence also democratic America supported the apartheid regime of South Africa and its allies in the civil war in Angola against the Cuban- and Soviet-backed government.

Cohen's geo-political region is defined as a sub-division of the geo-strategic;

> It expresses the unity of geographic features. Because it is derived directly from geographic regions, this unit can provide a framework for common political and economic actions. Contiguity of location and complementarity of resources are particularly distinguishing marks of the geopolitical region. (Cohen 1963: 62)

So the Trade Dependent Maritime World is divided into Europe and the Mahgreb; Africa minus Egypt, Sudan and Ethiopia (part of the Middle East Shatter Belt); North America; South America; and Australia along with New Zealand and Oceania. The Eurasian region is divided into the USSR and China. In this scheme, South Asia is distinctive and unique: Cohen classifies it as an independent geo-political region, the only such on earth, not contained within either of the geo-strategic regions. It is big enough to be a sub-continent in its own right; it has been and is guarded from the Eurasian power(s) by the massive wall of the Himalayas, from the Middle East by the Hindu Kush and other mountains of the Northwest Frontier, and from Myanmar and Indo-China by lower, but heavily forested, jagged mountain ranges. All the imperial phases before the British had created land-empires, which at some point more or less embraced the whole geopolitical region, and the British too had acknowledged this when their capital was moved to Delhi in 1911. If united, South Asia has clear lines of defence, and it has the options of self-sufficiency or access to the Trade Dependent Maritime World. But if divided, the scenario changes.

Antagonists and Protagonists since 1947: The Actors

When the British left India shortly after the end of the Second World War it was uppermost in their minds and an abiding passion with Mountbatten that the defence of South Asia was unitary. To that extent, having conceded Pakistan, nevertheless they also bequeathed a joint defence council for the two new neighbouring Dominions. Unhappily it did not survive long, because of the hostilities over Kashmir. From the Kashmir conflict onwards, South Asia was split open. Internal dissent laid the region prey to outside powers in a way which otherwise would not have been possible: for hostility within South Asia immediately meant that the rivals would be seeking outside support, if not alliances. The major world powers that emerged at the end of the war were the Soviet Union and the United States – the two poles of the two geo-strategic regions; but another major regional

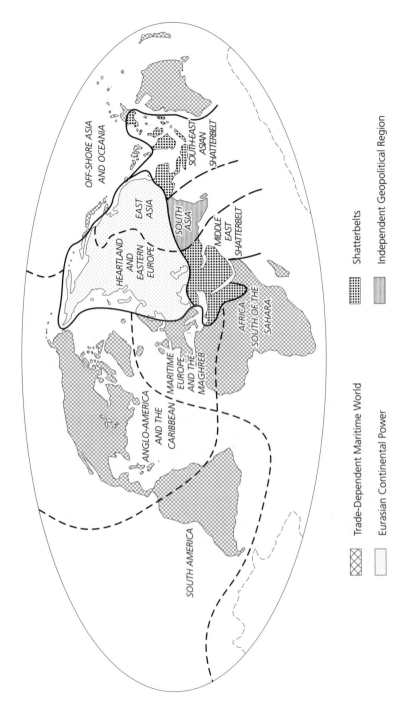

Figure 14.3 Cohen's Geopolitical Vision

Source: Cohen (1963)

Legend:

- Trade-Dependent Maritime World
- Eurasian Continental Power
- Shatterbelts
- Independent Geopolitical Region

Map labels:

OFF-SHORE ASIA AND OCEANIA

SOUTH-EAST ASIAN SHATTERBELT

EAST ASIA

HEARTLAND AND EASTERN EUROPE

SOUTH ASIA

MIDDLE EAST SHATTERBELT

AFRICA SOUTH OF THE SAHARA

MARITIME EUROPE AND THE MAGHREB

ANGLO-AMERICA AND THE CARIBBEAN

SOUTH AMERICA

power also emerged in Asia, with the communist victory in China in 1949. The alignments between India, Pakistan and these three outside powers are the focus of this chapter. What is remarkable is the extent to which the same pattern of alignment for long survived many changing circumstances. The reasons for this I do not think difficult to demonstrate, but the methods of my arguments are not those typical of most political scientists. I see less freedom of choice facing national leaders than I suspect they do. After the al-Qaida attacks of 2001 and the resulting American "War on Terror' in Afghanistan and Iraq and elsewhere, perhaps we are seeing new alignments. On the other hand the resurgence of a self-assertive and combative Russia has perhaps reflected the continuation of Mackinder's imperatives.

The Soviet Union/Russia

The Soviet Union was relatively slow in the years after the Second World War to develop a policy towards the emerging states of Asia. It was pre-occupied with events in Europe, and with its own reconstruction and rearmament as the Cold War enveloped East–West relationships. In India, the Communist Party was in avowed opposition to Congress, and in the early fifties sponsored a terrorist revolution in the Telengana region of Andhra. Ideologically the USSR 'should' have been supportive of this nascent revolution, but the ease with which it was suppressed confirmed for the Soviet bystanders what they had suspected also for ideological reasons, that it was not possible to jump from feudalism to communism without first going through the class-consolidating stage of capitalism. The Soviet Union therefore began to adopt a more pragmatic approach by allowing itself to extend economic and other aid to the Congress government, even though the latter was élitist and bourgeois and only partly socialist. Ideological doubts about Pakistan could not be so easily overcome, for this was a state founded on communal grounds. It was an emergent Muslim State not so far from its own southern Muslim Republics.

In the late 1950s the Chinese began to split from the Soviets. Their hostility to the imperialistic style of the USSR's management of East Europe and the USSR's reluctance to let China have access to nuclear technology, and a host of other issues, brought the two communist giants into open opposition. In 1969 they were involved in armed border skirmishes along their Amur and Ussuri river borders.

On the southern flank of their empire, the Soviets became more heavily involved in events in Afghanistan. They were clearly implicated in the overthrow of the King Md. Zahir Shah in 1973, and the establishment of a left-leaning republic. Their defeat at the end of the Afghan War, 1979–89, described more fully in the next section, helped bring down communism.

With the collapse of the USSR in 1990/91, the constituent republics became independent in name at least. But what has resulted from this is, in some sense, a more transparent acknowledgement of the true map of power within the Eurasian heartland. Russia probably remains the dominant power. Both through its association with states such as Kazakstan and Tajikistan in the Commonwealth of

Independent States (CIS), and through specific military treaties, Russia continues to project real power throughout the CIS. In Tajikistan, for example, a large-scale Russian military presence helps patrol the southern borders, which divide Muslim fundamentalist Tajiks in Afghanistan from the less radical Muslim groups of Tajikistan.

Foreign policy cannot, however, continue exactly as before. The Warsaw Pact has collapsed, and the outright confrontation with the West has been replaced by more diffuse and complex security concerns over the encroaching proximity of NATO; yet the nuclear weapons of the Cold War adversaries are no longer targeted at each other. To mollify Russian concerns, in 1997 a NATO–Russia Permanent Joint Council was established under a Founding Act on Mutual Security. The economic decline of Russia during the painful years of the transition from a command to a demand economy also had an impact on foreign policy, not just its own, but the foreign policy of former allies as well. Russia is now more prepared to sell military expertise for realistic prices to the Chinese and other bidders, rather than subsidise the export of military hardware and expertise for political purposes. This has meant, for example, that the cost to India of maintaining its Soviet weaponry has escalated rapidly, while the source has become far less reliable. President Putin (in office 2000–2008) reasserted Russian power, partly strengthened by higher gas and oil prices in an energy-short world.

Since the ending of Russian involvement in the Afghan war of the 1980s, Russia has fought two more wars against separatists in Islamic Chechnya. The territory, on the north flank of the Caucasus mountains and adjacent to the Caspian oilfields, was conquered by Czarist troops in the 19th century. Armed struggles to regain its independence have occurred at intervals since then. It is suspected that separatists (to use a term which is neither 'terrorist' nor 'freedom fighter') are behind bombings in Moscow and elsewhere in Russia. What is certain is that Chechen separatists have joined hands with extremists in Afghanistan.

The USA

The USA saw its prime role after the war as the containment of communism, and to that effect to assume, wherever possible, the mantle of global defence that the British were fast casting off. Ideologically, they were of course sympathetic to the world's newest and largest democracy: but such sympathy would only grow if India aligned itself with the crusade against what the Americans saw as democracy's greatest danger – the global Communist threat. Pakistan too was, in name, a democracy in the early 1950s, so there was no ideological difficulty in maintaining close diplomatic links with both countries.

The position in the early 1950s was therefore one of many open opportunities, but as yet few binding commitments. The first steps to change this were taken both by the Pakistanis and the Americans. The former desperately needed arms and outside support to survive what it perceived as the Indian threat. It played on its anti-Communist outlook with considerable force. The Americans wanted partners in their scheme to surround the USSR and its allies on all sides with military alliances, and were the founding force in the creation of NATO, CENTO[1] and SEATO (the North Atlantic Treaty Organisation, the Central Treaty Organisation, and the Southeast Asia Treaty Organisation respectively). These alliances completed a ring from Europe through Turkey and the Shah's Iran to South Asia, a 'rimland' containment of the Southern flank of Mackinder's Eurasian heartland. Pakistan signed a military aid agreement with the USA in 1954, and later joined both CENTO and SEATO. For the Americans, this meant that they had moved closer to the smaller of the two states of South Asia. While this might appear the lesser prize, the US had secured the co-operation of the state that guarded the North-West Frontier – and the relevance of that move became explicit in the aftermath of the Soviet invasion of Afghanistan in 1979. It also gave the Americans the opportunity in pre-satellite days to fly its U2 spy planes from Pakistan over Soviet territory to the UK and Norway and back the other way. In return Pakistan received military aid on a large scale, and civil aid which exceeded that given to India in per capita terms by a large amount – although in absolute terms India did receive more.

Indian reaction to the US-Pakistani arms deals was not warm; but it did not cause India to jump immediately into a pact with the USSR. Nehru's non-aligned ideals, the source of great irritation to the Americans, were genuine; but so was his commitment to some kind of watered-down socialism, even if it was only one of more five-year central planning and more nationalised industry within a mixed economy. In 1955, in the warming relationship between Moscow and Delhi, an aid agreement was signed for the construction of a large steel mill in India; Nehru visited Moscow and was the first non-communist leader to address the Soviet people; and Kruschev and Bulganin were given a warm reception when they visited India. In the post-Cold War world, US policy has been increasingly to attempt to mould Third World states to its own vision of market capitalism, open markets, and 'good governance', but in pursuit of these goals it was reluctant to intervene militarily unless either it saw a direct threat to a vital national interest, as over the Kuwait crisis, or it saw what appeared to be a low cost 'policing' operation, as in the case of Somalia. The US led NATO assault on the Taleban in Afghanistan after al-Qaida's attacks on New York neatly fits that pattern. The later invasion of Iraq does not, and is an aberration than can only be explained in terms of President George Bush's psychology and that of his cabinet. US concern over the possibilities of nuclear proliferation has deepened, and it is alarmed by the nuclear capabilities of Pakistan and India, and the refusal of either state to sign the nuclear non-proliferation treaty – though that has not stopped it trying to build new bridges with India.

There is a debate within the US military machine over the idea of Revolutionary Military Affairs (RMA). The nature of threats to national security and the nature of warfare changes – with revolutionary developments like the missile-carrying submarines. The current revolution encompasses a step-change in the accuracy of

---

1 CENTO started as the Baghdad Pact.

pilotless drones and airborne ordnance, with a shift in military doctrine to use this power to support allied land forces – exactly as in Afghanistan in 2001/2002. But there are also concerns about other revolutions, for example in bio-warfare and information-warfare, that may have great appeal to insurgent groups engaged in 'asymmetrical warfare'. The asymmetry is because a great power can be humbled by groups prepared to use methods (e.g. concealment within a civilian population) and targets (e,g, civilian populations) that conventional forces are supposed to avoid. From the point of view of insurgents, asymmetrical warfare is a logical answer to asymmetrical democracy – that is the ultimate control of weaponry that dominates the whole world residing in the hands of the electorate of but one country. At the global scale America's democracy is an oligarchy.

China, Tibet, and the Himalayan War

The biggest confrontation between China and India occurred in 1962 over the border between what in India was then known as the Northeast Frontier Agency (now Arunachal Pradesh) and the Chinese province of Tibet. This dispute, over the validity of the McMahon Line, has been considered in Chapter 12. However, this war was not fought just in the eastern Himalayas. In Figures 9.3 and 11.2, to the northeast of the Ladakh region of Kashmir, there is a salient of territory far beyond the Himalayan crest, known as Aksai Chin – also as the Soda Plains (of Aksai Chin). It is an arid desert at 4,000 metres. This too was claimed by India in accordance with the maps left by the British. When the frontier tensions between India and China escalated, following Nehru's 'Forward Policy' troops were despatched to occupy the boundary. No-one from India had been in this area for years, perhaps decades. To the few Chinese road engineers or Tibetans who had been there, there was not the slightest suspicion that they might have been on Indian territory. Nevertheless, India deployed troops to forward camps, well within China's claimed territory, where the Chinese had proper logistic support. When the shooting war started in earnest in the Northeast, the small, isolated pickets in Akai Chin were simply eradicated. The defeat of the Indians was predictable and rapid.

Other parts of the 'colonial' line between Kashmir and China along the Karakoram mountain range had been 'inherited' by Pakistan. The latter and China negotiated a settlement in 1963, which the Indians have always repudiated, since they do not recognise Pakistan's occupation of part of Kashmir. The area ceded by Pakistan is demarcated in Figure 9.3. Since Pakistan claims that Kashmir is still independent, it might seem at first as if this is contradictory to its own stance. However, the area is in that part of the tributary Gilgit and Baltistan Agencies, now claimed as part of Pakistan proper rather than tributary to Jammu and Kashmir.

Pakistan

The Muslim League had not quite the same opportunity to plan the future foreign policy of Pakistan as Congress had for independent India. The proclamation of the ambition to achieve an independent Pakistan came late – in 1940 – and the acceptance of that aim only six weeks before Independence itself. When the state was born, it was party to a defence council with India, and yet found itself shortly at war with India over Kashmir, and its fertile farms the subject of a water embargo. The first foreign policy aim of Pakistan was therefore survival; and in that aim it has been successful, even if 'dismembered' into two. In surviving it has seen, rightly or wrongly, that it has had to counter-balance the power of India, which it has always suspected of not being reconciled to Pakistan's existence. In the early years it is true that India assumed that Pakistan would not last long; but that does not mean that India wished to take steps to hasten its demise.

The alliances which Pakistan would naturally seek out were those with the great powers that were not well disposed to India. This included China, which had disputes over its border with India, and the United States, which was affronted by India's determined non-aligned stance. The 1962 war between China and India left the latter weak and demoralised, and then Nehru died in 1964, with no readily groomed successor. Ayub Khan, President (and dictator) of Pakistan was presented with an opportunity to even scores with India while its defences were still suspect. In 1965 he instigated the war over border disputes in the Rann of Kutch, which lead to mediation by the UK, but immediately followed it by a thrust on Kashmir. In this sector, and in the Punjab, the Pakistanis failed to attain their objectives: this was part of the despair of East Bengal which we noted earlier. The USA cut off supplies to both sides. The USSR arranged talks in Tashkent between the two protagonists, during which the Indian Prime Minster Shastri (Nehru's successor) died. The talks resulted in what passed as an agreement, the return to the *status quo ante*. It was the beginning of the end for Ayub, and the beginning of India's search for new arms partners. She began to purchase more from the British and the French, but above all turned to the Soviet Union.

The simple truth that my enemy's enemy is my friend began to work its inexorable logic. Pakistan began to forge an ever-closer relationship with China. The old Silk Road across the Karakoram between Pakistan and Tibet/China was developed as the Karakoram Highway, and Pakistan began to acquire Chinese arms as well. In Indo-China Chinese backing for the Soviet-backed Vietnamese began to waver, later to break out into open hostility and warfare. Although, for ideological reasons, it seemed highly unlikely, there were clearly strong grounds for the USA to find some accommodation with China, now openly hostile to the USSR and aligned with America's protégé Pakistan. In 1971 Pakistan made the arrangements for Kissinger's first secret trips to Beijing, and within a few years this new understanding was public.

1971 also saw the disintegration of Pakistan, and the emergence of Bangladesh. Although the USA tried for a time to maintain an even-handed approach, it did not manage to sustain impartiality between India and Pakistan for long. It 'tilted' towards Pakistan, and supported Pakistan diplomatically in the UN, despite the protest of the US diplomats in Bengal who warned Washington that genocide had started. It also deployed one of its fleets complete with an aircraft carrier in the

Bay of Bengal. From the Indian point of view, Washington's actions amounted to open hostility, and anti-American sentiment, already strong, became stronger. Neither were the Chinese uninvolved: troop movements and sabre-rattling on the Himalayan frontier provoked increasing concern over India's security. In 1971 India and the USSR signed a treaty of peace and friendship.

After the collapse of Ayub's military government in Pakistan, Zulfikar[2] Ali Bhutto's populist government was superficially anti-imperialist and anti-American. Bhutto was also strongly pro-Chinese, and used to wear his version (but well tailored in pure silk) of the Mao suit. But that was not enough to sever the external links upon which Pakistan depended: he was too pragmatic for that. The US therefore largely put up with the public denouncements, knowing exactly the extent to which Pakistan depended on its support. When Bhutto was executed by General Zia after the 1977 coup, American support for Pakistan began to slacken. It was worried by the resurgence of Islamic fundamentalism, and it was by then in confrontation with the Ayatollah Khomeini's new revolutionary Iran – with which Pakistan maintained cordial relations. And American public opinion was also increasingly hostile to Third World dictatorships. As Zia began to be pushed adrift, in 1979 the Soviet Union invaded Afghanistan. From that moment, the increasingly fundamentalist Zia was seen in a completely new light, as an ally in the fight against communism. The old priorities reasserted themselves, and American public opinion approved a massive new influx of arms, both to Pakistan, and through Pakistan to the Mujaheddin of Afghanistan.

From 1988 to 1999 the Government of Pakistan was overtly democratic, though the power of the army would always be somewhere hovering in the background. Indeed it is questionable how much Nawaz Sharif, the then Prime Minister, even knew about, or could control, the army's incursion over the cease-fire line in Kashmir in 1999. When the recent President, General Musharraf, seized power later in 1999, Pakistan was pushed by the international community further into isolation. Then, from being a pariah, in September 2001 after the attacks on the World Trade Centre in New York, overnight he became a gentleman-hero of the American coalition against terrorism. Pakistan was needed as the nearest land-base from which to attack the Taleban government of Afghanistan, a government which it had done much to help install in the first place. Though India had been quick to offer bases to the US, no Pakistani could have accepted overflights of any kind of military aircraft from India across Pakistani air space.

Pakistan has joined the (Middle Eastern) Regional Co-operation for Development and the Organisation of the Islamic Conference. This reflects both contemporary needs and historical past associations. It is thought that at least some of the finance for Pakistan's atomic weapons programme came from Libya, and on occasion Pakistan's bomb has been referred to in the Western Press as 'the Islamic bomb.'

---

2 Father to Benazir Bhutto who was assassinated in 2007.

The atomic weapons programme of Pakistan is an attempt to level the playing field with India. It is of course hugely expensive, but Zulfikar Bhutto asserted the Pakistani people would prefer to 'eat grass' if necessary to gain parity with India. Because of India's overwhelming superiority in conventional weapons, Pakistan has refused a 'no-first-use of nuclear weapons' treaty but in 2008 the new President, Asif Ali Zardari has reopened the possibilities.

Afghanistan

The Great Game between Russia and Britain to dominate Afghanistan waned in the 1920s and 1930s, because post-Revolutionary Russia was engaged in its own civil wars and economic de- and re-construction. During the Second World War defence of the Motherland against German Nazi invasion absorbed nearly all of its efforts. But in the Cold War and after Independence in India, the pattern of the Great Game re-emerged. A traditional and relatively quiet Afghanistan received aid from both the USSR and the inheritors of the British world-policeman role, the USA. In 1978 a coup in Afghanistan installed a communist government, but the army fractured and the country descended into civil war. For a while the government clung to power in Kabul with Soviet help. The Russians claim that they were simply forestalling a commitment by Afghanistan to the Americans, which would have opened up the possibility of penetration into the oilfields of the Caspian and the USSR's empire in central Asia. The new Afghan government was clearly about to fall when the Soviets launched a full-scale invasion in 1979. The internal opposition, the Mujahadeen, was a pro-Islamic coalition of different tribes, united against the common foe. In this situation the United States had no hesitation in recognising the struggles of a people, who, like the Americans, were God-fearing, in the face of atheist communist aggression. Through Pakistan they armed the Mujahadeen on a massive scale, and help in the training of Afghans and foreign volunteers, amongst them Osama Bin Laden, leader of al-Qaida (in Arabic this means 'The Base'), from Saudi Arabia. By 1989 the Soviets had had enough, and withdrew. Their 'Vietnam' was implicated in the fall of the USSR and the end of communism.

What the war had done to Afghanistan was truly awful. Soviet forces reputedly killed 1.3 million people and forced five and a half million Afghans (a third of the pre-war population) to leave the country as refugees. Another two million Afghans were forced to migrate within the country. On a proportionate basis, the Soviet Union inflicted more suffering on Afghanistan than Germany inflicted on the Soviet Union during the Second World War.

The US, which was also guilty of prolonging this conflict, and other western powers, walked away, and left the ravaged country to its own devices, in an act of cynicism which has returned to haunt the world. Afghanistan again became a country of warlords and petty fiefdoms, but with each group better armed than before. Its economy became (as it often has been throughout history) heavily dependent on the cultivation and sale of narcotics, particularly opium and its derivative, heroin. Some fiefdoms had all the trappings of independent states. Then Pakistan's ISI (Inter-Services Intelligence) helped support a new fundamentalist Islamic movement amongst the Pathans led by Mullah Omar. The movement has the name Taleban, meaning 'students', but perhaps in this context better translated as 'disciples'. The journalist Anita Pratap visited Balkh, the kingdom mentioned in Chapter 5, in 1996. 'While reporting the Taleban invasion from Kabul, I realised it was important to go across and meet Dostum because it wasn't yet clear whose side he was on, the Taleban's, or the ousted regime of Burhanuddin Rabbani. But Ariana, the national airline of Afghanistan, did not operate between Kabul and Mazar-i-Sharif. The safest and quickest way to get to Mazar was to go to Peshawar and take a plane from there. In Peshawar we realised there was an airline called Balkh Air that could take us to Mazar.' (Pratap, 2001). She arrived in Mazar to be refused 'entry' by an official who told her that her Afghan visa was of no use, since she was now in Balkhsthan. 'General' Dostum's fiefdom had its own flag, army, bureaucracy, airline, laws and currency.

Mullah Omar's fundamentalism expressed itself in many ways. In March 2001 he destroyed the huge rock Buddhas in Bamiyan, two of the world's great cultural monuments, in an act of contempt which ranks alongside Ghori's destruction of the temple at Somnath some one thousand years before. The vicious war to 'unite' Afghanistan under the rule of the Taleban, involving terrible massacres in, amongst many other places, Mazar-i-Sharif, was almost completed when Osama Bin Laden, the head of al-Qaida launched the worst of his attacks to date against the USA. The American response was to use a combination of air-power and Afghan ground forces from the Northern Alliance (that is a combination which included, for example, General Dostum and his Uzbek forces from Balkhsthan) to overthrow the Taleban. By 2002 they had managed to install a care-taker government in Kabul, but the majority of the Taleban and al-Qaida leadership melted across the mountain and disappeared in Pakistan, whence they have resurfaced, to challenge the new government. The war against them and in support of Karzai's government in Kabul has dragged on now for more than seven years, and is putting NATO's political unity under severe stress. Just as during the Afghan War of 1979–89 (and indeed the British-Afghan wars of the 19th century described in Chapter 5) the violence is spilling over into Pakistan's Northwest Frontier Province and even into Punjab and Sind.

The Americans also built bases in Kyrgystan, Uzbekistan, and Tajikistan, and so the maritime powers have penetrated Mackinder's heartland for the first time ever, although the Uzbeks terminated their agreement in 2005. It is not conceivable, particularly given the unwillingness of the USA to expose their forces to long-term risk, that this is the end of the Great Game.

There is an enigma the two decades since the end of the Soviet Afghan war. It is almost inconceivable that the American CIA did not know about, and did not connive in, Pakistani support for the Taleban. The conspiracy theorists believe that such covert support would have been because any government that could dominate the whole of Afghanistan was preferable to chaos, in order that pipelines could be built by major oil companies from the Caspian to the nearly 1.5 billion people of oil-deficient South Asia. The USA has had no problems supporting an autocratic regime in Saudi Arabia, which has, so far, delivered political stability rather than democracy, and stable oil supplies in consequence. If this was the simple aim in Afghanistan, it was obviously compromised by al-Qaida's attack from its Afghan bases, and so Pakistan had to be courted again, armed again, and set to work as a regional policeman, a role it has been slow to fulfill as the whole of the northwest has again been radicalised.

Kashmir

The festering sore of Kashmir has never been healed. Though to some extent the state became more tranquil after 1972, and by the mid-1980s tourism, so important to its economy, was soaring again, the possibility of either direct conflict between India and Pakistan or a new local insurgency has always been present.

As we have noted above, the Mujahadeen triumphed in 1989 when the Soviet Union withdrew from Afghamistan. But through history, the opprtunities and instabilities wash across the Northwest from Afghanistan to Kashmir. It was across this terrain that the tribesmen struck in 1948, beginning the Kashmir crisis. In 1989 victorious Jihadists were certainly some of those involved in switching their attention to the unfinished business of 'liberating' Kashmir. In 1989 the insurgency in Kashmir erupted, no doubt with some strong local support as well. It provoked an intensive and unfortunately very insensitive crackdown by Indian security forces, so escalating the spiral of violence. There are many groups now involved, not all striving for the same ends. Some groups want union with, and are supported by, Pakistan; and other more numerous groups are calling for Independence. Several times in 1995, 1996 and 1999 the Indian Government has tried to hold elections to return the state to its own administration (within federal India) and 'normality', but has been thwarted by violence and intimidation. Just like Northern Ireland for the British, Kashmir seems an insoluble problem, unless there are major concessions by all sides. Pressure from the outside – the United States, the USSR or Russia, and even the British – on this issue has at times proven counter-productive. This is an issue which touches the raw nerves of national identity and sovereignty for both states.

The knock-on effect across the Northwest continue. When the Taleban were kicked out of Afghanistan in 2002 by the US-led NATO coalition, India claimed that terrorist infiltration across the Line of Control had increased, even possibly involving al-Qaida. There were strong hints that President Musharraf was not sufficiently in control of Pakistan's ISI, and that, from India's point of view, Pakistan was sponsoring terrorism. The military build-up by both sides along the border seemed to presage another war between India and Pakistan (which many observers feared could lead to an exchange of nuclear weapons). Clearly it suited the leadership of al-Qaida and the Taleban to tie down Pakistani forces on the Indian rather than the Afghan border. But in yet a further twist, American pressure forced Pakistan to suspend the greater part of its support for Kashmiri separatist terrorists.

Bangladesh

Bangladesh was born one of the poorest countries on earth, with very little infrastructure, less than 10 per cent urbanisation, and a largely illiterate peasantry. It is surrounded on most sides by India, and its borders are essentially indefensible. A short stretch of border abuts Myanmar (Burma). Its foreign policy from the beginning has been dictated by the need for survival, and it has played its role as a major client of international aid agencies well. Although formally the Ganges water dispute appears resolved, it is quite possible the Treaty will prove unworkable. This and other resource problems have the potential for friction even in the near future. It also has had difficulties over finding a home for 500,000 Urdu-speaking Biharis, who left Eastern India for East Pakistan, and were not accepted by (West) Pakistan post-1972 as 'Pakistani', although in the early 1990s 'repatriation' of these refugees to Pakistan did start. Persecution of Muslims by Myanmar has resulted in an influx of more the 250,000 refugees into the southern coastal region and Hill Tracts of Bangladesh. This area is already the scene of anti-government insurgency, led by the local Chakmas who are mainly Buddhist. Thus the southeastern border with Myanmar becomes Bangladesh's second great foreign policy concern.

India

India's experience of two centuries of domination by a European power left it with an undeveloped economy which supplied raw materials in exchange for manufactures, and a mostly illiterate and mostly rural workforce. It had been party to two World Wars, neither of which was of its own making. In the first of these no national interest of its own had been at stake. In the second, the issue is more complex. Some leaders of the Indian National Congress believed that it could negotiate diplomatically with the Japanese, and that the Japanese planned invasion of India was simply and only to eliminate British military power. However, the reading of Japanese behaviour in China, S.E. Asia, Indonesia and the Philippines suggests that Congress' assessment might have been a bit naïve.

But from the beginning Nehru, who retained the Foreign Policy portfolio as well as that of Prime Minister, determined that India wanted complete independence, from external military alliances and from external economic dependency. It wanted to become a leader of what would become to be known as the Third World, and to champion the cause of other colonial peoples – both in S.E. Asia and in Africa. It became a founder member of the 'Non-Aligned-Movement', together with Egypt and Yugoslavia. To Nehru this meant both a practical policy of non-involvement in the Cold War, and a moral standpoint, which curiously permitted a pragmatic approach to issues, siding with major powers over specific issues if their actions were deemed worthy. To begin with Nehru courted China, in the belief that they could both share this new leadership in Asia and Africa; but the reality of China's uncompromising competitive hostility was revealed in the invasion of Tibet and the 1962 Himalayan War.

Economically India became autarkic – throwing up high tariff barriers against the rest of the world, and devising its own Five-Year Plans to promote massive heavy industrialisation – all reminiscent of the achievements of the Soviet Union. The combined effects of foreign policy neutrality, international moralising, and economic socialism and protectionism, became quite offensive to the USA. In 1956 the Chairman of Republican Policy Committee in Congress, Styles Bridges, observed:

> I lose patience with those nations which are not only neutralist in their military position, but insist on neutralism in their moral position. I know of no worse offender in this regard than Nehru, who proclaims himself the moralist of Asia. I know of no instance of Nehru having openly and sincerely taken the side of freedom and democracy. I know only of weasel words and idle pretension. (Cited in Singh, 1966: 109)

India's moralising and hostility to the USA probably reached a peak over the Vietnam War. And India had many points on its side too. The USA had fought its own war of independence against the British, and had then withdrawn from any imperial defence system. It had indeed preached isolationism, had joined the First World War late, and after the shock of that had retreated back into isolationism until Pearl Harbour had compelled it to join in the Second World War. India, an ex-colony, simply preached its own version of isolationism, at a time when the USA, the new world hegemon, had become paranoid about the communist revolutionary threat.

India's non-aligned status did not prevent it receiving substantial aid inflows from the West. The USA was for decades the largest donor, and India one of its largest non-military aid recipients. Pakistan always got less aid in total terms, though much more in per capita terms and, later, much more indeed in military terms. In 1971 India signed a 20-year Treaty of Peace, Friendship and Co-operation with the USSR, and by 1980 the USSR was India's largest external economic partner as well as being militarily important. The move by the USSR into Afghanistan worried India, both because it felt the act was unjustified and its legitimacy suspect, but also perhaps because of fears over the Northwest Frontier. But Delhi did not publicly condemn Moscow; the relationship was too close for that. And it co-operated with the communist governments of Afghanistan, far preferring to see one such survive than a take-over by militant Islam.

India has seen its role within South Asia since Independence to be the regional hegemonic power – in a sense to replace the strong centre that the British and the Moguls had represented. This means that she still sees the Northwest Frontier as much an Indian security problem as a Pakistani one. After the Russian invasion of Afghanistan in 1979, even though Mrs Gandhi was not publicly hostile to it, it was clear that Indian sensitivities had been upset. In the early 1980s India and Pakistan held dialogues to see if they could come to a common defence agreement on the issue, but in the end nothing transpired.

Independent India has intervened at some stage in most of the neighbouring states of South Asia. In 1961 she launched an armed invasion of the Portuguese colony of Goa, and annexed the territory. Nehru had lost patience with the Portuguese over their retention of what was 'naturally' part of India. But there had been no plebiscite of the Goan population, who, after four centuries of Portuguese Catholicism, had a very distinctive culture of their own. The justification was slim indeed, and the act in contravention of international law. The US objected to this treatment of its NATO ally Portugal. In 1971 India was instrumental in the birth of Bangladesh. In 1975 the Kingdom of Sikkim – whose exact status until then as a protectorate of India's may be disputed – was absorbed as a constituent state of India, officially after a full poll on the issue. This annexation has not been accepted by China. In 1987 the connection between the Tamil separatist war in Sri Lanka and political support in Tamil Nadu persuaded India to send its Indian Peace Keeping Force into Sri Lanka. This move resulted finally in military ignominy, and the assassination of Rajiv Gandhi. In 1988 India dropped parachutists on the Maldives to re-instate the overthrown government of President Gayoom. With respect to Nepal, India has been prepared to use trade embargoes to pressure Nepalese policy.

The years of turmoil have provoked India into becoming a major arms producer of considerable sophistication. Having feared Chinese nuclear weapons, she has also developed her own nuclear programme. Its exact status is no longer a close secret. Her explosion in 1974 of an atomic bomb underground in the Rajasthan desert near Pakistan's border, 'for peaceful purposes', left no-one in doubt of her technical mastery. But it did not provoke quite the same hostile reaction from the international community that the series of tests in 1998 caused. The 1974 test was one-off, and at that time India did not possess an appropriate delivery system. Nevertheless, India did ignore pressure to sign the non-proliferation pact, preferring instead to keep all options open. In India's view, universal nuclear disarmament is the only viable option. The 1998 tests caused a major uproar; they were an early act of the incoming Hindu nationalist Bharatiya Janata Party. This time Pakistan responded, and this time it seems that both sides have delivery systems capable of hitting each other's cities. Finally, having felt the power of US naval deployment in the Bay of Bengal in 1971, India is committed to developing a 'blue water' navy, including aircraft carriers and nuclear -powered submarines. This means a navy capable of strategic action within the Indian Ocean and not just coastal defence. There is some logic to this, in that many important cities and installations, including most of her atomic power stations, are within range of submarine-launched missiles, not necessarily intercontinental missiles like Polaris, but small cruise missiles launched from a submerged or surfaced submarine. These are now within Pakistan's capability. In other words, India now recognises that the defence of the seas is as important as the defence of the land borders. The heir to the Moguls and the British appreciates India's position as a unique geopolitical region, which must face both the potential of land power and sea power.

Since the collapse of Soviet power and the subsequent economic problems of the new Russia, India, though still maintaining very cordial relations with Moscow, has overtly courted closer ties with the USA. The overtures have been reciprocated, and in 2005 America proposed a treaty for closer nuclear co-operation and the sharing of technical information. To its irritation, this was blocked by the Communist partners in India's Congress-led coalition government. At the same time, short-term American interests in Afghanistan have fortuitously pushed Pakistan back into America's favour. India is anxious to be one of the big players on the international scene, and would like a permanent seat on the UN Security Council, just like China.

The involvement of outside powers in South Asian affairs has grown considerably in the last four decades, primarily because of the split within South Asia between India and Pakistan. It forced the USA, for example, to choose – the famous 'tilt' of 1971. If India and Pakistan had maintained friendly relations and a joint security pact, then perhaps they would have had that far greater autonomy that Nehru always had hoped for, and smaller defence budgets. Belatedly, there are moves afoot to bring them closer together. Realisation of the common interest is growing and has found expression in the founding of SAARC, the South Asian Association for Regional Co-operation.

Sri Lanka, Nepal and Bhutan

I do not propose to say much about these other nations, because they have featured little in the international geo-politics of South Asia. Given the horrible civil war in Sri Lanka between the LTTE (the Liberation Tigers of Tamil Eelam – Eelam being the new state they want in the North and East of Sri Lanka) and the Sinhala-dominated Government that has dragged on since the 1970s, this may seem a little odd. However, as I pointed out at the end Chapter 5, on Afghanistan and the Northwest Frontier, apart from India, no other state has had a stake in Sri Lanka's conflict.

In the British period, Ceylon (Sri Lanka) was a separate colony within the Empire, but a very significant part of the Imperial defence system. At Trincomalee on the eastern coast there is one of the finest natural harbours in the bay of Bengal – the east coast of India from Tamil Nadu to Kolkata is devoid of any equivalent. It was a base of great strategic importance to the British Navy, rather like Singapore. It protected the Bay of Bengal's maritime trade, and could be used to support the Middle East too. After the fall of Singapore to the Japanese, in 1942 they attacked Colombo and Trincomalee with a carrier task force – but for a number of reasons never managed to follow this up with a proper invasion, which the British greatly feared.

From India's perspective, a friendly government in Sri Lanka that denies any other nation access to these facilities, is all she currently requires; although the admirals building India's new Blue Water navy must from time to time think of Trincomalee's value in regional defence, and the navies of both countries continue to cooperate in action against the Tigers. India's other problem with the Sri Lankan war is the efflux of refugees to its own Tamil south – one of the reasons prompting its ill-fated intervention between 1987 and 1990, when the Tigers gave the Indian army a bloody nose.

Nepal has one of the world's lowest per capita incomes, and on most other development measures, such as infant child mortality, fares badly. It is ethnically diverse, with a population scattered in mountain valleys and on hill-sides, with few or no roads, and on the plains of the Terai. Most of its trade is with or through India, and Kolkata operates as its principle transit port for bonded goods. Over the years quantities of imported goods have been smuggled back across the border into India, evading Indian tariffs. India has from time to time flexed its muscles in consequence, imposing trade embargos. Nepal does of course have another neighbour – China, via Tibet, and the Nepalese have from time to time tried to balance the power of India with overtures to China. But the reality of its geopolitical position means that India has the upper hand. The country was until the late 1950s an absolutist Hindu monarchy, but with slowly evolving local democracy from then until the first stages of parliamentary democracy in 1990. Some political leaders were not satisfied with the quality of these 'developments', and from the early 1990s there was a struggle between the Royal Family (many of whose members and the then King were murdered in bizarre circumstances in 2001), parliament, the Royal Nepalese Army and a Maoist People's War – led by self-style Maoists, but who have not been supported by China. The significance of this from India's point of view is that Naxalites and other insurgent groups in Bihar and North India have cooperated with the Maoists – so that security concerns on both sides of the border are linked. In early 2008 the Maoists took part in new elections, and took the largest number of seats in a new parliament. They have since formed a government and abolished the monarchy.

Bhutan was effectively a protected state during the British period – internally autonomous but in external treaty obligation to the British, who were happy to maintain its isolation. It was a largely Buddhist state with a Buddhist monarchy and a society administered through a network of monasteries, but it has a significant number of immigrant workers in the lower plains, particularly from Nepal. India has taken over the external British role – and the state has slowly moved to admit outside influences, although it has a small quota for high-priced tourist visas, thus avoiding some of the problems that Nepal's hippy tourism induced. Television was introduced in 1999, and the King introduced democracy, with the first elections held in 2007. The country already exports hydroelectric power to India, and it seems likely that pressure from India for more dams will grow. India is also involved in the problems of ethnic tensions which spill over between Nepal, Sikkim, Bhutan and Assam and Bengal.

SAARC, SAPTA and SAFTA

A favourite idea of President Ziaur Rahman of Bangladesh, SAARC (the South Asia Association for Regional Co-operation) was inaugurated at a meeting in Dhaka Bangladesh in December 1985. Its founding membership was India, Pakistan, Nepal, Bhutan, Bangladesh, Sri Lanka and the Maldives. Since 2005 Afghanistan has also been a member, and there are several other nations with Observer status. Such an Association had been mooted for some time, and Prime Minister Callaghan of the UK and President Carter of the USA had at an earlier date offered financial help if such a movement got going. But India was suspicious, that the result of such a group would be a forum in which the smaller countries ganged up on the one undisputed regional power. The framework in which it has finally been established is therefore restrictive. The Standing Committee and the annual summit are thus far to consider only general issues of concern to all; no bi-lateral issues can be put on the agenda. These are still for individual countries to resolve, and not to be submitted to any regional pressure or arbitration.

Members of SAARC like to see a model of their future in ASEAN, the successful grouping of Southeast Asian states that is edging towards a common-market status. The late President Zia of Pakistan stated publicly that the success of ASEAN had

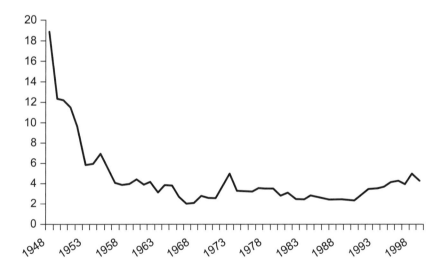

Figure 14.4   Intra-regional Trade as a Percentage of International Trade in South Asia, from Partition till 1999

Source: World Bank (2004)

Table 14.1   South Asian Trade as a Percentage of Total Trade, by Country

	1981	1990	1995	1998
India	1.8	1.4	2.7	3.2
Pakistan	3.1	2.7	2.2	3.6
Bangladesh	5.4	5.8	12.7	12.4
Sri Lanka	6.5	5.6	7.5	8.2
Nepal	47.4	11.9	15.0	32.8
Maldives	9.4	9.2	6.7	9.4
Bhutan	NIA	9.7	73.5	71.8
South Asia	3.2	2.4	4.1	4.9

Source: World Bank (2004)

stemmed from the fact that the one 'giant', Indonesia, had kept a low profile. His implication was that India should do the same to make SAARC successful. So far, this has run counter to the common perception of India's past actions (the invasion of Goa, the absorption of Sikkim) and recent actions, in particular its involvement in Sri Lanka, its continued military and naval build-up, and its former support for pro-Soviet Afghanistan, which India would have liked to see admitted to SAARC. But, at a lower level, there are signs of slow progress. Co-operation is mounting in diverse fields such as agriculture, meteorology, sports, terrorism, drugs, and tourism, and there are schemes to provide for linkages and exchanges between industrialists and universities. A number of senior businessmen now have open visas to all SAARC countries. Cross-border transport links are being improved – in 2008 rail links between India and Bangladesh were re-opened or the first time since 1947.

But there is a long way to go. After 1947, intra-regional trade collapsed (comments on the trade war between India and Pakistan after their currencies diverged in 1949 are to be found in Chapter 10), and have shown few signs of revival (see Figure 14.4 and Table 14.1). What regional trade there is, is dominated by exports from India. These, in 2002–2003, accounted for about three quarters of total regional trade. Most of India's regional exports go to Bangladesh and Sri Lanka, both of which run large trade deficits with India. Nepal and Bhutan normally run trade surpluses with India, but given the relative sizes of the countries and their economies, the total volumes of this trade is very small from India's point of view, and a trade restriction on Nepal by India causes little national pain.

In 1993 SAARC adopted a South Asian Preferential Trading Arrangement (SAPTA), which provided the framework for bilateral reductions on tariffs and quotas. In practice this has proved complicated, and has affected trade little. In 2004 SAARC moved one step further, when members signed a treaty for a South Asian Free Trade Area (SAFTA) which began to reduce tariffs across the board between member states from 2006. The treaty allows the lesser developed states (defined as Nepal, Bhutan, Bangladesh and the Maldives) to reduce their tariffs more slowly than India, Pakistan and Sri Lanka, but in theory all should be eliminated by 2016. Given the emerging global patterns of trade noted above, it will be interesting to see what new patterns and benefits emerge – and indeed if full implementation is achieved.

Why intra-regional trade should be so small can be explained in part historically, by India's and Pakistan's adoption of autarchic development policies, their political hostility and their trade embargoes. But that is not the whole picture. The classical picture of trade between developing countries, which are suppliers of raw materials, and the developed countries, which supply industrial goods, can explain another part of the trade patterns of the smaller states of South Asia. As a further clue, in the last few decades trade has grown faster between the developed industrial countries than within any other group. This seems curious to begin with, in that they could be thought to have similar and not complementary economies. However, the growth of international corporations with distributed manufacturing, and the growth of consumer demand in aggregate and for consumer diversity of choice within product categories, suggests powerful reasons why this should occur. The conclusion to this is that if there were complete free trade, both within SAFTA and with the rest of the world, then intra-regional trade would grow as *a consequence* of development, rather than be the cause of development. ASEANS's experience of free trade is that it has not increased trade between members states much. Rather, one external power, Japan, has been increasingly involved in investment with each member state. In the case of India, the trade patterns revealed by its latest statistics show that trade is greatest with developed nations (Europe, North America and Japan etc), with newly industrialising countries (China and Hong Kong – the latter reported separately), and with sources of energy – Nigeria, the Gulf States etc.

Yet the states do have in common the heritage of greater India, and cultural and linguistic continuities across their borders. And the smaller states have many problems which inevitably involve India, whether it be Sri Lanka's Tamil problem, or water in Bangladesh. Creeping in here is the idea of the New Security Agenda.

The New Security Agenda

The United Nations Development Programme in 1997 outlined the seven areas of new security; economic, nutritional, health, environmental, personal, community and political. Since then the more crystallised term 'New Security Agenda' has been pushing its way into debates in international relations. It challenges the comprehensiveness and utility of the 'Old security Agenda.' Whether there ever was quite such a simple animal as 'old security' is uncertain; but for the purposes of this chapter we can agree that the theorists of the New Security see the Old Security as essentially the preservation of any given Westphalian State by conventional military and diplomatic means. India and Pakistan and the other states have defined territories to defend, by the providing and equipping adequate military forces to deter or defeat perceived military threats. Their governments are the dominant actors in forging supporting alliances.

By contrast, in the post-Cold War world, the new agenda is growing in prominence because international relations theorists, diplomats, many governments, and agencies of the United nations are increasingly aware of the extent to which the internationalisation of crime, terrorism, the trade in narcotics, and environmental issues, all impinge, sometimes collectively, on the security of both states and their constituent publics. More recently the most basic of all securities – access to food – has again become an international security issue. The issues, then, may involve linkages which cut across states or which crystallise below the level of states. Essentially, New Security relates to people rather than to nations. Duffield (2005) proposes the idea of biopolitical security, in which the principal concern is not securing states but lives. 'Geopolitics, the security of states, and biopolitics, the security of population, are not mutually exclusive; they are complementary, interdependent and work together to a lesser or greater degree.'

In July 2004 the Foreign and Commonwealth Office of the UK sponsored a conference to discuss the new security agenda in South Asia (Jacques, 2004). Delegates were initially presented with the following questions as jumping off points:

Is competition over water a potential source for conflict?
Can environmental cooperation reduce tensions?
What role is there for regional cooperation in the management of natural or man-made disasters?
How can South Asia's energy demands be met?
The regulation of South Asian trans-border migration
Can trans-border drug trafficking and crime be tackled through a regional response?
The challenges of HIV/AIDS to security.

Terrorism was not included because within seconds of opening such a debate there was a danger that the dispute over Kashmir and 'freedom fighters' (aka 'terrorists') and 'occupying forces' (aka 'liberators') would stymie any discussion on any other issues. In the best traditions of peace negotiation, confidence has to be built by dealing first with tractable rather than intractable issues.

Many of these issues have cropped up in the preceding chapters. The partition of Punjab and its impact on water management was dealt with in Chapter 13, in which we also discussed the issue of the Farakka barrage, on the Ganges just upstream of Bangladesh, and the embryonic plans for the massive River Linkage Programme of India – which would redefine sub-continental hydrology to fit an arbitrary Westphalian sovereignty. In Chapter 12, on India's Northeast, I commented on the level of illegal migration continuing from Bangladesh to India, and its destabilising effects (see Hazarika, 2000, for extensive field-research on this topic). Sometimes these issues are linked: the raising of the Ganges by seven metres at Farakka has destabilised it both upstream (and downstream) of Farakka. Upstream in Maldah district of West Bengal the river has shifted 10 kms east in 30 years. (It may soon outflank Farakka barrage.) It has eroded villages, schools, roads, houses, post-offices – the whole investment of modern life – and displaced hundreds of thousands of people, while throwing up new *char* then *diaria* lands in its wake, the size of a small English county (Rudra 2003). It has become a driver for some of the cross-border migration. Neither the state of West Bengal nor Jharkhand will accept the responsibility to administer resettlement on the char – which therefore operates under *goonda* control, and no public services are established. The Farakka scheme is also implicated in the millennium flood of 2000 that destroyed the houses of 20 million people in West Bengal and 3 million in Bangladesh, with attendant huge losses in agricultural equipment and in livestock (Chapman and Rudra, 2007). This is insecurity which repeatedly wipes out any small accumulations of capital which poor people may have contrived.

Among the major states of India, Bihar has the lowest per capita income, the highest levels of corruption, the highest levels of violence, and the least development of infrastructure, the districts north of the Ganges being even more backward than the southern part of the state. Some of its poverty has been ascribed to the capricious behaviour of the River Kosi, which, in debouching from Nepal onto the plains, has built a massive inland deltaic fan, across which it wanders in destructive floods. Attempts within India to control it by embankments have failed. Whether the super dams in Nepal discussed in Chapter 13 would, if built, reduce flood damage, is a matter of speculation. However, what is not speculative is that the border between Nepal and Bihar is highly porous, and that the Maoists of Nepal cross it relatively freely. They have contacts with nearly all other insurgency groups in India such as the People's War Group and the Naxalites (Hutt 2004). Cairn Energy, a UK company which is investing in India's energy sector since liberalisation and which has found oil in Rajasthan, has the concessions to drill in north Bihar. But presently it cannot not do so, because of the security situation. Development in this troubled part of the northeast demands security: security demands development.

So, whereas the previous section stressed the low-levels of economic interaction between the states of South Asia, this section has found other ways in which their destiny is still regional. Unfortunately, the Mumbai massacres of 2008 underline this simple fact, and the frequent impotence of Old Security.

The Politics of Triangles

Although in the foregoing each country has had its relationships looked at somewhat independently of each other, it is clear that these relationships form an interconnected web. Within that web some relationships emerge that some protagonists do not really want, and some persist that one or other party would like to change but cannot. For example, during the Zulfikar Ali Bhutto years, Pakistan would have liked a more anti-US policy but did not manage to get very far with it. Both the US and India would often have liked a closer relationship with each other. Russia at times would have liked a closer relationship with Pakistan. In this section I look at a way of conceptualising the web of relationships that offers some insight into how they are formed and why they persist. I do not claim that this analysis, which is simplistic in some senses, explains everything, nor that it can predict everything.

The inspiration for this analysis derives from Bhaneja (1973) *The Politics of Triangles*. We start with the simplest case of one triangle. Suppose Pakistan, the USA and India form a triangle, and that each of the edges is either positive or negative. If they are positive it means the two countries basically have a common point of view and co-operate. If negative, it means they have different interests, do not co-operate, and in some sense compete. There are four basic arrangements of negative and positive values on three sides. These are + + +, + + -, + - -, and - - -. If these plusses and minuses are added up, then overall the value of these four possible sets of values becomes +, -, +, -. The two which come out positive are called stable patterns, and the two which come out negative are called unstable. To see why, we can plot them and look at the implications (Figure 14.5). In the first case, Pakistan, India and the USA all like each other, all co-operate, so the pattern is stable. Counting the number of positive sides gives an index of Harmony, in this case H= 3. In the second case Pakistan and India dislike each other intensely. The USA tries to like both, but at some stage India may turn to the USA and say 'You are helping my enemy, you are either against my enemy, or I will reject your friendship.' Pakistan will do the same. In essence, the Americans are forced to choose. A likely outcome is therefore for the USA and India to fall out. Thus a triangle of Harmony 2 which is unstable becomes a triangle of Harmony 1 which is stable. This is also of course the third state. It is also possible for the unstable triangle of Harmony 2 to reach the stable form of Harmony 3, if the two warring factions can be persuaded to bury the hatchet.

The last state is the unstable state of three negatives. History is riddled with examples of this, and the outcome: Churchill forming a pact with the 'Devil himself' – Stalin – to defeat Hitler, and more recently the Croats and the Bosnians who had fought each other nevertheless formed an alliance against the greater common enemy, the Serbs, in former Yugoslavia. This behaviour is predicted in the Arthashastra, a treatise on statecraft written in India about c. 250 BCE for Kautilya, the Mauryan Emperor. In it is to be found the dictum 'My enemy's enemy is my friend'. Figure 14.5 suggests further implications. The arrows show possible shifts from unstable to stable states, hence there is no arrow from 14.5b to a more harmonious state – either 14.5c or 14.5a. To get from 14.5b to 15.5a requires that the first shift is to the unstable state 14.5c. This will require an act of will, 'making friends with my friend's enemy'. Clearly it is a risk-bearing strategy.

Although for much of the time international relations are ambivalent, crisis tends to clarify them. In the triangle USA, Argentina, UK during the Falklands crisis, in the end the USA decided to supply advanced sidewinder missiles to the

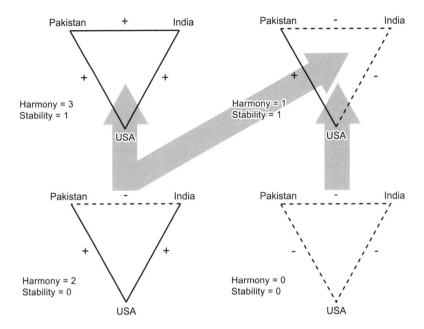

Figure 14.5   Triangular Relationships

UK, not Argentina. In his reaction to the attacks of al-Qaida in 2001, President Bush of the USA declaimed, 'you are either with us or against us'. It is, however, conceptually easy to develop more sophisticated and realistic models with both positive and negative values on multiples of the same links. For example, we could draw a positive link between Pakistan and India at the level of common cultures, another at the level of cognate languages, while having a negative link for the governmental level. Neither is there anything to stop one assigning different significance to different edges: the USA–Russia edge dominated all other edges in the thinking of the United States until very recently.

This basic idea is first generalised to the case of six states which form a small network of relationships. The six states are Afghanistan, Russia (or, its former guise the Soviet Union), China, India, the USA, and Pakistan. These six states can be simply drawn as a hexagon, in which it is possible to draw a line between all pairs of countries. (Figure 14.6) The countries are called nodes, and the relationships edges. There are then 6x5/2 = 15 possible such edges. Many subsets of three edges also make triangles – and in this six sided figure the 15 edges define (6x5x4)/(3x2x1)= 20 possible triangles. So instead of dealing with Stability and Harmony in one triangle, we now have to consider Stability and Harmony for 20

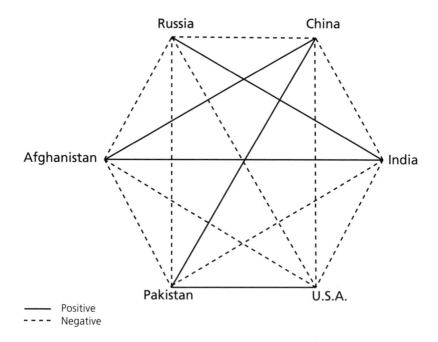

Figure 14.6 Triangular Relationships: South Asia and External Powers

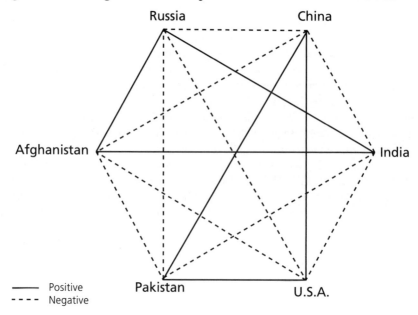

Figure 14.7 Triangular Relationships: South Asia and External Powers: New Alliances

triangles and 15 edges. An interactive computer programme is used to explore the relationships in this network.[3]

I sketch the situation at the end of the 1960s in Figure 14.6. Bigger actors and their relationships can form one key, but so also can particular local associations. In this case I have taken the USA–USSR hostility to be the key to the post-Second World War world, and I have taken Indian-Pakistani hostility to be a regional key. The USA would like to have friendly relations with both India and Pakistan, but that produces an unstable triangle. The USA has to choose while local hostility between India and Pakistan remains. The USSR initially is not well disposed towards India, a bourgeois state, but certainly does not like the theocratic state of Pakistan. In the USA–USSR–Pakistan–India complex the warming of US-Pakistani and USSR-India ties will be mutually reinforcing, and productive of a stable pattern of alliances. The USSR–India ties will reinforce China's rivalry with India, now seen from Beijing as a partially client state of the USSR (a fate China struggled to avoid). US and Chinese relations are hostile, partially over Vietnam. The last line drawn is between Afghanistan, before the Russian invasion, and China, and is assumed to be positive, as part of Chinese containment of Russia. This pattern has low ranking Harmony (5 edges positive out of 15) and moderate Stability (12 triangles out of 20). There are therefore 8 triangles remaining which are unstable, which will have forces pushing them to find a more stable state.

Table 14.2a   The Starting Relationships in South Asia

	AF	RU	CH	IN	US	PK
AF	0	-1	1	1	-1	-1
RU	-1	0	-1	1	-1	-1
CH	1	-1	0	-1	-1	1
IN	1	1	-1	0	-1	-1
US	-1	-1	-1	-1	0	1
PK	-1	-1	1	-1	1	0

Table 14.2b   Number of Unstable Triangles on Each Edge

	AF	RU	CH	IN	US	PK
AF		3	2	2	1	2
RU			1	1	2	1
CH				2	3	2
IN					1	0
US						1

---

3 The 1st and 2nd editions of this book contain an Appendix which goes deeper into this modelling approach.

Table 14.2a shows this pattern of 'starting relationships', and Table 14.2b the number of unstable triangles associated with each edge. There is only one edge which is completely associated with stability – the Pakistan–Indian edge. This is an unpredicted but very deep structural property of the relationships shown in Figure 14.6. In simulating dynamic change, I postulate the most unstable edge will change first. This is either the Afghan–Russia edge, or the Chinese–US edge. If the US–Chinese edge is changed positive first (Nixon goes to Beijing in 1972), then the pattern of instabilities change, and it is the –Afghan edge which has most instability. This has to be turned from positive to negative – so Harmony is going down while Stability increases. The next change is the Russian–Afghan edge – which can be made positive by invading Afghanistan and sustaining a pro-Soviet government. The result of this is shown in Figure 14.7. The final result is universal Stability 20/20 – but Harmony stays at the fairly dismal level 6/15. There are two competing alliances US–China–Pakistan versus India–USSR–Afghanistan – i.e. competition at the individual state level has been transformed into competition at a higher hierarchical level. The point about the pattern that has been achieved is that it is very difficult to change any one part of it without changing other parts. The pressure of knock-on effects is immediate: or, to put it another way, the knock on effects are so difficult to achieve that the system will push back to its original stable form. The picture overall is one of great stability, and one which cannot be changed without ramifications on other axes. It is for this reason that changes of regime do not necessarily make such radical differences. So long as Bhutto remained anti-Indian and pro-Chinese, Pakistan's pro-American stance is difficult to change. It also has a certain graphic dramatic appeal, in that the crossing point between two antipathetic axes – China+Pakistan+USA vs USSR+Afghanistan+India – is in the centre of the world, the land of mythical Mt Meru, somewhere in the Hindu Kush, Pamir, Karakoram mountains. The competing alliances resulted in the wars over regime change in Afghanistan.

One of the striking aspects of the analysis is that given a particular structure, the actual historical sequence is not necessarily important. In other words, the Soviets can invade Afghanistan before Nixon makes up with China – the same alliances emerge.

The analysis also says much about the difficulties of the Himalayan States and Bangladesh (not shown). For example, Nepal might like to strike a balanced friendship with both Delhi and Beijing, but it cannot do so too strongly, since then an unstable triangle will emerge. Certainly India does not want a close ally of China south of the Himalayas. Bangladesh finds it difficult to have close ties with both India and Pakistan, and simultaneously with India and China. If the China-Bangladesh–Pakistan triangle ever got very strong, India would feel frozen out and threatened in Bengal.

The West and post-Soviet Russia have achieved some better accommodation – although the relationship is fragile. China and Russia have mended some fences. If (and I have to emphasise the condition) the USA and Russia ever came really close, and if Indian-Pakistani hostility remains, then the pressure is on the USA and Russia both to choose the same ally in South Asia. India is the more likely candidate, because of its size, potential market, and regional dominance. For the moment, the war against the Taleban of Afghanistan has again pushed the USA into working with Pakistan, but the triangle USA–India–Pakistan is unstable. The key to making a Pakistan–India–US alliance work is clearly solving the deepest structural impediment in the matrix, the hostility between India and Pakistan over disputed Kashmir.

Concluding Remarks

This chapter has looked at South Asia in its international context at some different moments in time, and from a variety of perspectives. The arguments which these perspectives support are both deductive and inductive. The major deductive viewpoint is that provided by Mackinder and Cohen, that South Asia is an independent geopolitical region, strategically placed as one of the rim-land regions flanking the central Eurasian heartland. The actual calculation of stability and harmony in the web of triangular relationships and the prediction of change is deductive, but the starting relationships in the webs are modelled empirically however crude the binary relationships may appear.

The four most significant countries which this account has focused upon are the USA, Russia, China and India. In terms of the ability to project power, the USA has had greater capacity than the USSR/Russia, and China greater than India. The reason for the latter differential can be found at least in part in the fact that China has inherited, by coercion if necessary, the whole of its own geo-political region (with the exception of off-shore Taiwan/Formosa, arguably not part of the mainland region anyway), whereas India has not inherited the whole of its equivalent region, because it was made to accept the principle of self-determination in 1947. The antagonism of the two successor states in South Asia has a bitterness about it which can only spring from fraternal civil war, the threat of which precipitated the division, and the reality of which accompanied its birth. Kashmir has become the symbol of all that the two nations stand for, and the resolution of the conflict one way or the other would seem to represent a judgement of history – between the Indian claim that the creation of Pakistan was essentially unnecessary, and the Pakistani claim that it was the only way of protecting the rights of the Muslim minority. The judgement of history will probably be made from the victor's viewpoint, hence either side will go to any lengths to protect its claim. Both will turn for support to outside powers, the weaker of the two having aligned itself with the greater super-power and with the greater Asian regional power.

With the demarcation of new state borders, the space-economy and the mentality of the citizens begin to adjust to the new lines. Over the course of history there have been periods when much of what is now Pakistan was separate from the powers of the Ganges valley, and more closely related to Afghanistan and perhaps Persia. These periods add up to a small proportion of historical time, but they do occur. Perhaps such a period is being or is about to be repeated. However, the gains to be made from settling the quarrel between India and Pakistan are great, and both sides know this. At least the dialogue can continue, inside SAARC, inside the Commonwealth (which Pakistan left but rejoined – although as of 1999 its membership has been suspended after the military coup of Pervez Musharraf) and with direct bilateral meetings. But the states are taking on their new characteristics – and perhaps what was not so true in 1947 is becoming truer now – a more Hindu and less secular India, and a more Islamic and less secular Pakistan. It is a conclusion which, sadly, fits with Huntington's (1997) *Clash of Civilizations*, in which he sees civilisation/cultural differences as the underpinning of the new world order.

Finally, many people wonder whether Mackinder's century is over. This is for two main reasons – one being the simple truth that the USA managed to establish bases in central Asia to support the war in Afghanistan, despite profound Russian disquiet. The second reason is that global geography is changing. The Arctic has been warming for the last half century, and the area of sea ice has shrunk by half. There is speculation that soon the North-west Passage, skirting northern Canada, will be open to shipping, hugely shortening trade routes between say Europe and Japan. But it may also open up Russia's Arctic shoreline as well, connecting the world heartland with the maritime periphery. It will also enable the Arctic ocean to be exploited for oil, gas and minerals, and already there is a major political struggle between Denmark, Canada, Norway and Russia to maximise their respective claims. The NATO, CENTO, SEATO thesis of containment is then old-hat not just because of failed or hostile states around the Persian Gulf, but also because such profound shifts in global geography will change South Asia's relevance. Indeed, in geopolitical strategic terms, it might well become far less relevant than it is today. Its interest for the rest of the world will be in its population, and hence market, in its competitive industrialisation and consequent competition for resources.

PART IV
Conclusions

Chapter 15

States and Region in South Asia

Introduction: Nature Proposes

The aphorism, 'nature proposes, man disposes' is applicable to South Asia. Nature – the result in this case of the interplay between plate tectonics and meteorology – has proposed a geo-political region. Within that region, the size of the population, the undogmatic qualities of its ancient theologies, and the time elapsed between external human shocks, has resulted in a unique and evolving continuity of culture despite the detailed complexity of society and regional variation. The details and variation are important, but to use a ceramic analogy, they form the finely patterned cracks in the glaze of an old, cracked, plate – yet the plate exists.

Thus, until the advent of the million-plus cities, India can best be described as divided into numerous 'pays', rather like those described and defined in France by Vidal De La Blache. There are numerous ways in which such 'pays' can be identified. Schwartzberg (1985) uses variations in local language and local perception to demarcate the folk regions of northwest India for example. A regional geography such as that by R.L.Singh (1971) (India only) or O.H.K. Spate and A. Learmonth (1967) (India, Pakistan and Ceylon), provides descriptions of the economy and to some extent the society for some 240 or so such local territories, often with changes in the physical resource base marking the boundaries. There is a close correlation between ecosystem and 'pays'. There is however one important distinction between the 'pays' of De La Blache, and the 'pays' we note here.

The 'pays' of De La Blache existed within a well-defined and centrist state, whose laws were made centrally and recognised universally. In India there was no such centrist tradition – partly for reasons of scale. Given the early technologies there were plenty of costs but little economic advantage in the integration of large areas of India. More significantly it was because such functions as maintaining the social order were organised within caste, each having a tribunal (panchayat) for its own members. Inter-caste matters would be settled by the dominant caste of any one area, often by invoking the panchayat of lesser castes to take action against its members where necessary. (In contemporary India, some observers suggest that the lack of secessionist movements in states such as Bihar is a result of the caste tensions within the state, which replicate similar tensions elsewhere in other states, and which prevent the emergence of a unified 'state' view.) In such a society the concept of King or Monarch had a very specific connotation. The Raja, usually a Kshatriya and ritually inferior to the Brahmins, might be rich, but his wealth had, beside his own gratification, two major functions (Bayly, 1983). One was for pomp and ceremony which was for public consumption, the other was that of a

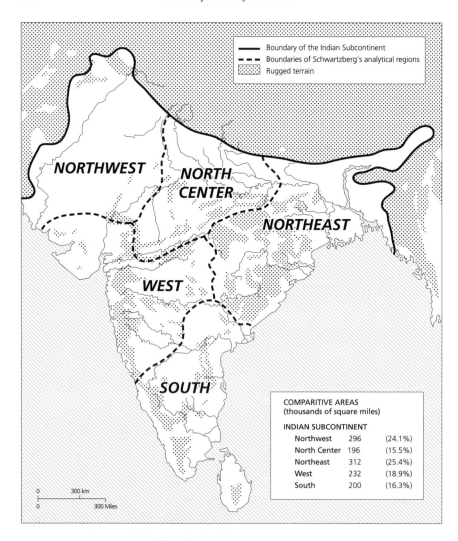

Figure 15.1   Basic Regions in South Asia
Source: Schwartzberg (1992)

General, or, Minister of Defence. In other words, the interpretation of customary law was the preserve of the Brahmins, his was the defence or aggrandisement of the territory. The geopolitical history of South Asia may, in many ways, then be seen as the fluctuating ways in which such monarchs have assembled the *pays* into larger units.

Historically, it seems that after the periods of building and integration, dissolution has followed. Thus at any one time empirical analysis might see the units as defined either by integration of smaller units, or as the remnants of disintegration of larger ones. If understanding the patterns is approached as an

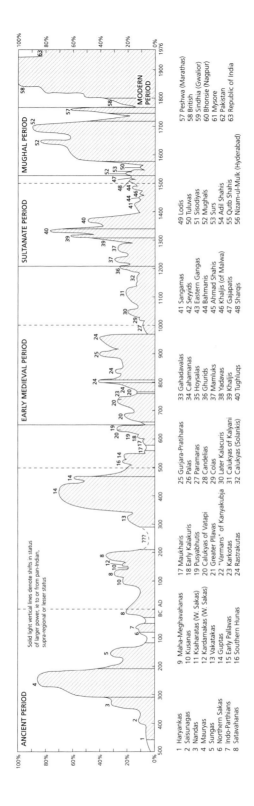

Figure 15.2   Pan-Indian Power over Time

Source: Schwartzberg (1992)

exercise in deductive analysis, there are the same two ways in which we may approach such a regionalisation. But the two differ in what they suggest about the boundaries that may emerge. If we build up bigger units by aggregating the smaller ones, the boundaries of the bigger units should reflect some of the boundaries of the constituent parts. If we divide the whole, to get progressively smaller units, in this case if the territory divides 'naturally' along the boundaries of the pays, then again the boundaries of the units will be made out of some pays boundaries. But if for whatever reason the division of territory ignores such lines, then new boundaries may be etched in the landscape, unrelated to the cleavages of the pays. In Africa the colonial powers were notorious for ignoring local ecology and culture in the construction of boundaries. The arbitrary divides using latitude and longitude in North America are also well known.

Schwartzberg (1992) concludes his Historical Atlas of South Asia with an analysis of the evolution of regional power configurations in the Indian sub-continent, and a geopolitical synopsis, using a variation of the top-down analysis. He starts his analysis by dividing South Asia into five regions (Figure 15.1) – of a scale and size which are reminiscent of John Bright's 19th century imagination of India after British withdrawal. The basis of these units is not actually explained, but they do seem to conform to 'natural' and repeated boundaries. The overall outline of South Asia which he uses is also of interest. Assam is included, but not the other six states of the Northeast. In the Northwest, the tribal areas around Quetta which harbour much of the current insurgency against the Pakistani government, are not included as part of South Asia. At both borders, the 'problems' are at least in part caused by a South Asian state incorporating that which is not South Asian.

The basic regions which Schwartzberg has drawn are, by implication, large assemblages of 'pays' which are frequently attained throughout history. He then defines a pan-Indian power as one which incorporates significant portions of at least four of these five regions, and a supra-regional power, as one which incorporates significant portions of at least two. By these definitions it follows that a pan-Indian regional power cannot easily coexist with supra- regional powers, but only with one regional power, and that at most only two supra-regional powers can coexist.

His analysis produces a diagram (Figure 15.2) which shows how pan-Indian integration has fluctuated in time – and he discusses the circumstances which would have favoured such periods. He further produces a map (Figure 15.3) of the core areas of the pan-Indian States, the vast majority of which are within the Ganges Valley. The three exceptions are: the upper Indus Valley which was important in the early Graeco-Buddhist cultures, and which is now the capital area of Pakistan; Malwa, the area around Indore and Bhopal; and the Mahratta area around Pune. Although the latter might, in some sense be said to be partly 'southern' (in this case meaning south of the Satpura range and the Narmada river) in language and culture it is still 'northern'. Schwartzberg's conclusion is that the resources of the north are necessary to sustain the military forces which have integrated the rest of South Asia, and that the north will contain the core areas.

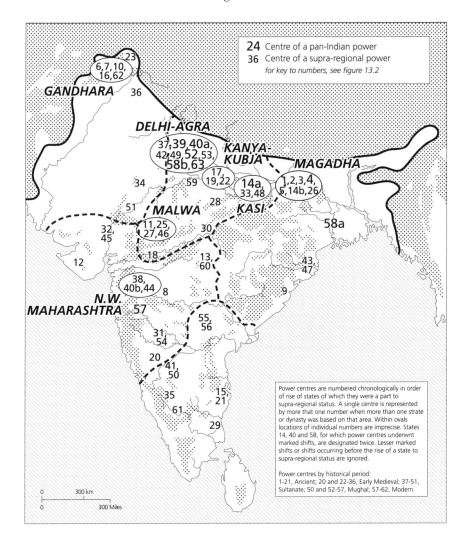

Figure 15.3   Core Areas in South Asia

Source: Schwartzberg (1992)

Malik provides a similar kind of diagram, Figure 15.4. He reduces the two dimensional map of India's regions to a single dimension, so time may be added as the second axis. Note that, instead of Schwartzberg's five regions, Malik proposes ten. The diagram shows how the different phases of imperial integration have come and gone – but he does not define nor establish in the diagram an equivalent of Schwartzberg's Supra-regional or Regional powers. The coexistence of such states with an empire is not an issue for him. His diagram does however stress the importance of the same core areas that Schwartzberg derived, and also shows how

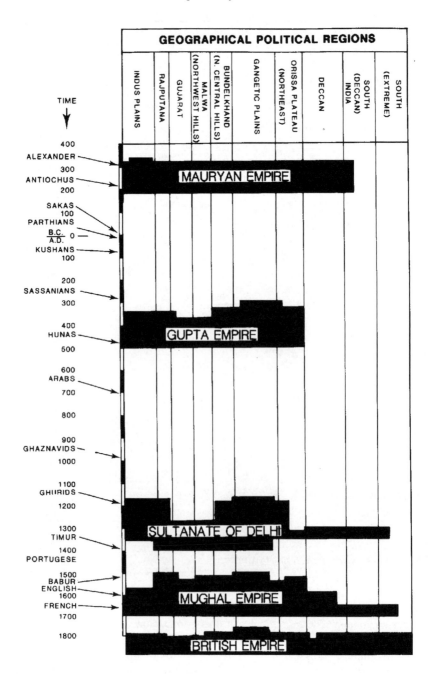

Figure 15.4   Pan-Indian Power over Time

Source: Malik (1968)

the Deccan and far south is hardly ever incorporated in these empires – not until
the final achievement of hegemony by the British in the early 19th century. If a
simple probabilistic reading of history were ever to establish anything, it is that
such a diagram would cause one to wonder whether or not the south might again
go its own way. Seemingly, from this evidence, this is a much more likely scenario
than the partition of Bengal or Punjab.

What this section has established is that although many patterns of division or
integration in South Asia are possible, nature has established the most likely core
areas, and that integration into much larger units – possibly the whole subcontinent
– is, with the passage of time, increasingly likely.

Humankind Disposes

The implicit model of the 'pays' of South Asia being integrated into larger states
works quite well not only for the ancient and mediaeval periods, but also for the
Islamic periods of the Sultanate and the Mogul Empire, even despite the fact that
Islam introduced an immiscible dogmatic theology into the subcontinent. Islamic
society in India was and is recognisably South Asian, and gave birth to its own
much-valued South Asian language, Urdu. In India we have already noted the
complexity of social groups that Hinduism spawned. When some of these groups
were converted to Islam, they did not abandon their origins overnight, no more than
someone today by proclaiming himself a Christian could therefore expect to change
his job tomorrow. Islam may prescribe the equality of man, but it does not command
that people marry at random. Within Islam-i-Hind, therefore, in significant ways
caste persists, defined not so much by pollution rules, as family marriage rules. In
Pakistan the network of families, bound within a marriage group known as a *biradri*,
is fundamental to all social and political life. Further, the acceptance of Islam and the
recitation of the Koran in Arabic does not deprive a man of his native tongue – so
that a Bengali Muslim is as a rule first and foremost a Bengali, yet also a Muslim.
So, within Islam as within Hinduism, local regional cultures persisted. Usually the
same regional culture pervaded both religions in one place. The major difference
between the two major religions was, and is, that for Muslims the common and exact
reference point of a dogmatic revealed and egalitarian theology could be established
with Muslims from different areas, and indeed internationally, whereas for Hindus
such common references were (and still are in many ways) much harder to establish
and were always confounded by caste.

In sum, though religion and culture may overlap, they cannot be seen as the
same thing and these different facets of their world view may be drawn out of
people, or repressed, augmented or diluted, according to circumstances. These
circumstances fluctuated, sometimes with a dogmatist in authority, sometimes with
a more accommodating ruler. This was the nature of the fluid mental landscape in
which the British merchants and imperialists became immersed, as a truly radical
new kind of 'circumstance'. They may have wished for no more than power, law

and order, in order to trade, but the impact of their new infrastructure and the turning inside out of the economy, and above all the gradual osmosis of their native democratic ideals and institutions, meant that the relationship between religion and culture in individual world-views would be changed and would be a nerve that could be tweaked by all manner of agents with all manner of intentions. Before the British, to a large extent emperors and nature usually disposed together. By the time they left, it was much more the masses that disposed; in the last months leading, as much as following, the politicians.

The distinction between the élites and the masses is useful in elaborating all three of the forces of integration. An élite may have identitive bonds in common, although the subject peoples do not. These bonds can then form the cement of integration and, so long as the masses are divided, they cannot combine to eject the élite. In the case of utilitarian bonds, these may be perceived more easily by the élite than by the masses. In the case of the use of coercive force to achieve integration, this almost by definition – but not quite completely – has to be controlled by an élite. This is because the use of force by the masses against other masses is more likely to lead to anarchy, genocide, and disintegration than to integration.

The identitive bonds are those mutually recognised by a people as the symbols of their community. They are usually associated with language and religion, but they may also be associated with territory. Where they are strong, utilitarian integration can also follow if the technology permits. Utilitarian bonds are those of economic self-interest. The British now know that they are bound economically to Europe and that to break away would be injurious, no matter if they do not 'feel' European. Coercion is expensive, and its fundamental premise is the threat of destruction. So after the costs of an invasion, which may instantaneously be met by plunder, a period of accommodation and reconstruction has to occur.

The British integration of South Asia relied on all aspects of the three forces we have mentioned, but not equally at élite and mass levels. They used superior technology as the basis of coercion where necessary. They relied on the identitive bonds of the British as 'British' to cement the rulers of Empire, and they were forbidden to become landed gentry. The civil service was early on Europeanised at the highest levels, and English was instituted as the language of Government, supplanting Persian. But the new rulers came from a country which had a rudimentary parliamentary democracy. At some stage they would have to confront questions about the legitimacy of their rule, and the exclusive proprietorship by their group of that right. Although, for the majority of the rural masses they did nothing, they nevertheless fostered utilitarian integration by the development of the railways, and by the development in many areas of major irrigation schemes. They established large new and cosmopolitan cities, and also founded new universities in which a new middle-class intelligentsia studied in English and incidentally discovered its own Indian-ness. They were the harbingers of the demise of the 'pays', just as industrialisation was in de la Blache's France.

What the British did not do was deliberately foster the identitive bonds of the Indian masses as 'Indians.' Partly, it was not in their own self-interest; partly they believed that India was a sub-continent of many races and tongues, an empire in itself, not just because that was what history said, but also because it suited their own propaganda. Identitive integration at the level of the masses was the last thing they either believed in or wanted to believe in. C. Rahmat Ali's rejection of the 'All-Indian-ness' of Congress was close to the British understanding of South Asia, and that tacit understanding may have played an unacknowledged role in the failure of the British to leave behind a unitary successor state.

With the filtering of Western liberal ideas into India's small emergent middle class, demands for representation were first conceded by elections on very small franchises to town boards, while true power resided with the Viceroy and his Council and the Provincial Governors and their Councils. But it was the beginning of a long (and still continuing – more quickly in India than either in Pakistan or Bangladesh) road of bottom-up encroachment on power, which grew on a wider and wider mass base as Gandhi spread the message of Congress. In order to spread this message, it had to be phrased in a way which the masses could understand; it could not be handed down in an uncompromised fashion. Thus although Congress had always sought to be secular and multi-communal, its behaviour locally was often more parochial and partisan. The new leaders might well have been nationalistic: but the masses were sunk still in local perceptions. When they had to be enlisted in the struggle for power, they were told it was for self-determination. But who or what was self? To a Tamil it is Tamil, or perhaps Tamil Brahmin or Tamil non-Brahmin. In Bengal self meant one's own community, here very clearly either Hindu or Muslim. The Bengali part went unspoken – taken-for-granted as the starting point. With classic myopia local differences seemed large, distant ones less important. Thus, later, Jinnah could and did appeal to Muslim Bengalis as Muslims, to join his Pakistan movement. Playing on the different religious and cultural facets of the world views of the masses was the key to power in the new South Asia. The question was whether any representative kind of federal constitution could conserve the sovereignty of the whole, while devolving maximal self-rule in the parts.

Here we see the point behind the remark made above about coercion by the masses rather than the élites. Jinnah held very few cards, which was one of the reasons that he was given so little credit by Congress. The Muslim communities were the minority, and not strong in the institutions of the new society, not strong in trade or banking, nor strong in the civil service. But the masses could be awoken, and what Jinnah could threaten was, simply, anarchy. As law and order collapsed and genocide began, so did all alternatives to partition, and any time for debate, nuance and subtlety in the settlement evaporated. The logic of partition was applied forthwith, not only by separating province from province, but by the vivisection of Punjab and Bengal too. The new lines drawn on the map were not 'normal' or 'natural' re-using ancient 'pays' boundaries. This was a partition of minds, and the homes that those minds lived in. This was not a partition that reflected local ecology, the basis of 'pays', nor the communications that ran within and between the 'pays'. Even at the largest scale what was produced was 'unusual' and 'odd'.

Though in the past the Indus Valley has often been the core of a regional power (but see Schwartzberg's map which includes Rajasthan within this region), until now no Indian power ever included the South and Assam, and only half of Punjab, and half of Bengal.

The tragedy of the hundreds of thousands of deaths and twelve million refugees is testimony to the unpreparedness or incapacity of either British, Congress or Muslim League élites to deal with the forces they had in part unleashed. Four and a half decades later, at the end of the communist imperium in Yugoslavia, murder, genocide and ethnic cleansing and migration reproduced in Europe a similar chain of events. The tigers unleashed by the community leaders becoming harder and harder to ride, though with the twist that here 'secularity' was proclaimed by the 'Muslim' government of Bosnia. In Yugoslavia the divisions and fighting sucked in external powers. In South Asia, although not in the same way as in Yugoslavia, the internal divisions of the sub-continent and the open hostility between the two largest successor states has also brought in the armaments and interests of external powers. By the end of Empire in 1947 pan-Indian identitive bonds among the masses were not strong enough to maintain the integration of South Asia. For this, the Indians may blame the British who divided to rule, keeping Princely states apart, and acquiescing in separate electorates for the Muslims. But the Muslims blamed Congress – for not honouring their secular pretensions at local levels. Casting blame to one side, it is clear now that to have expected this sub-continent of creeds and castes, at that time still largely illiterate, and a veritable linguistic Tower of Babel, to have formed a national identity at that moment in history was to expect the impossible. As an alternative to identity, coercion was not possible either – except arguably at the fringe over issues like Kashmir. For independence was about self-determination and the rejection of imperial coercion; and reluctantly Congress had in 1947 concluded that negotiated independence for Pakistan was best, at least in the short term. That way they inherited a stronger Centre in the new India, still the largest self-governing unit the sub-continent has ever seen, and the world's most populous democracy.

There is plenty of evidence to support the contention that India saw Pakistan in 1947 as a temporary aberration of the political map. In accepting the plan for partition, Congress issued a most blatant geographical and ideological statement:

> Geography and the mountains and the seas fashioned India as she is, and no human agency can change that shape or come in the way of her final destiny.... [w]hen the present passions have subsided, India's problems will be viewed from their proper perspective and the false doctrine of two nations in India will be discredited and discarded by all. (cited in Mansergh 1978)

Since then the Government of India has stated publicly that it has no claims on Pakistan; but there are still regional imperatives that interlock the destinies of the two countries. Both countries, and Bangladesh too, may well wish to keep internal matters to themselves and not to interfere in their neighbours' affairs; but precisely because many issues are regional in a pan-South Asian sense, it is impossible to isolate many issues as purely domestic. There are still a few Sikhs in Pakistani Punjab, who can offer transborder support to their brethren. There are Bengali Muslims and Bengali Hindus in both Bengals. The Indus river basin is shared by Pakistan and India. The Ganges-Brahmaputra is shared by India and Bangladesh (as well as Nepal, Bhutan and China). And there are Tamils in both India and Sri Lanka, some of the latter being locked in a civil war with the Sinhala majority.

Pakistan has cohesion when exposed to external threat by India; but left to its own devices is riven by regional dissent and the fight between the Mohajirs and indigenous groups, which has brought virtual civil war to Karachi. Thus, the most remarkable feat of its history is that, apart from the loss of Bangladesh, it has survived. It has not done so through identitive bonds: though perhaps there is a new generation that accepts Pakistan as a natural sovereign state, although clearly there are fewer such people in the tribal tracts of Northwest Frontier Province.. It has done so partly by a policy similar to India's: protected industrialisation which has created a new middle class with vested interests. But this class has been smaller and more concentrated than in India, and it does not reach far into the mountains. Pakistan has done so also through coercion, through several periods of army rule, by an army hugely supported by outside funds from the USA.

The point is this. Though no-one has ambitions to absorb Pakistan, if the country itself fell to pieces, the little bits could soon enter into different arrangements with neighbours, as the Czech Republic and Slovakia have gone their separate ways since the end of the Communist Imperium in East Europe.

Besides defence, the rationale of a unified South Asia had other merits. In 1946 the Raj governed a sub-continent with a uniform currency, uniform external tariffs, unified postal service, and a commercial and civil legal code which had many common elements. In specified spheres, such as agriculture, considerable powers were delegated to the provinces. The overall impression is of the kind of common market which Europe is now trying to achieve – except that at its core was the coercion of empire. Since 1947, South Asia has de-common-marketed itself. But the regional imperative cannot simply go away, and recognition of this has resulted in the foundation of the South Asian Association for Regional Co-operation (SAARC). This is years, if not decades, away from re-building a common market; but if one day it is achieved it will hopefully do so without coercion from the centre.

States of Development

This book is not about the process of 'development' in the successor states of South Asia, but it is useful to make at least some passing reference to what has happened in South Asia since 1947. Pakistani and Indian governments since Independence have accepted that they have a role and responsibility in the process of economic growth, even if the commitment to development has varied in time and varied in policy. The nature of the commitment is part of the contract between State and people. The trauma of Partition in 1947 left a real fear of the possibility of further Balkanisation and secession. Therefore, since 1947, the central Government of India has had as an absolute priority the instillation into the minds of each and everyone of the citizenry of the republic, that they are first and foremost Indian. The Prime Minister, Jawarhalal Nehru, viewed the India of 1947 as sunk in backwardness, ignorance, poverty and superstition. India had been colonised by a small, but powerful, industrial country, which had treated it as a source of materials and as a vast market, and which had left India's own level of industrialisation, even simple industrialisation, abysmally low (at Independence in 1947 India produced no bicycles indigenously). The pursuit of Indian-ness, of international economic independence, and of industrialisation and modernisation, therefore, all fitted together. Nehru believed that for the country to progress, planning by the Government was essential, and that 'planning was science in action' (Vasudeva and Chakravarty, 1989: 417). In short, the Government would seek to 'develop' India, and it readily acknowledged its place among the 'developing nations'. It took inspiration from the rapid development (and in the early years it was rapid) of the Soviet Union, and developed its own idea of five-year planning, and public sector dominance in the commanding heights of the economy.

India's share of world trade collapsed dramatically, and import substitution became a central goal. Public sector investment was poured into heavy industry; but the rural sector was not forgotten. Here the effort was as much administrative as financial. In sweeping reforms in the late 1950s, villages were grouped into new Blocks, about 100 villages at a time, to implement a programme known as Community Development. Each Block was headed by a Block Development Officer who oversaw Village Level Workers. Some efforts, often abortive, were made in land reform, to eliminate the excesses of feudal landlordism, and to give more equal holdings to the peasantry. Nor did technology pass rural areas by. India accelerated the development of large- scale irrigation, and began the construction of huge new dams – the temples of modern India as Nehru called them. The point is that the word 'development' intruded virtually everywhere that society and government intermeshed.

India's rate of growth in the years 1947 to 1990 was disappointingly slow. Regional disparity, which Congress had hoped to reduce, increased, so that there were, and still are, two Indias now; a 'high-speed' India in the West, more urban, and including Mumbai, Gujarat, and the national capital area, and a 'slow-speed' India in the East, more rural and including much of Uttar Pradesh, Bihar and Orissa. (See Chapman, 1992 for an elaboration of this.) This slow-speed India retains the world's biggest concentration of absolute poverty. In the late 1960s this area saw again true famine – something which Congress was supposed to have eradicated. Urbanisation increased from 15 per cent to 28 per cent, so that India remained predominantly rural.

Slow-speed development, new awareness of the external world, and the collapse of the socialist model, all conspired to end Nehru's India, and to prepare India for again opening its economy to outside forces. In the end, a foreign exchange crisis also forced the government's hand, and in the 1990s it adopted the orthodox IMF medicine of liberalisation, though the process was enacted with deliberation and some finesse, avoiding the worst side effects this policy produced elsewhere. India's exports have grown fast, and its economy boomed. In the process more people have been lifted out of poverty than during the years of planning. As regions and sectors of the economy have grown, so has national self-confidence. This led to the common utterance of a simple phrase: 'India shining'.

The new post-socialist and self-confident India saw a rise in Hindu assertiveness, marked by the victory of the Bharatiya Janata Party in the 1998 general election, which led a coalition Union Government until 2004. The advent of the BJP and the relative decline of Congress can in part be seen as the sloughing off of the British shadow. Nehru's vision of secularity was in part an acceptance of plurality. His India never promulgated a uniform civil code, and Hindu and Muslim customary law prevailed within the family life of each community. The BJP's view of secularity is the adoption of Hindu codes and the reaffirmation of the cultural distinctiveness of India with which this book began.

The newspaper editors of Mumbai have identified their new élite readership as 'post-nationalist'. This term does not mean internationalist and non-Indian, but rather it identifies those in major cities who take for granted that India now exists and has coherence, that their part of this India has developed, that they are now aware of the outside world, and are keen to exchange with it. But it would be false to suppose that this means that they are international in the sense of being supra-national in outlook, part of an undifferentiated world cosmopolitan class. Amongst these people will be found some of the fiercest critics of Western hegemony, and, although there may be a general awareness of international forces, pragmatically, issues at home make the news.

The image of India in the outside world has changed dramatically – its growth rates and competitiveness are to be 'feared' by the developed world, and its industrial conglomerates have started buying up 'first-world' industry. As something of a corrective to this view, I list the 12 largest economies in the world in Table 15.1, plus three others from South Asia, where the size of the GDP and the size of the population (in 2007 and 2008 respectively) are expressed as multiples of India's. The US has an economy nearly 13 times bigger, with a population just one quarter of India's. China has an economy 3 times bigger. The UK has an economy 2.5 times the size of India's with a population of less than 5 per cent of India's. These are huge disparities. On the other hand, in South Asia, Pakistan, Bangladesh and Sri Lanka's economies are 13 per cent, 7 per cent and 3 per cent of India's, and their populations 15 per cent, 13 per cent, and 2 per cent. India is a regional power, but still far from a global power.

Pakistan, too, has been through similar development fashions. But it is also fashioning a 'nation state', which does not have the same historical depth as India. It embraced import substitution, which produced a high cost industrialisation in many sectors. It embraced a degree of planning and, in the Ayub years tried, too

Table 15.1   India's Relative Economic and Population Size

World Rank		2007 IMF Nominal GDP US$ million	Ratio of GDP to India	Population in millions 2008	Ratio of Population to India
1	United States	13,843,825	12.60	304	0.26
2	Japan	4,383,762	3.99	127	0.11
3	Germany	3,322,147	3.02	82	0.07
4	China	3,250,827	2.96	1330	1.16
5	United Kingdom	2,772,570	2.52	61	0.05
6	France	2,560,255	2.33	64	0.06
7	Italy	2,104,666	1.92	58	0.05
8	Spain	1,438,959	1.31	40	0.03
9	Canada	1,432,140	1.30	33	0.03
10	Brazil	1,313,590	1.20	192	0.17
11	Russia	1,289,582	1.17	141	0.12
12	India	1,098,945	1.00	1148	1.00
47	Pakistan	143,766	0.13	168	0.15
58	Bangladesh	72,424	0.07	154	0.13
78	Sri Lanka	30,012	0.03	21	0.02

Source: Reworked from http://www.photius.com/rankings/

late, to use this to help reduce the inequality with East Pakistan. It has urbanised too. But it has done little in terms of land reform, and has left a feudal rural aristocracy fairly intact. It has a poorer record in education than India, and one of the world's highest population growth rates. Its population has overtaken Bangladesh's, and in the next century it will probably become the world's third most populous state. It too is now trying to open its economy more to the outside world. Some of its structural distortions are greater than India's, and the pain and political consequence of adjustment are not to be downplayed. In both India and Pakistan the era of autarchic economic policy has probably strengthened utilitarian integration within each state. However, it would be easy to overstate the political significance of this. Regional inequality has accompanied autarchy, already threatens stability, and could even worsen with liberalisation.

In both India and Pakistan, defence has laid claim to a significant part of public expenditure (Table 15.2). Defence expenditure in the world as a whole is 2.5 per cent of GDP, and in the UK it is also 2.5 per cent. In 1997 Pakistan's defence expenditure was 6 per cent of GDP, which was an appalling indictment of its priorities. This percentage has decreased significantly, as the Pakistani economy has also grown in recent years. (I have no proof of my other conjecture: that the 'improvement' is a result of massively increased US military aid during the 'War on Terror'.)

Table 15.2   Defence Expenditure and Human Development in South Asia

Table 15.1 Defence Expenditure and the Economies of South Asia

	India	Pakistan	Bangladesh	Bhutan	Maldives	Nepal	Sri Lanka
Population (millions) 2007	1169.0	163.9	158.7	1.6	0.4	28.2	20.1
Est. popn. Annual Growth Rate per cent 2007	1.5	1.8	1.7	1.6	1.7	2	1.1
GNI per capita 2006	820	770	480	1410	2680	290	1300
Human Development Index	0.61	0.53	0.53	0.54	0.74	0.53	0.76
Defence Expenditure as per cent of GDP 1997	3.00	6.00	2.00			1.00	6.00
Defence Expenditure as per cent of GDP 2007	2.80	3.50	1.00			2.10	2.60
Armed forces (personnel) 2003	1205000	610000	126000	5000		46000	115000
Armed forces per 100,000 population	103	372	79	313		163	572
Doctors per 100,000	60	74	26	5	92	21	55
Defence Expenditure US $m 2006	23933	4752	692			160	616
Defence Expenditure US $m 1997	9900	3200	602			161	616

Sources: Stockholm International Peace Research Institute, World Bank

Independent Bangladesh started life as an international begging bowl. Despite nepotistic governments, corruption, periods of dictatorship, and natural calamities, its economy has begun to grow. Dhaka may be Asia's fastest growing city – much of the growth coming on the back of a cotton goods industry that has emerged spontaneously. New sources of rural production have emerged – shrimp farming plays a large part in its export trade. Natural gas fields have been discovered – which India covets, but so far there is no sign that Bangladesh will permits its export by pipeline. Though Bangladesh still has appalling problems of poverty, the idea that change and improvement is possible has taken root. Jinnah's rural slum is now noticeably less rural, and is modernising. Though Islamic, it is clearly aware of its Bengali and South Asian identity. That is not an issue.

Nature, Culture and Civilisation

In his book *The Clash of Civilizations and the Remaking of World Order* (1997) Samuel Huntington observes that there have been two (truly global) World Orders – the European (Western) domination from the Renaissance to World War II, and the Cold War, from World War II to the collapse of the Warsaw Pact in the 1990s. What will be the nature of the next World Order? In this age of globalisation, industrialisation and urbanisation, it might be tempting to think in terms of one culture and one hegemony – essentially the Westernisation of the globe. But Huntington's lengthy analysis suggests otherwise. He suggests it is possible for modernisation to occur without Westernisation, and that – to take an example – Islamisation may strengthen in Pakistan at the same time as that nation develops shopping malls and atomic bombs.

The same point is made by Khilnani with respect to India:

> Nehru wished to modernise India, to insert it into what he understood as the movement of universal history. Yet the India created by this ambition has come increasingly to stand in an ironic, deviant relationship to the trajectories of western modernity that inspired it … [T]he 'garb of modernity' has not proved uniform, and Indians have found many and ingenious ways of wearing it. (Khilnani, 1997: 8)

Huntington does not claim that his model is the only model of international relations, nor the best in any objective sense. He sees it as the most revealing in understanding the contemporary world. He believes the primary divisions of the world are civilisational; which other authors may refer to in terms of high-level cultures. He sees civilisations/cultures as the highest levels at which individuals understand their own identity – and these are the highest levels to which they can appeal in understanding the worth of their own behaviour. This is the highest level which the super-ego can incorporate when a new 'self' is being formed. In Huntington's view, the clashes (often grumbling 'fault-line' wars like those of the Balkans) in the new world order are and will be the clashes between civilisations.

The primary civilisations are Western, Orthodox (i.e. Eastern Christian Orthodox Churches and predominantly Slavic areas), Islamic, Hindu, Sinic, Japanese, and Buddhist with emergent roles for African and Latin American. He concludes that it is imperative for the West to realise that its civilisation is but one of several, that it is neither universal nor universalising, and that the best hope for a more peaceful future is not for the imposition of one civilisation's values on the others, but for dialogue to find and reinforce whatever inter-civilisational values exist, and for understanding of alternative perceptions. To understand the nature of Islamic states it is necessary to realise that in them government of a Western secular style will always be weak – since pan-Islam is prescriptive and embracing, and there is little authority left to the State, except perhaps in defence. There are of course many reasons for political instability in Pakistan since 1947, but surely one of them has to be the difficulty of building a secular Islamic state, and another may be (the same thing seen another way) the inappropriateness for Pakistan's regional cultures of the Western Sovereign State model (the 'European system of states'). By definition, Huntington's cultural and civilisational analysis is less environmentally determinist than the geo-political one with which this chapter opened. Looked at from this viewpoint, the whole of Pakistan is India's North-West Frontier Agency (not a comforting thought) and an individually weak component of a greater Islamic realm. Certainly the 'freedom' with which Islamic militants from other territories have joined the affray in Kashmir could suggest just that. Here, in other words, is one of the world's fault lines, within South Asia. Bharat/ Hindustan confronts an Islamic realm stretching to the Middle East. But:

> … culturally and not least in security terms, Pakistan could not fully free itself from its South Asian moorings even if it wanted to.' (Talbot, 1999: 34)

In fact, the 1998 BJP government made strong attempts to build human bridges with Pakistan, and the succeeding Congress government has followed the same policy of engagement. The opening of road and rail routes from India across both Pakistani and Bangladeshi borders (and including a route between the two Kashmirs), even if they are perhaps for the moment limited and somewhat token, and the contacts between concerned citizens on both sides, are all testimony to the pull of South Asian cultural continuity and shared history.

The Politics of Reaction

There is a school of thought that says that the history of South Asia since the Mutiny of 1857 has been a history of reactionary politics: that the British gave the nationalists too little and too late – for had they been given Dominion Status at the end of the First World War, there would have been no calls for Pakistan. The Congress is likewise accused of offering Jinnah too little, too late, and even as late as 1946 Pakistan was not inevitable even though the British no longer accepted the responsibility for foisting unity on India. And it is said West Pakistan offered the East too little, too late. But this is a little simple: counterfactual history cannot prove that a great Indian Federation in 1919 would have survived.

History has shown that there have been no spontaneous grass roots movements for sub-continental integration until this century. Before that, the sheer scale and complexity of South Asia has meant that only coercive empire could in any sense (usually slight) unite it. What the 20th century has given the public at large is some conception of the right to self-determination. This 'self' has usually had fairly narrow confines, and regionalism remains a serious threat to both Pakistan and India. But wider nationalisms, post-nationalisms and internationalisms are emerging in the 21st-century age of mass-communication and industrialisation. Whether the concept of the strong centrist state will survive long in the new century is open to question.

The scale of problems, and the advent of large-scale technologies, combine to propel whatever forms of state may survive to negotiate with each other those common resource problems (particularly in river basin management, and trading complementarities where possible) which they must accommodate to mutual advantage. Apart from resource issues, and the continuing sore of Kashmir, there are few causes for continuing outright hostility between the current states of South Asia. Nearly all boundary problems are resolved, and from the nadir of trade embargoes one assumes that things can only get better. However, the size of India so outweighs the other states that they must have their anxieties quietened by Indian diplomacy. This will require that the newly self-confident India acts less unilaterally in its own interests, and more multi-laterally in terms of regional leadership in the New Security Agenda.

What is the future for South Asia? This book has been an attempt to explain the broad history of this geo-political region, a story which can be followed in the press every day. I proclaim my inability to make any predictions about the future. So, the most I can do, is hazard the haziest of guesses. On the pessimistic side India and Pakistan are often singled out as the most likely combatants in the world's next nuclear war. Neither appears to have a well-worked out nuclear doctrine, there is no hot-line between Delhi and Islamabad, Pakistan does not have a second-strike capability and there are also doubts that India does. This, coupled with the short distances and very fast flight-times to targets, means that there could be a provocation to a pre-emptive strike. In 2002 nearly all OECD countries advised their nationals to leave the two countries because of the fear that fighting in Kashmir could escalate to a nuclear exchange.[1]

---

1 On 5 June 2002, President Musharraf said he would he would not renounce Pakistan's right to use nuclear weapons first. In one of those unanticipated shocks, like the tsunami of 2004, on 6 June an asteroid exploded over the eastern Mediterranean with a force equal to that of the Hiroshima bomb. Given the latitude, had it happened four hours earlier, this would have been over Kashmir, and might have prompted both sides to 'retaliate'.

If India and Pakistan do nothing to settle this issue, we are likely again to see the politics of reaction, the politics of too little accommodation too late, and yet more war and carnage. What is needed now is the politics of pro-action. Both Pakistan and India know this. Any international assistance in settling this dispute from within SAARC or from outside South Asia must be available when needed – and should be readied with quiet diplomacy that provokes neither side into even more entrenched positions. Supposing this issue is resolved, then it seems to me that the other imperatives promoting closer co-operation among the nations will prevail. Whatever the states of South Asia may then be, they will come to form a more integrated whole, with perhaps less strong sovereign governments. Then South Asia will have refound itself, as a subcontinent within its own secure borders. It could even contemplate its own South Asian Treaty Organisation, banishing Hedley Bull's vision of anarchy, with which this book started, beyond its regional borders.

References and Bibliography

Introductory Note

As I said in the Foreword, this book is intended to be introductory. I hope readers will wish to go into different periods and subjects in much more detail. This is getting easier – there is more material available all the time – but that means there is more to have to choose from.

Below is the list of cited references, only one of which (the second) is avowedly a work of fiction – but such a compelling read – and the quote I have taken from it is 'real' in that it comes from the Foreword by a then-living soldier. There is plenty of other good fiction to get the sense of a place or period, such as J.G.Farrell's *The Seige of Krishnapur* about the 'Mutiny' of 1857 or Humayun Kabir's *Men and Rivers*, about rural life in Bengal somewhere about the 17th century, or more recently Kiran Desai's *The Inheritance of Loss*, which gives a ground level insight into the ethnicities and politics of contemporary Sikkim.

The list contains cited references, and also other bibliography which I think useful. I might also add that in the last stages of preparing this text I have checked facts and rumours with sites on the World Wide Web. A key-word like Kashmir throws up both facts and also the ongoing polemic of groups contending not only the present and future, but also the ownership of history. This is both alarming and also healthy.

Abbas, B.M. (1982), *The Ganges Water Dispute*, New Delhi: Vikas.
Afghan (pseudonym) (1922), *Exploits of Asaf Khan*, with an introduction by Sir George Younghusband, KCMG, London: Herbert Jenkins.
Ahmad, Kabbir Uddin (1972), *Breakup of Pakistan*, London: The Social Science Publishers.
Aitchison, C. (1892), *Life of Lord Lawrence*, London.
Ali, Imran (1988), *The Punjab under Imperialism*, Princeton: Princeton University Press.
Allchin, B. and Allchin R. (1982), *The Rise of Civilization in India and Pakistan*, Cambridge: Cambridge University Press.
Ambedkar, B.R. (1940), *Pakistan: Or Partition of India*, Bombay: Thacker and Co.
Ambirajan, S. (1978), *Classical Political Economy and British Policy in India*, Cambridge: Cambridge University Press.

Anon (1839), *Assam: A Sketch of its History, Soil, and Productions with the Discovery of the Tea Plant and the Countries adjoining Assam*, London: Smith Elder and Co.

Aziz, K.K. (1967), *The Making of Pakistan: A Study in Nationalism*, London: Chatto and Windus.

Aziz, K.K. (1974), *Britain and Pakistan: A Study of British Attitude towards the East Pakistan Crisis of 1971*, Islamabad: Islamabad University Press.

Bareh, H. (1967), *The History and Culture of the Khasi People*, Calcutta.

Barnds, W.J. (1972), *India, Pakistan and the Great Powers*, New York: Praeger.

Baruah, A.K. (2003), 'Tribal Traditions and Crises of Governance in Northeast India, with special reference to Meghalaya', *Working Paper 22. Crisis States Programme.* Development Studies Institute, London School of Economics.

Baruah, S. (1999), *India against Itself: Assam and the Politics of Nationality* Philadelphia: University of Pennsylvania Press.

Baruah, S. (2003), 'Confronting Constructionism: Ending India's Naga War', *Journal of Peace Research*, Vol. 40 No 3, pp. 321–38.

Basham, A.L. (1954 and later), *The Wonder That Was India: A Survey of the History and Culture of the Indian Sub-continent Before the Coming of the Muslims*, London: Sidgwick & Jackson.

Bayly, C.A. (1983), *Rulers, Townsmen and Bazaars: North Indian Society in the Age of British Expansion 1770–1880*, Cambridge: Cambridge University Press.

Bayly, C.A. (1985), 'The pre-history of Communalism: Religious Conflict in India 1700–1860', in *Modern Asian Studies*, Vol. 19, No 2, pp. 177–203.

Bayly, C. and Harper, T. (2004), *Forgotten Armies: The Fall of British Asia 1941–1945*, Harmondsworth: Penguin, Allen Lane.

Berry, B.J.L. (1968), 'Essays in the Spatial Structure of the Indian Economy', Chicago, Department of Geography, *Research Paper No 111.*

Bhagabati, A.K., Bora, A.K. and Binal, K.K. (eds) (2002), *Geography of Assam*, New Delhi: Rajesh Publications.

Bhaneja, B. (1973), *The Politics of Triangles*, New Delhi: Research Publications.

Bhattacharyya, B. (1995), *The Troubled Border*, Guwahati: Lawyer's Book Stall.

Bhaumik, S. (2005), 'Guns, drugs and rebels', in *Gateway to the East: A Symposium on Northeast India and the Look East Policy*, www.india-seminar.com#550.

Bilham, R. and England, P. (2001), 'Plateau pop-up during the great 1897 Assam earthquake', *Nature (Lond)*, 410, 806–809.

Bora, D. (1942), 'A short sketch of the northeast frontier policy of the Great Mughals', *Journal of Assam Research Society*, Vol. IX, Nos 3 and 4, pp. 78–84.

Bose, A. (1973), *Studies in India's Urbanization 1901–1971*, Institute of Economic Growth, New Delhi: McGraw Hill.

Bouglé, C. (1971), *Essays on the Caste System*, Translated with an Introduction by D.F. Pocock, Cambridge: Cambridge University Press.

Bowman, I. (1921), *The New World: Problems in Political Geography*, New York: World Book Co.

Brammer, H. (1990a), 'Floods in Bangladesh. 1. Geographical Background to the 1987 and 1988 Floods', in *The Geographical Journal*, Vol. 156, No. 1, pp. 12–22.

Brammer, H. (1990b), 'Floods in Bangladesh. 2. Flood Mitigation and Environmental Aspects', in *The Geographical Journal*.

Brass, P.R. (1994), *The Politics of India since Independence*, New Cambridge History of India, Cambridge: Cambridge University Press.

Brush, J.E. (1949), 'The Distribution of Religious Communities in India', in *The Annals of the Association of American Geographers*, Vol. 39, pp. 81–98.

Buckley, R. (1880), *The Irrigation Works of India and their Financial Results*, London: W.H. Allen.

Buckley, R.B. (1893), *Irrigation Works In India And Egypt*, London.

Buckley, R.B. (1905), *The Irrigation Works of India*, London.

Bull, H. (1977), *The Anarchical Society: A Study of Order in World Politics*, London: Macmillan.

Carnegie Endowment for International Peace (1988), *Nuclear Weapons and South Asian Security: Report of the Carnegie Task Force on Non-proliferation and South Asian Security.*

Caroe, O. (1962), *The Pathans 550 B.C.–A.D. 1957*, London: Macmillan.

Chakravarty, S.R. and Narain, V. (eds) (1986), *Bangladesh: History and Culture*, New Delhi: South Asian Publishers.

Chakravorty, B.C. (1964), *British Relations with the Hill Tribes of Assam since 1858*, Calcutta.

Chanda, A. (1965), *Federalism in India*, London: George Allen and Unwin.

Chapman, G.P. (1983), 'The Folklore of the Perceived Environment in Bihar', *Environment and Planning A*, Vol. 15, pp. 945–68.

Chapman, G.P. (1990), 'Religious v. Regional Determinism: India, Pakistan and Bangladesh as Inheritors of Empire', Ch. 6, pp. 106–34, in Chisholm, M. and Smith, D. (eds), *Shared Space: Divided Space*, London: Unwin Hyman.

Chapman, G.P. (1992), 'Change in the South Asian Core: Patterns of Growth and Stagnation in India', Chapter 2, pp. 10–43, in Chapman and Baker (eds).

Chapman, G.P. (1996), 'Environmental Myth as International Politics: The Problems of the Bengal Delta', Ch. 10, pp. 163–86, in Chapman G.P. and Thompson, M. (eds).

Chapman, G.P. (2001), '"Other" cultures, "other" environments, and the mass media' Ch. 10, pp. 127–50 in J. Smith (ed.) *The Daily Globe: Environmental Change, the Public and the Media*, London: Earthscan.

Chapman, G.P. (2002), 'Changing Places: the Roles of Science and Social Science in the Development of Large Scale Irrigation in South Asia', Ch. 4, pp. 52–78, in R. Bradnock and G. Williams (eds), *South Asia in a Globalising World: A Reconstructed Regional Geography*, London: Prentice Hall.

Chapman, G.P. and Baker, K.M. (eds) (1992), *The Changing Geography of Asia*, London: Routledge.

Chapman, G.P. and Rudra, K. (2007), 'Water as Foe, Water as Friend: Lessons from Bengal's Millennium Flood', *Journal of South Asian Development*, Vol. 2, 1: 19–49.

Chapman, G.P. and Thompson, M. (eds) (1996), *Water and the Quest for Sustainable Development in the Ganges Valley*, London: Mansell.

Chapman, G.P. and Wanmali, S. (1981), 'Urban-rural relationships in India: a Macro-scale approach using Population Potentials', *Geoforum*, Vol. 12, No 1, pp. 19–44.

Chapman, G.P., Kumar, K., Fraser, C. and Gaber, I. (1997), *Environmentalism and the Mass Media: The North South Divide*, London: Routledge.

Charlesworth, N. (1982), *British Rule and the Indian Economy 1800–1914*, London: Macmillan.

Chatterjee, S.P. (1947), *The Partition of Bengal: A Geographical Study*, Calcutta Geographical Society Pub. No 8. 1947.

Chaudri, M.A. (1967), *The Emergence of Pakistan*, New York: Columbia.

Chauhan, B.R. (1992), *Settlement of International and Inter-state Water Disputes in India*, New Delhi: Indian Law Institute.

Chen, L. (ed.) (1973), *Disaster in Bangladesh: Health Crisis in a Developing Nation*, New York.

Chesney, G. (1894), *Indian Polity: A View of the System of Administration in India*, London: Longman, Green & Co.

Choudhury, P.C. (1959), *The History of the People of Assam to the Twelfth Century A.D.*, Government of Assam, Dept. of Historical and Antiquarian Studies, Gauhati.

Chowdhury, M. (1988), *Pakistan – Its Politics and Bureaucracy*, New Delhi: Associated Publishing House.

Clark, S. (1864), *Practical Observations on the Hygiene of the Army in India*, London: Cox.

Cohen, S.B. (1963), *Geography and Politics in a Divided World*, London: Methuen.

Cohn, B.S. (1971), *India: The Sociology of a Civilization*, Englewood Cliffs NJ: Prentice Hall.

Collins, L. and LaPierre, D. (1975), *Freedom at Midnight*, New York: Simon & Schuster.

Cotton, A. (1866), *The Famine in India*, Lecture at the Social Science Congress, Manchester.

Cotton, A. (1872), *1000 Short Teloogoo Sentences*, Madras.

Cotton, A. (1875), *Study of Living Languages for Colloquial Purposes*, Dorking.

Cotton, A. (1877), *The Madras Famine*, with appendix containing a letter from Miss Florence Nightingale.

Cotton, A. and Hope E.R. (1900), *General Sir Arthur Cotton: His Life and Work*, London: Hodder and Stoughton.

Coupland, R. (1943), *Report on the Constitutional Problem in India*, London, Pt I The Indian Problem 1833–1935, Pt II Indian Politics 1936–1942, Pt III The Future of India, Oxford: Oxford University Press.
Coupland, R. (1945), *India: A Restatement*, Oxford: Oxford University Press.
Cranner, M. (ed.) (1994), *The True Cost of Conflict*, London: Earthscan.
Crow, B., Lindquist, A. and Wilson, D. (1995), *Sharing the Ganges: The Politics and Technology of River Development*, New Delhi: Sage.
Dalvi, J.P. (1969), *Himalayan Blunder*, Delhi.
Davies, C. Collin (1959 [1953]), *An Historical Atlas of the Indian Peninsula*, 2nd edn, Oxford: Oxford University Press.
Day, W.M. (1949), 'Relative Permanence of Former Boundaries in India', *Scottish Geographical Magazine*, pp. 113–22.
De Blij, H. (1967), *Systematic Political Geography*, New York: Wiley.
de Reuck, A. and Knight, J. (1967), *Caste and Race: Comparative Approaches*, London: J.A. Churchill.
Deakin, A. (1893), *Irrigated India*, London: Thacker and Co.
Deloche, J. (1993), Vol. 1 (1994), Vol. II, *Transport and Communications in India Prior to Steam Locomotion, Vol. 1 Land Transport, Vol. II Water Transport*, French Studies in South Asian Culture and Society VII trans. from the French by James Walker, Delhi: Oxford University Press, first published as *La Circulation en Inde avant la Révolution des Transports*, Paris: Publications de l'Ecole Française d'Extrême Orient (1980).
Derbyshire, I. (1985), *Opening up the Interior: The Impact of Railways on the North Indian Economy and Society 1860–1914*, Unpublished PhD Dissertation, Cambridge, 1985.
Desai, A.R. (1966), *Social Background to Indian Nationalism*, 4th edn, Bombay: Popular Prakashan.
Digby, W. (1901), *Prosperous British India*, London: T. Fisher Unwin.
Dubois, Abbe J.A. (1906), *Hindu Manners, Customs and Ceremonies*, Trans. H.K. Beauchamp. Oxford: Clarendon Press.
Duffield, M. (2005), 'Getting Savages to Fight Barbarians: Development, Security and the Colonial Present', *Conflict, Development and Society*, 5 (2).
Dumont, L. (1970), *Homo Hierarchicus: The Caste System and Its Implications*, Translated from the French by M. Sainsbury, London.
Dunbar, G. (1943), *A History of India from the Earliest Times to the Present Day*. London: Nicholson and Watson.
Elwin, V. (ed.) (1959), *India's North-east Frontier in the Nineteenth Century*, Oxford: Oxford University Press.
Etzioni, A. (1968), *The Active Society: A Theory of Societal Political Processes*, London: Collier-Macmillan.
Fowler, F.J. (1950), 'Some Problems of Water Distribution between East and West Punjab', in *Geographical Review*, Vol. 40, pp. 583–99.
Gadgil, M. and Guha, R. (1992), *This Fissured Land*, Delhi: Oxford University Press.

Gait, E. (1905, 1926), *A History of Assam*, Calcutta: Thacker Spink and Co.

Golant, W. (1975), *The Long Afternoon: British India 1601–1947*, London: Hamish Hamilton.

Gole, S. (1988), *Maps of Mughal India*, Kegan Paul International, London.

Goudie, A.S., Allchin, B. and Hegde, K.T.M. (1973), 'Former Extensions of the Great Indian Sand Desert', *Geographical Journal* Vol. 139, pp. 243–55.

Governor of Assam (1998), 'Report on Illegal Migration into Assam', Raj Bhavan, Guwahati (available through www.satp.org – South Asian terrorism portal).

Greenberger, A.J. (1969), *The British Image of India: A Study in the Literature of Imperialism*, London: Oxford University Press.

Griffin, K. and Khan, A.R. (eds) (1972), *Growth and Inequality*, London: Macmillan.

Griffiths, P. (1952), *The British Impact on India*, London: Macdonald.

Guha, R. (1989), *The Unquiet Woods: Ecological Change and Peasant Resistance in the Himalaya*, New Delhi and Oxford: Oxford University Press.

Guillaume, A. (1954), *Islam*, Harmondsworth: Penguin.

Habib, I. (1982), *An Atlas of the Mughal Empire*, New Delhi and Oxford: Oxford University Press.

Hardgrave, R.L. (1975), *India: Government and Politics in a Developing Nation*, 2nd Edn. New York.

Hardy, P. (1972), *The Muslims of British India*, Cambridge: Cambridge University Press.

Hasan, Mushirul (1997), *Legacy of a Divided Nation: India's Muslims from Independence to Ayodhya*, London: C. Hurst & Co.

Hazarika, S. (2000), *Rites of Passage: Border Crossings, Imagined Homelands, India's East and Bangladesh*, Harmondsworth: Penguin.

Hazarika, S. (2004), 'Land, conflict, identity in India's north-east: negotiating the future.' *Futures*, Vol. 36, 6–7, pp. 771–80.

Hettne, B. (1978), *The Political Economy of Indirect Rule: Mysore 1881–1947*, London and Malmo: Curzon Press.

Hewitt, V. (2002), *Towards the Future? Jammu and Kashmir in the 21st Century*, Cambridge: Granta Editions.

Hodson, H.V. (1969), *The Great Divide: Britain–India–Pakistan*, London: Hutchinson.

Hunter, W.W. (1871), *The Indian Musalmans: Are They Bound in Conscience to Rebel Against the Queen?*, London: Trubner and Co.

Huntington, S. (1997), *The Clash of Civilizations and the Remaking of World Order*, New York: Simon & Schuster.

Hussain, A. (1979), *Politics in an Ideological State: The Case of Pakistan*, Folkstone: Dawson.

Inayatullah (1964), 'Basic Democracies, District Administration and Development', *Research Publication Series* No 9, Academy for Rural Development, Peshawar.

Indian Government (1902), *Land Revenue Policy*, Calcutta.

Indian Government Secretary of State for India (1906/1908), *The Imperial Gazeteer of India. The Indian Empire Vol. III Economic*, New Edition, Oxford.

Islam, M.R. (1987), *Ganges Water Dispute: Its International Legal Aspects*, Dhaka: University Press.

Islam, N. (1992), 'Indo-Bangladesh Common Rivers: the Impact on Bangladesh', *Contemporary South Asia*, Vol. 1, No 2, pp. 203–25.

Jaffrelot, C. (1996), *The Hindu Nationalist Movement and Indian Politics, 1925–1990s*, London: C. Hurst & Co.

Jannuzi, P. Tomasson and Peach, J.T. (1980), *The Agrarian Structure of Bangladesh*, Boulder, CO: Westview Press.

Jeffrey, Robin (1974), 'The Punjab Boundary Force and the Problem of Order, August 1947', *Modern Asian Studies*, Vol. 8, pp. 491–520.

Jennings, I. (1957), *Constitutional Problems in Pakistan*, Cambridge: Cambridge University Press.

Karan, P.P. (1961), 'Dividing the Water', *Professional Geographer*, pp. 6–10.

Khan, A.R. (1972), *The Economy of Bangladesh*, London: Macmillan.

Khilnani, S. (1997), *The Idea of India*, New York: Farrar, Straus and Giroux.

Khubchandani, L.M. (1983), *Plural Languages, Plural Cultures: Communication, Identity and Sociopolitical Change in Contemporary India*, University of Hawaii Press: East West Centre.

Khullar, D. (1999), *When Generals Failed: The Chinese Invasion*, New Delhi: Manas Publications.

King, A.D. (1976), *Colonial Urban Development: Culture, Social Power and Environment*, London: Routledge, Kegan and Paul.

King, H. (1875), *Madras Manual of Health*, Madras: Government Printer.

King, L.C. (1983), *Wandering Continents and Spreading Sea Floors on an Expanding Earth*, Chichester: John Wiley.

Kirk, W. (1975), 'The Role of India in the Diffusion of Early Cultures', *Geographical Journal*, Vol. 141, pp. 19–34.

Klein, I. (1974), 'Population and Agriculture in Northern India 1872–1970', *Modern References and Bibliography 313 Asian Studies*, Vol. 8, pp. 191–216.

Knowles, L.C.A. (1924), *The Economic Development of the British Overseas Empire*, London: George Routledge and Sons.

Kumar, D. and Desai, M. (eds) (1983), *The Cambridge Economic History of India Vol. 2, c. 1757–1970*, Cambridge: Cambridge University Press.

Lamb, A. (1966), *Crisis in Kashmir, 1947–1966*, London: Routledge, Kegan Paul.

Lamb, A. (1968), *Asian Frontiers*, London: Pall Mall Press.

Lapidus, D.F. (1987), *The Facts on File Dictionary of Geology and Physics*, New York.

Law, B.C. (1954), *Historical Geography of Ancient India*, Paris: Société Asiatique de Paris.

Lewis, O. (1958), *Village Life in Northern India*, New York: A. Knopf.

Macfarlane, C. (1857), *History of British India from the Earliest English Intercourse*, 3rd edn, London 1857.

Mackenzie, A. (1884), *History of the Relations of the Government with the Hill Tribes of the North-east Frontier of Bengal*, Calcutta: Home Department Press.

Mackinder, H.J. (1904), 'The Geographical Pivot of History', *Geographical Journal*, Vol. XXIII, pp. 421–44.

Mackinder, H.J. (1922), 'The Sub-Continent of India', Chapter 1, pp. 1–36, in *The Cambridge History of India: Volume 1, Ancient India*, E.J. Rapson, (ed.), Cambridge: Cambridge University Press.

Maine, H.S. (1871), *Village Communities in the East and West*, London: John Murray.

Maity, S.K. (1970), *Economic Life in Northern India in the Gupta Period*, Varanasi: Motilal Banarsidass.

Majumdar, D.N. (1961), *Races and Cultures of India*, 4th edn, London: Asian Publishing House.

Malik, S.C. (1968), *Indian Civilisation: The Formative Period*, Indian Institute of Advanced Study, Shimla.

Malik, S.C. (1989), *Modern Civilisation: A Crisis of Fragmentation*, New Delhi: Abhiner Publications.

Mansergh, N. (1978), 'The Prelude to Partition: Concepts and Aims in Ireland and India', *The 1976 Commonwealth Lecture*, Cambridge: Cambridge University Press.

Mansergh, N. (ed.), *The Transfer of Power, 1942–47*, Vols 1–12, London: HMSO.

Marshall, P.J. (1990), *Bengal: the British Bridgehead in Eastern India 1740–1828*, *New Cambridge History of India*, Cambridge: Cambridge University Press.

Marshman, J.C. (1867), *The History of India from the Earliest Period to the Close of Lord Dalhousie's Administration*, Three Volumes, London.

Maxwell, N. (1970), *India's China War*, London: Jonathan Cape.

McIntosh, J. (2001), *A Peaceful Realm: The Rise and Fall of the Indus Civilization*, Boulder, CO: Westview Press.

Menon, V.P. (1956,), *The Story of the Integration of the Indian States*, Calcutta and London: Orient Longman.

Metcalf, T.R. (1994), *Ideologies of the Raj*, *New Cambridge History of India*, Cambridge: Cambridge University Press.

Meyer, F.V. (1948), *Britain's Colonies in World Trade*, Oxford: Oxford University Press.

Michel, A.A. (1967), *The Indus Rivers: A Study in The Effects of Partition*, New Haven and London: Yale University Press, p. 595.

Minorsky,V. (1982), *Regions of the World: Translated from Hudu-Alam: a Persian Geography 982 A.D.*, London 1937, 1970, Reprinted 1982.

Moon, Sir P. (1961), *Divide and Quit*, London: Chatto and Windus.

Moreland, W.H. (1923), *From Akbar to Aurangzeb: A Study in Indian Economic History*, London: Macmillan and Co.

Naik, J.P. (1963), *Selections from Educational Records of the Government of India. Vol. 2: Development of University Education, 1860–1887*, New Delhi: National Archives of India.

National Water Development Agency (2007), *National Perspective Plan*, Ministry of Water Resources, New Delhi: available at http://nwda.gov.in/index2.asp?su blinkid=48&langid=1.

Nehru, J. (1961), *The Discovery of India*, New Delhi: Asia Publishing House.

Nock, O.S. (1978), *Railways of the World No 5: Railways of Asia and the Far East*, London: A. & C. Black.

Noman, O. (1988), *The Political Economy of Pakistan 1947–1985*, London: KPI Ltd.

Owen, S. (1872), *India on the Eve of the British Conquest*, London.

Philips, C.H. and Wainwright, M.D. (1970), *The Partition of India: Policies and Perspectives 1935–1947*, London: Macmillan,

Prasad, P. (1967), 'The Emergence of the Demand for India's Partition', *International Studies*, Vol. 9.

Prasad, R.N. (1998), *Public Administration in North-east India*, New Delhi: Vikas Publishing.

Pratap, A. (2001), *Island of Blood*, New Delhi: Penguin Books,

Prendergast, W.H. (1944), 'A sleepy tea and jute railway', *Journal of Assam Research*, Vol. XI, Nos 1 and 2, pp. 11–31.

Rahman, M.A. (1968), 'East and West Pakistan: A Problem in the Political Economy of Regional Planning', Harvard University Centre for International Affairs, *Occasional Paper No 20*.

Rahman, M. (1978), *Bangladesh Today: An Indictment and a Lament*, London.

Rahmat Ali, C. (1942), *What does the Pakistan Movement stand for?*, pamphlet, Cambridge.

Rahula, W. (1967), *What the Buddha Taught*, London: Gordon Fraser.

Raychaudhuri, T. and Habib, I. (eds) (1982), *The Cambridge Economic History of India Vol. 1 c. 1200–c. 1750*, Cambridge: Cambridge University Press.

Razvi, M. (1971), *Frontiers of Pakistan*, Karachi: National Publishing House.

Richards, J.F. (1993), *The Moghul Empire, New Cambridge History of India*, Cambridge: Cambridge University Press.

Risley, H.H. (1915), *The People of India*, London.

Roberts, P.E. (1952), *History of British India under the Company and the Crown*, Oxford: Oxford University Press.

Robinson, A. (1974), 'The Economics of Malthusia', *Modern Asian Studies*, Vol. 8, pp. 521–34.

Robinson, F. (1982), *An Atlas of the Islamic World since 1500*, Oxford: Equinox.

Robinson, F. (ed.) (1989), *Cambridge Encyclopaedia of India, Pakistan, Bangladesh and Sri Lanka*, Cambridge: Cambridge University Press.

Rose, L.E. and Fisher, M.W. (1967), The North-east Frontier Agency of India, *Office of External Research, Department of State Publication 8288, Near and Middle Eastern Series 76*, Washington, D.C.

Rothermund, D. (1993), *An Economic History of India from Pre-Colonial Times to 1991*, 2nd edn, London: Routledge.

Rudolph, L.I. and S.H. (1980), *The Regional Imperative*, New Delhi: Concept.

Rushbrook Williams, L.F. (1925), *India in 1924–25*, Calcutta 1925.

Rushbrook Williams, L.F. (1972), *The East Pakistan Tragedy*, London: Tom Stacey.

Rustomji, N. (1983), *Imperilled Frontiers: India's North-eastern Borderlands*, New Delhi: Oxford University Press.

Said, E.W. (1991), *Orientalism: Western Concepts of the Orient*, London: Penguin.

Sarkar, B.K. (1925), *Inland Transport and Communication in Mediaeval India*, Calcutta: Calcutta University Press.

Sarkar, S. (1983), *Modern India 1885–1947*, London: Macmillan.

Sarma, S.S. (1986), *Farakka – A Gordian Knot: Problems on Sharing the Ganga Waters*, Calcutta: Asit Sen.

Satow, M. and D.R. (1980), *Railways of the Raj*, London: Scholar Press.

Saxena, D.P. (1968), 'Indian Agriculture During the Vedic Period', *Proceedings of Symposium on Land Use in Developing Countries*, 21st International Geographical Congress, Aligarh.

Schofield, V. (1996), *Kashmir in the Crossfire*, London: I.B. Tauris.

Schofield, V. (1999), *Kashmir in Conflict*, London: I.B. Tauris.

Schwartzberg, J.E. (1966), 'Prolegomena to the Study of South Asian Regions and Regionalism', pp. 89–107, in *Regions and Regionalism in South Asian Studies: An Exploratory Study* (ed.) Crane, R.I., Durham, NC: Duke University Press.

Schwartzberg, J.E. (1985), 'Factors in the Linguistic Reorganization of Indian States' in P. Wallace (ed.), *Region and Nation In India*, New Delhi.

Schwartzberg, J.E. (1985), 'Folk Regions in Northwestern India' in A.B. Mukerji and A. Ahmad (eds), *India: Culture, Society and Economy. Essays in Honour of Prof. Asok Mitra*, New Delhi.

Schwartzberg, J.E. (1992), *An Historical Atlas of South Asia*, Chicago: Chicago University Press.

Schwartzberg, J.E. (1995), 'An American Perspective II', *Asian Affairs: An American Review*, Special Issue 'Kashmir', Vol. 22, No. 1, pp. 71–87.

Schwartzberg, J.E. (1997), 'Who are the Kashmiri People? Self-identification as a vehicle for self-determination', *Environment and Planning A*, Vol. 29, pp. 2237–56.

Sen, K.M. (1961), *Hinduism*, Harmondsworth: Penguin.

Sharan, P. (1983), *Government and Politics of Pakistan*, New Delhi: Metropolitan.

Singh, G. (2000), *Ethnic Conflict in India*, London: Macmillan.

Singh, M.R. (1972), *A Critical Study of the Geographical Data in the Early Puranas*, Calcutta: Punthi Pustak.

Singh, P. (1966), *India and the Future of Asia*, London: Faber & Faber.

Singh, R.L. (1971), *A Regional Geography of India*, New York: A. Knopf.

Sonwalker, P. (2004), 'Mediating otherness: India's English-language press and the Northeast', *Contemporary South Asia* 13(4), 389–402.

Spate, O.H.K. (1947), 'The Partition of Punjab and Bengal', *Geographical Journal* Vol. 110, pp. 205–218.

Spate, O.H.K. (1948), 'The Boundary Award in Punjab', *Asiatic Review*, Vol. XLIV, pp. 1–8.

Spate, O.H.K. (1948), 'The Partition of India and the Prospects of Pakistan', *Geographical Review*, Vol. 38, p. 5–29.

Spate, O.H.K. and Learmonth, A.T.A (1967), *India and Pakistan: A General and Regional Geography*, London: Methuen and Co.

Spear, P. (1965), *A History of India: 2*, Harmondsworth: Penguin.

Srinavas, M.N. (1962), *Caste in Modern India and Other Essays*, Bombay: Asian Publishing House.

Stepanek, J.F. (1979), *Bangladesh: Equitable Growth?*, New York: Pergamon.

Stephens, I.M. (1967), *Pakistan*, London: Benn.

Stockhom International Peace Research Institute (2002), http://www.sipri.se.

Stone, I. (1984), *Canal Irrigation in British India: Perspectives on Technological Change in a Peasant Economy*, Cambridge: Cambridge University Press.

Swinson, A. (1967), *The North-West Frontier*, London: Hutchinson and Co.

Talbot, I. and Singh, G. (eds) (2000), *Region and Partition*, New York: Oxford University Press.

Talbot, I. (1998), *Pakistan: A Modern History*, London: Hurst and Co.

Talboys, Wheeler, J. (1973), *India of the Brahmanic Age*, Reprint, New Delhi: Cosmo Publications.

Tapponnier, P., Peltser, G. and Armijo, R. (1986), 'On the Mechanics of the Collision Between India and Asia', pp. 115–57 in *Collision Tectonics* (Geological Society Special Publication No. 19), Coward, M.P. and Ries, A.C. (eds), Oxford: Blackwell Scientific.

Taylor, D. and Yapp, M. (eds) (1979), *Political Identity in South Asia*, London: Curzon Press.

Tayyeb, A. (1966), *Pakistan: A Political Geography*, Oxford: Oxford University Press, London.

Temple, Sir R. (1880), *India in 1880*, London: John Murray.

The Telegraph (Kolkata) (2007), 'Want to grab land? Send jumbos – Naga hoodlums let loose wild herds to scare away Assam villagers', 21 July, available at: http://www.telegraphindia.com/1070721/asp/northeast/story_8084986.asp.

Thapar, R. (1966), *A History of India: 1*, Harmondsworth: Penguin.

Thapar, R. (1992), *Interpreting Early India*, Oxford University Press, Oxford

Tomlinson, B.R. (1979), *The British Raj 1914–1947: The Economics of Decolonization in India*, London: Macmillan.

Tomlinson, B.R. (1993), *The Economy of Modern India 1860–1970*, New Cambridge History of India, Cambridge: Cambridge University Press.

Vakil, C.N. (1950), *The Economic Consequences of Divided India*, Bombay: Vora and Co.

Vakil, C.N. and Raghava Rao, G. (1968), *Economic Relations between India and Pakistan*, Bombay: Vora and Co.

Verghese, B.G. (1994), *Winning the Future: From Bhakra to Rajasthan Canal*, New Delhi: Konark Publishers.

Wallace, P. (ed.) (1985), *Region and Nation In India*, New Delhi: Oxford and IBH.

Waqif, A.A. (1997), 'Economic Cooperation in South Asia: A Review', *EIAS Briefing Papers*, No. 97/08, European Institute for Asian Studies, Brussels.

Ward, F.K. (1930), *Plant Hunting on the Edge of the World*, London: Victor Gollancz.

Westwood, J.N. (1974), *Railways of India*, Newton Abbot: David and Charles.

Whitcombe, E. (1972), *Agrarian Conditions in Northern India Vol. 1 The U.P. under British Rule 1860–1900*, Berkeley: University of California Press.

Wilcox, W. A. (1963), *Pakistan: The Consolidation of a Nation*, New York: Columbia University Press.

Williams, G. (1966), *The Expansion of Europe in the Eighteenth Century*, London.

Wilson, A.J. and Dalton, D. (eds) (1982), *The States of South Asia*, London: C. Hurst and Co.

Wirsing, R.G. (1994), *India, Pakistan and the Kashmir Dispute: On Regional Conflict and its Resolution*, London: Macmillan.

Wiser, W.H. (1936), *The Hindu Jajmani System*, Lucknow Publishing House, Lucknow, reprinted 1969, 1988. New Delhi: Munshiram Monaharlal.

World Bank (1989), *Bangladesh Action Plan for Flood Control*, Washington, DC.

World Bank (2004), 'Trade Policies in South Asia: An Overview', *Report No 29949, Poverty Reduction and Economic Management Unit*, South Asia Region, Washington DC.

Wylly, H.C. (1912), *From the Black Mountain to Waziristan*, London: Macmillan and Co.

Zaman, M. et al. (eds) (1983), *River Basin Development*, Dublin: Tycooly Publishing.

Index

Laurasia 3
Lawrence, Henry 142
Learmonth, A. 317
Lebanon 39
Lhasa 240, 248, 251ff
Liberal Party (India) 154
Libya 294
lingua franca 226
Linlithgow, Lord 158
Lithuanian language 9
Lodi Gardens 49
Lodi, Bahlul 49
Lodi, Ibrahim 50
London 69, 75ff, 83, 103, 144ff, 157, 171, 178ff, 208, 252, 282
Lower Bari Doab Canal 129ff, 187, 263, 265
Lucknow 82
Lucknow Pact 1916 151
Lyallpur 131
Lytton, Lord 144

Macauley, T.B. 80
Mackenzie 242
Mackinder, H. 22, 89, 97, 283, 285, 289, 291, 313, 314
Macmillan, Harold 143
Madhopur 129, 187
Madhya Pradesh 6, 161, 226, 253
Madras xvii, 70ff, 89, 115, 123ff, 134, 175, 223, 250
Mahaban 99
Mahabharat range 3
Mahabharata 13, 154
Mahanadi river 8
Maharashtar 162
Maharashtra 56, 58, 205, 224, 314
Mahatma; *see* Gandhi
Mahavira 15
Mahmud of Ghazni 45
Makran coast 17
Malabar 60
malaria 120, 125, 129, 135
Malaya 158, 178
Malda 190
Maldives 300
Malik, S.C. 322
Malthus, T. 128

Manasowar lake 16
Manchuria 285
Mandal Commission 37
Mangla dam 131
Mangrol 184ff.
Manipur 239, 243ff, 281
Mansergh, N. 326
Manu Syayambhu 15, 16
Manu, Laws of 31
Marala 131, 266
Marartha Confederacy 79
Marathas 58, 70, 73, 79
Marshman 95
Marx, Karl 11
Marxism 43
Masulipatam 76
Mauryan empire 11, 13, 20, 23, 102, 308
Maxwell 252
maya 35
McMahon Line 251
McMahon, Henry 103, 251
McNaghten 94
Mecca 39
Medina 40
Mediterranean 68
Meerut 82
Meghalaya 5
Meghalaya 239
Meghna river 278
Meiji Restoration 12
Menon, V.P. 179
Meru, Mt. 16, 196, 312
Mesopotamia 16
Miani 96
Miani 96
Michel, A.A. 129
Midnapore 127
Minto, Lord 91, 150
Mir Jaffa 75
Mirza, Iskander 232ff
Mizoram 239
Mughal Army 54
Mughal currency 53
Mughal Empire xii, xvii, 18ff, 48ff, 71, 73ff, 90, 91, 113, 119, 138ff, 151, 195, 235, 241, 283
Mohajirs 202, 207, 236, 238, 327
Mohammad, Ghulam 231, 232, 266

Yeatts 175
Younghusband, Colonel 252
yuga, kali 28
Yugoslavia xiii, 217, 298, 308, 326

Zahir Shah, King 109, 289
zakat – almsgiving 41
Zamindars/Zamindari 78, 128
Zimbabwe 20, 52, 129